New Trends for Hamiltonian Systems and Celestial Mechanics

ADVANCED SERIES IN NONLINEAR DYNAMICS*

Editor-in-Chief: R. S. MacKay *(Univ. Warwick)*

Published

Vol. 27 *Simulation and Control of Chaotic Nonequilibrium Systems*
by W. G. Hoover & C. G. Hoover

Vol. 26 *Geometry of Nonholonomically Constrained Systems*
by R. H. Cushman, J. Śniatycki & H. Duistermaat

Vol. 25 *Smooth Particle Applied Mechanics – The State of the Art*
W. G. Hoover

Vol. 24 *Microscopic Chaos, Fractals and Transport in Nonequilibrium Statistical Mechanics*
R. Klages

Vol. 23 *Geometrical Theory of Dynamical Systems and Fluid Flows (Revised Edition)*
T. Kambe

Vol. 22 *Energy Localisation and Transfer*
eds. T. Dauxois, A. Litvak-Hinenzon, R. MacKay & A. Spanoudaki

Vol. 21 *Spatio-Temporal Chaos & Vacuum Fluctuations of Quantized Fields*
C. Beck

Vol. 20 *The Mathematical Theory of Permanent Progressive Water-Waves*
H. Okamoto & M. Shoji

Vol. 19 *Integrability and Nonintegrability of Dynamical Systems*
A. Goriely

Vol. 18 *Symplectic Twist Maps*
C. Gole

Vol. 17 *Smooth Dynamical Systems*
M. C. Irwin

Vol. 16 *Positive Transfer Operators and Decay of Correlations*
V. Baladi

Vol. 15 *Methods in Equivariant Bifurcations and Dynamical Systems*
P. Chossat & R. Lauterbach

Vol. 14 *Topics in Nonlinear Time Series Analysis – With Implications for EEG Analysis*
A. Galka

Vol. 13 *Time Reversibility, Computer Simulation, Algorithms, Chaos (Second Edition)*
W. G. Hoover & C. G. Hoover

Vol. 12 *Localization and Solitary Waves in Solid Mechanics*
A. R. Champneys, G. W. Hunt & J. M. T. Thompson

*For the complete list of titles in this series, please visit
http://www.worldscientific.com/series/asnd

ADVANCED SERIES IN NONLINEAR DYNAMICS
VOLUME 8

New Trends for Hamiltonian Systems and Celestial Mechanics

Mexico
13–17 September 1994

Editors

Ernesto A Lacomba
Universidad Autonoma Metropolitana - Iztapalapa, Mexico

Jaume Llibre
Universitat Autonoma de Barcelona, Bellaterra, Barcelona, Spain

World Scientific
Singapore • New Jersey • London • Hong Kong

Published by

World Scientific Publishing Co. Pte. Ltd.
5 Toh Tuck Link, Singapore 596224
USA office: 27 Warren Street, Suite 401-402, Hackensack, NJ 07601
UK office: 57 Shelton Street, Covent Garden, London WC2H 9HE

British Library Cataloguing-in-Publication Data
A catalogue record for this book is available from the British Library.

Advanced Series in Nonlinear Dynamics — Vol. 8
NEW TRENDS FOR HAMILTONIAN SYSTEMS AND CELESTIAL MECHANICS

ISBN-13 978-981-02-2645-9
ISBN-10 981-02-2645-4

PREFACE

The papers in this volume are an outgrowth of the second "International Symposium on Hamiltonian Systems and Celestial Mechanics" which was held in Cocoyoc, Morelos, México from September 13 to 17, 1994. The proceedings of the first symposium were published in Volume 4 of this series under the title "Hamiltonian Systems and Celestial Mechanics". The Scientific Organizers of the Conference were Ernesto A. Lacomba and Jaume Llibre and the members of the Technical Organizing Committee were Joaquín Delgado, Ernesto Pérez-Chavela, from the Universidad Autónoma Metropolitana, México and Kurt B. Wolf on the part of CIFMA (Centro Internacional de Física y Matemáticas Aplicadas, Cuernavaca, México). Special thanks go to J. Delgado, for his effort and the time spent in helping with the edition of these proceedings.

The second symposium gathered together 46 researches (14 from México and 32 from abroad) in addition to 10 graduate students, 38 lectures were given during 32 session hours. The emphasis of all the talks was on Hamiltonian Dynamics and Celestial Mechanics and their relationship to several aspects of topology, mechanics and dynamical systems in general. All speakers were invited to submit a contribution to these proceedings. All papers that were received went through the refereeing process typical of a mathematical research journal and those that were accepted form the present volume.

This symposium would not have been realized without the initial support of the CONACYT (México) grant 1772-E9210. The organization and realization of the meeting was possible due to a lot of work on the part of a number of people. Special thanks are due to K. B. Wolf and Petra Seligman from the CIFMA who took care with great enthusiasm of the logistic administration of the meeting and many others things. We also thank the economical support provided by the following institutions. From México the Universidad Autónoma Metropolitana; from Spain CIRIT and DGICYT; and from the international agency Organization of American States. And last but not least, the participants at the conference must be thanked for offering everyone a week-long intense and highly interesting mathematical experience.

Ernesto A. Lacomba and Jaume Llibre

CONTENTS

THE CURVATURE CRITERION AND THE DYNAMICS OF A ROLLING ELASTIC CYLINDER

C. M. ARIZMENDI
Universidad Nacional de Mar del Plata, Argentina
arizmendi%unmdp.edu.ar@secyt.gov.ar

R. CARRETERO-GONZÁLEZ
Queen Mary and Westfield College, UK
Mile End Road, London E1 4NS UK
r.carretero@qmw.ac.uk

H. N. NÚÑEZ-YÉPEZ
Departamento de Física
Universidad Autónoma Metropolitana-Iztapalapa, México
nyhn@xanum.uam.mx

A. L. SALAS-BRITO
Laboratorio de Sistemas Dinámicos
Universidad Autónoma Metropolitana-Azcapotzalco
Apartado Postal 21-726, Coyoacán 04000 DF, México
salas@lsd.uam.mx

ABSTRACT

The motion of an expanding and contracting cylinder rolling on a curved surface is analysed. We establish the existence of autoparametric instabilities for certain values of the cylinder parameters and also found its integrable cases. We next use what we have called the curvature criterion to predict an order-chaos-order transitions whenever the system is chaotic. The Poincaré surface of section method is used to check some of the predictions with good results. The curvature criterion and some of its implications are also briefly discussed.

1. Introduction

An elastic cylinder rolling without slipping on a cylindrical surface may be thought of as a simplistic model for an elastic roller bearing. But, despite its possible uses as a model for something, we show here that it can be a system worth of study by its own sake (Wilms and Cohen 1991, Arizmendi *et al.* 1995). We exhibit that the rolling elastic cylinder is describable using a two-degree-of-freedom Hamiltonian which becomes integrable at two different values of the system parameters. We also show that the system is prone to unstable behaviour at certain values of its parrameters and, after applying what we call the *curvature criterion for the regularity of motion* or CC, we guess that for other parameter values it may exhibit ch ic motions. Furthermore, we predict that whenever the rolling elastic cylinder undergo chaotic oscillations it would also exhibit an order-chaos-order transition on increasing its energy. We argue that this behaviour may be understood as the result of geometric properties of the potential energy surface (PES) of the system. These predictions are checked against numerical computations of the cylinder behaviour. Whenever possible, we use the criterion to predict other features of the Hamiltonian transition to chaos.

2. The rolling elastic cylinder

Let us consider a thin elastic cylinder of radius $r_0 + r$ and mass m, rolling on a cylindrical surface of radius R (Fig. 1). The cylinder is expanding and contracting but never loses its circular profile. We call r_0 the undeformed constant radius and r the *change* in the radius as the oscillation proceeds. The normal v_n and the tangential v_t components of the velocity of point P in the plane of the paper (see Fig. 1) can be easily seen to be

$$\begin{aligned} v_n &= \dot{r} - (r_0 + r)\Omega \sin\phi - \dot{r}\cos\phi, \\ v_t &= \dot{r}\sin\phi + (R - r_0 - r)\dot{\theta} - (r_0 + r)\Omega\cos\phi. \end{aligned} \tag{1}$$

In these equations $\Omega = \dot{\phi}$ is the angular velocity of the cylinder, and the angles θ and ϕ are illustrated in Fig. 1. Furthermore, the rolling without slipping condition gives the relationship $(R - r_0 - r)\dot{\theta} = (r_0 + r)\Omega$ between the variables of the system. With the help of these equations we can get the lagrangian of the system

$$L = m\left(\dot{r}^2 + (r - r_0 - r)^2\dot{\theta}^2\right) - mg\left(R - r_0 - (R - r_0 - r)\cos\theta\right) + \frac{1}{2}kr^2; \tag{2}$$

where g is the acceleration of gravity, m the mass and k the elastic constant of the cylinder. It then follows that the Hamiltonian of the elastic cylinder is

$$H = \frac{1}{4m}\left(p_r^2 + \frac{p_\theta^2}{(a - r)^2} + mga(1 - \cos\theta)\right) + \frac{k}{2}r^2, \tag{3}$$

where we defined $a \equiv R - r_0$.

For the purposes of this work, it is convenient to introduce the dimensionless cartesian coordinates

$$y = (1 - \frac{r}{a}) \cos\theta, \quad \text{and} \quad x = (1 - \frac{r}{a}) \sin\theta, \tag{4}$$

for the description of the system. The origin of these coordinates (marked $\mathcal{O}$ in Fig. 1) is located at the geometrical centre of the outer cylindrical surface. The problem suggest polar coordinates, but cartesian coordinates are best suited for the numerical calculations since these reduce to normal mode coordinates in absence of the geometrical coupling. These coordinates are also important for the application of CC. The transformation to the new variables x and y can be easily extended to a canonical transformation, in this way we get the dimensionless Hamiltonian function

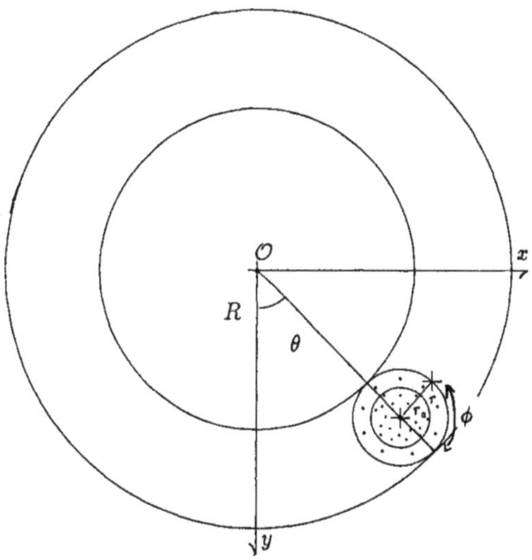

Figure 1. Schematic representation of the system.

$$H = \frac{1}{4}(p_x^2 + p_y^2) + c(1 - y) + \frac{1}{2}\left(\sqrt{x^2 + y^2} - 1\right)^2, \tag{5}$$

where we introduced the nonnegative, nondimensional number $c \equiv mg/ka$ parametrizing the behaviour of the system. This number c is the ratio of the weight of the cylinder to the elastic force caused by a deformation a, or may be taken as the ratio of the frequency of rolling back and forth in a pendulum-like fashion on the outer cylindrical surface to the frequency of elastic oscillations of the cylinder. The introduction of this nondimensional H, is equivalent to saying that we are scaling all energies by ka^2 and the times by $\sqrt{m/k}$.

Using the Hamiltonian (5), the equations of motion of the system can be expressed as the canonical system of equations for the phase space coordinates (q_1, q_2, p_1, p_2) of the cylinder

$$\dot{x} = p_x/2, \quad \dot{p}_x = -x + \frac{x}{\sqrt{x^2+y^2}},$$
$$\dot{y} = p_y/2, \quad \dot{p}_y = c - y + \frac{y}{\sqrt{x^2+y^2}}. \tag{6}$$

Notice that the special case of an elastic cylinder rolling on a flat surface corresponds to taking $c = 0$. The system is integrable in this particular case, since, when $c = 0$, the system becomes equivalent to a central field problem in the plane (Arizmendi *et al.* 1995). Furthermore, the system has two points of equilibrium located at $(0, c+1)$ and $(0, c-1)$, the first point corresponds to stable equilibrium whereas the second one corresponds to a non-stable equilibrium point, notice also the presence of a mild type of singularity at the origin. This singularity is only important when the cylinder fits closely into the external cylindrical surface; the most important consequence of the singularity is that the system becomes again integrable (this is equivalent to taking the $c \to \infty$ limit). This directly follows as the cylinder may only spin about its own axis in such a case. As we see in section 4 when we apply CC to the system, the shape of the potential energy function $V(x, y)$ carries important clues on the behavior of the rolling elastic cylinder. But for now, let us check on the low energy behaviour of it.

3. Parametric resonance

At low enough energies the Hamiltonian (5) can be approximated by

$$H \simeq \frac{1}{4}(p_x^2 + p_y^2) + c(1-y) + \frac{1}{2}\left(\frac{cx^2}{(1+c)} + y^2\right) + \frac{x^2 y}{6(1+c)^2}. \tag{7}$$

From this hamiltonian the low energy equations of motion for the cylinder are

$$\ddot{x} + \left(\frac{c}{2(1+c)} + \frac{y}{6(1+c)^2}\right)x = 0, \tag{8}$$

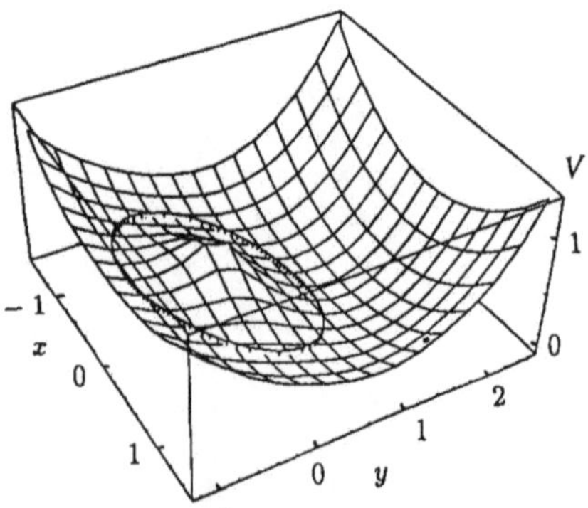

Figure 2. Potential energy surface of the rolling elastic cylinder, the "barrier" marks the zero curvature line.

$$\ddot{y} + \frac{y}{2} = -\frac{x^2}{12(1+c)^2}, \tag{9}$$

equation (8) describes a driven harmonic oscillator whereas equation (9) corresponds to a parametric oscillator. If we now assume that $x \ll y$, then we can approximately solve equation (9) with the simple solution $y(t) = A\cos(\omega_2 t)$. Substituting this solution in (8) we get

$$\frac{d^2x}{d\tau^2} + (b + 16p\cos(2\tau))\,x = 0, \tag{10}$$

where we defined $b \equiv 4c/(1+c)$, $12p \equiv A/(1+c)^2$, and $\tau \equiv t/2^{3/2}$. Notice that, with these identifications, we have $0 \le b \le 4$. Equation (10) has the standard form of a Mathieu equation and, therefore, to study the low energy behaviour of the system we can use the well known properties of it (McLachlan 1947). The stability properties of the Mathieu equation are summarized in the stability diagram of Fig. 3. As is well known, the solutions to Mathieu equation show unstable behavior at $b = n^2, \quad n = 1, 2, 3, \ldots$, but, to evaluate the parameter values at which one should expect the onset of autoparametric oscillations in the rolling elastic cylinder, we have to plot the behavior of it in the same stability chart. This is easily done since we can get for the cylinder

$$p = \frac{A}{192}(4-b)^2. \tag{11}$$

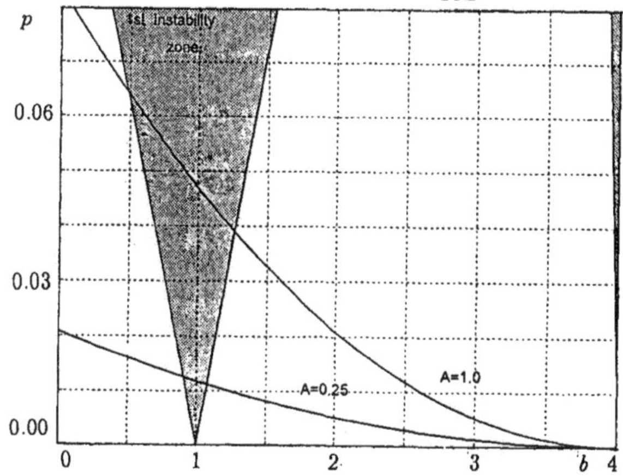

Figure 3. Stability diagram of the Mathieu equation with the behaviour of the cylinder at two A values superimposed.

This curve is plotted for two different A values in figure 3, notice that it always passes trough the point (4,0). On the other hand, since b cannot be greater than four, all instability zones starting at $b = n^2$ for $n \ge 2$ are forbidden to the cylinder, excepting for the point (4,0) — which corresponds to $c = \infty$ and hence to an integrable case— which just touches the apex of the second instability zone.

In this way, we have established that the cylinder may become parametrically unstable when $b = 1$, that is when $c = 1/3$. This is the only parameter value producing the onset of autoparametric oscillations.

4. CC and the cylinder dynamics

A criterion for the appearance of stochastic or chaotic behavior in Hamiltonian systems with two degrees of freedom, like the one under analysis, was suggested by Toda (1974) and later rediscovered by Brumer and Duff (1976); this was a variant of what we have called CC. This criterion was used for addressing problems in several fields (Bolotin *et al.* 1989, Núñez–Yépez 1990, Núñez–Yépez *et al.* 1990, Akhiezer *et al.* 1991, Carretero–González 1992) but it does not appear to be reliable enough (Núñez–Yépez *et al.* 1990) and thus, it is not much used now. We have recently analysed the matter and have found that this criterion is in fact reliable, but for studying the regularity, *not the chaotic*, properties of Hamiltonian systems (Arizmendi *et al.* 1995). CC is easy to describe; to analyse whether a two-dimensional Hamiltonian system shows only regular motions or not, it suffices to compute the Gaussian curvature, K, of the potential energy surface (PES). *If K is always greater than zero then the system cannot be chaotic.* This assertion is justified in the Appendix. The Gaussian curvature of a two-surface $z = V(x, y)$ is defined as ($V_{ij} \equiv \partial^2 V/\partial x_i \partial x_j$)

$$K = \frac{V_{xx}V_{yy} - V_{xy}^2}{\left(1 + V_{xx}^2 + V_{yy}^2\right)^2}, \tag{12}$$

K is the function we have to analyse for deciding whether a system is regular or not. If $K \geq 0$, like in the case of an harmonic oscillator, the system is got to be regular. The system under analysis has the possibility of being chaotic, only if there is a region with negative curvature on its PES. Obviously, there is no guarantee that the system is going to behave chaotically but, if the chaotic behaviour can be established by other means, then, we may use CC to predict other properties and to obtain estimates of energies pertaining to the transition. For example, if the PES of a chaotic system has got a confined and bounded region of negative curvature, at large energies the system would spend most of its time moving outside that region and the motion would be then regular again. In other words, a *chaotic system with a confined and bounded PES always shows an order-chaos-order transition as its energy is increased.* We may even use these ideas to estimate the energy value at which the transition to chaotic motion takes place. To this end we have to compute the set of points for which $K(x, y) = 0$, and determine the minimum energy, E_c, necessary for reaching this curve. E_c would play the role of the critical energy for the onset of chaos since, for any $E > E_c$, the phase trajectory of the system is able to get to the negative curvature region and the motion may thus become chaotic. We may also estimate the energy of maximum chaos (*i.e.* the energy at which the volume of phase space occupied by chaotic orbits has a maximum), E_{max}, as the maximum energy the system attains on the zero curvature line. It should be clear that this is only a rather poor lower bound estimate, as it is based on the assumption that at energies greater than E_{max} the phase-trajectories spend very little time traversing through the $K < 0$ region (Núñez–Yépez 1990, Carretero-González 1992).

To analyse the rolling elastic cylinder from the perspective of the CC we have to pay attention to the potential energy function (7). A plot of the PES of the cylinder is shown in Fig. 3. Notice that $-c^2/2$ is the minimum energy the cylinder may have. To check for the possible existence of chaos in this system we have to evaluate if there is a region of negative curvature in its PES. This is easily done since K is

$$K = \frac{(x^2+y^2)^{\frac{3}{2}} - (x^2+y^2)}{(x^2+y^2)^{3/2}} \tag{13}$$

which take negative values inside a circle of radius 1 centered at the origin (Fig. 2), thus we may expect chaotic behaviour. As Fig. 4 shows the system is indeed chaotic for $c \neq 0$. The existence of chaos already established, we may estimate the critical energy for the onset of chaos as the minimun energy for reaching the zero curvature line: $E_c = 0$. Thus if $c \neq 0$, for energies $E > 0$ there must exist phase space points traversed by chaotic orbits. Furthermore, as the negative curvature region is confined, the criterion predicts that at large enough energies the motions become regular again. The energy of maximum chaos can also

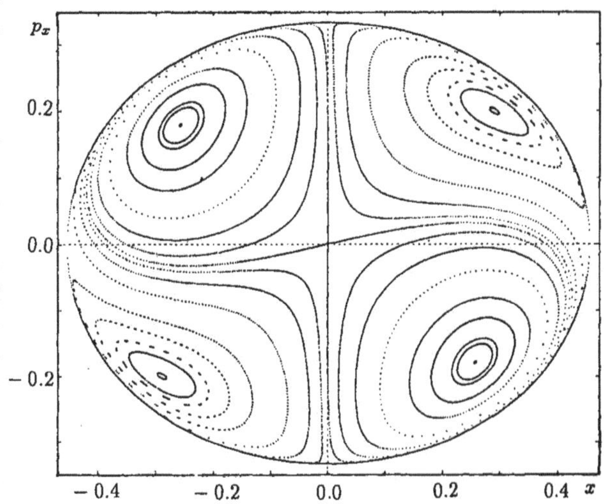

Figure 4a. Poincaré section for $c = 1/3$ at $E = -1/36$. Notice the regularity of the motions.

be easily estimated, it is $E_{max} = c$.

To test these conclusions, we have numerically integrated the canonical equations (6) and plotted the points where the phase orbit pierces the plane x-p_x when $p_y > 0$, $y = 1 + c$. In this way we construct Poincaré sections for several energies at $c = 1/3$. The onset of chaos as the energy is increased is apparent in Fig. 4. Note the existence of only regular orbits at negative energies and the presence of an increasingly noticeable chaotic splatter of points near the origin as the energy of the system grows beyond zero. Notice also the obvious inversion symmetry in the Poincaré sections, which reflects the fact that the cylinder oscillates back and forth in the course of its rolling motion but that it does not have enough energy for going around circling completelly the outer cylindrical surface. Also, as predicted by CC, chaotic motions disappear at energies greater than a certain energy $E_c' > E_{max}$. The breaking of the aforementioned symmetry

Figure 4b. Poincaré section for $c = 1/3$ at $E = 1/64$. As predicted by CC, chaos begin to appear.

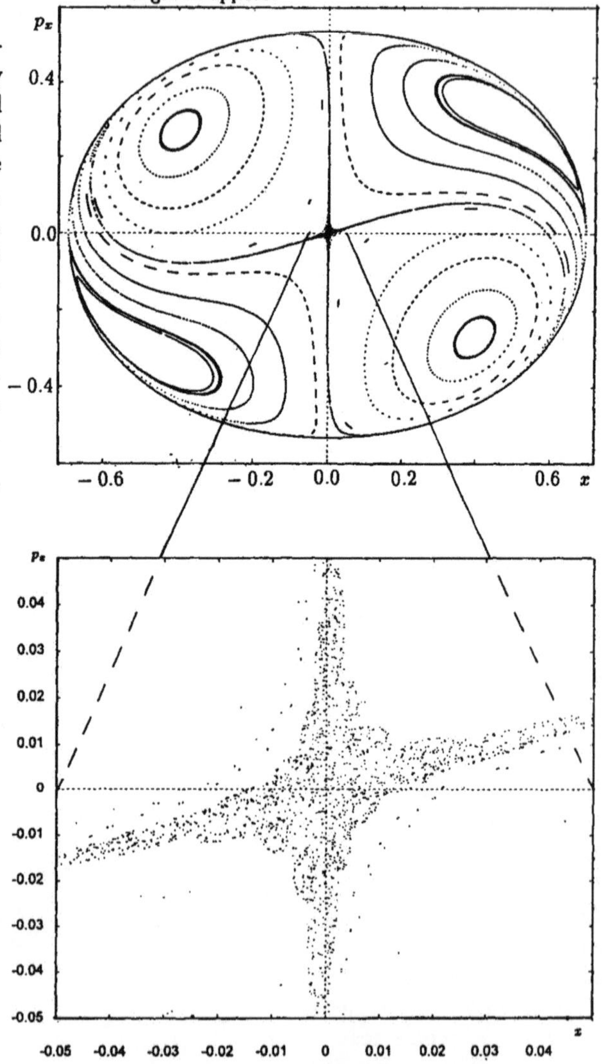

Figure 4c. Poincaré section for $c = 1/3$ at $E = 1/64$. Blow up of the chaotic region in 4b.

is also apparent in the plots at higher energies where the motion becomes regular again. This breakdown is related to the ocurrence of rotational motions in the cylinder which is thus capable of going in a preferred direction, either clockwise or counterclockwise, on the inner surface of the outer cylinder. This makes clear the fact that the chaotic oscillations of the cylinder are a result of the interaction of its radial elastic vibrations with the oscillations produced in the course of its rolling motion. The predictions we get from applying CC are completelly fulfilled by the numerically calculated dynamics of the system, but the quantitative agreement is so to speak fortituous. As you may check, at other c-values, $E_c = 0$ is only a lower bound estimate, the actual transition to chaos occurs at a higher energy. But the important point is, as long as the system is chaotic, an order-chaos-order transition is always found to occur.

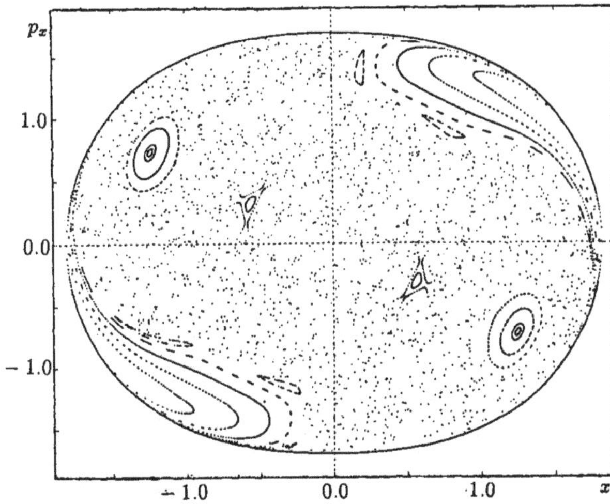

Figure 4d. Poincaré section for $c = 1/3$ at $E = 2/3$.

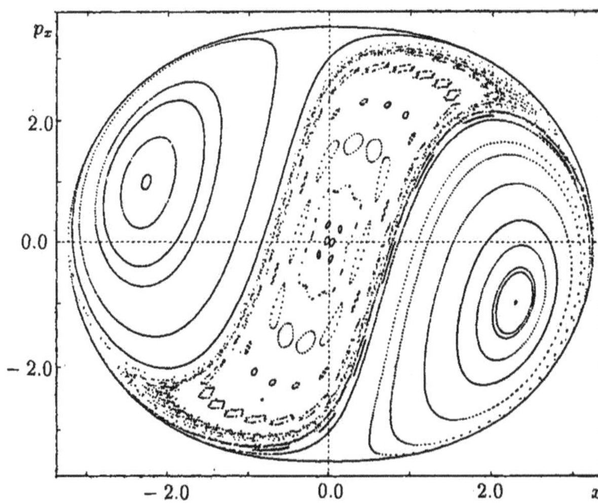

Figure 4e. Poincaré section for $c = 1/3$ at $E = 3$. The motions have become regular again.

5. Appendix. The curvature criterion and the order-chaos-order transition

To understand the mechanism behind what we have called the curvature criterion (Toda 1974, Núñez–Yépez *et al.* 1990, Akhiezer *et al.* 1991), let us consider two initially very close trajectories $\mathbf{z}(t) \equiv (x, y, p_x, p_y)$ and $\mathbf{z}'(t) = \mathbf{z}(t) + \boldsymbol{\xi}(t)$ in the phase-space of a two-dimensional Hamiltonian system. This means that, if we write the phase space distance between the two trajectories $\mathbf{z}$ and $\mathbf{z}'$ as

$$||\boldsymbol{\xi}(t)|| = d_t(\mathbf{z}', \mathbf{z}) \equiv \sqrt{\sum_{a=1}^{2}(x'_a(t) - x_a(t))^2 + (p'_a(t) - p_a(t))^2}, \tag{A1}$$

then, assuming that the system is chaotic, this distance should change as

$$d_t = d_0 \exp(\lambda t), \tag{A2}$$

where λ is a number, essentially a Liapunov exponent (Carretero-González *et al.* 1994), characterising the rate of exponential divergence at $t = 0$. It should be clear now that the vector $\boldsymbol{\xi}$ lives in the tangent space to the phase space at the point $\mathbf{z}(t)$. As we are interested in local properties of the motion, we have to analyse the Jacobi or variational equations for the deviations $\boldsymbol{\xi}$ —which may be obtained in an elegant way from a variational principle (Salas–Brito 1984). The Jacobi equations can be written as the non-autonomous linear system

$$\frac{d}{dt}\boldsymbol{\xi} = \mathcal{M}(t) \cdot \boldsymbol{\xi}, \tag{A3}$$

where the matrix $\mathcal{M}(t)$ is the product of the symplectic matrix (Arnold 1978)

$$\mathcal{J} = \begin{pmatrix} 0 & 0 & 1 & 0 \\ 0 & 0 & 0 & 1 \\ -1 & 0 & 0 & 0 \\ 0 & -1 & 0 & 0 \end{pmatrix} \tag{A4}$$

times the matrix of second derivatives of H (Cf. equation (5)) evaluated along the fiducial trajectory $\mathbf{z}(t)$:

$$\mathcal{M}(t) \equiv \left.\begin{pmatrix} 0 & 0 & 1/2 & 0 \\ 0 & 0 & 0 & 1/2 \\ -H_{xx} & -H_{xy} & 0 & 0 \\ -H_{yx} & -H_{yy} & 0 & 0 \end{pmatrix}\right|_{\mathbf{z}(t)}, \tag{A5}$$

where $H_{ij} \equiv \partial^2 H/\partial x^i \partial x^j$. The matrix $\mathcal{M}(t)$ can be transformed to a diagonal form with the help of a similarity transformation. Such transformation is crucial since the eigenvalues λ of this matrix characterize the nature of the motion. If one of these eigenvalues is real, then the trajectories diverge exponentially. But if λ is imaginary, the trajectories just oscillate one around the other and thus the motion is regular. Generally, however, the eigenvalues of $\mathcal{M}(t)$ changes with time; but we can make the approximation, good for small t and thus compatible with the local caracterization we are trying to obtain, of replacing the time dependent point on the trajectory $\mathbf{z}(t)$ by the phase space *coordinates* $\mathbf{z}$. With this approximation the problem is greatly simplified, the four eigenvalues of the now constant matrix $\mathcal{M}$ are also constant and are given by

$$\begin{aligned}
\lambda_1 &= \frac{1}{2}\left(-1+(1-\frac{4\alpha}{\beta^2})^{1/2}\right)^{1/2}, \\
\lambda_2 &= \frac{1}{2}\left(-1-(1-\frac{4\alpha}{\beta^2})^{1/2}\right)^{1/2}, \\
\lambda_3 &= -\frac{1}{2}\left(-1+(1-\frac{4\alpha}{\beta^2})^{1/2}\right)^{1/2}, \\
\lambda_4 &= -\frac{1}{2}\left(-1-(1-\frac{4\alpha}{\beta^2})^{1/2}\right)^{1/2},
\end{aligned} \qquad (A6)$$

where, for the sake of simplicity, we introduced $\alpha \equiv V_{xx}V_{yy} - V_{xy}^2$ and $\beta \equiv V_{xx} + V_{yy}$ and $V(x, y)$ is the potential energy function appearing in (5). Notice that the eigenvalues appear in pairs with the opposite sign and thus that their sum always vanishes: $\sum_{a=1}^{4} \lambda_a = 0$. These are expected properties in any 2-D Hamiltonian system like the cylinder under analysis (Lichtenberg and Lieberman 1992, Carretero–González *et al.* 1994).

A straightforward analysis of the eigenvalues (A6) show that they are real— that is, the system is chaotic— when $\alpha < 0$. Hence $\alpha < 0$ is a necessary consequence of the chaoticity of the system. The criterion can then be stated in a more easy to remember form by saying that if a two-dimensional Hamiltonian system is chaotic then there necessarily exists a zone of its potential energy surface (PES) where the Gaussian curvature is negative (Toda 1974, Brumer and Duff 1976, Bolotin *et al.* 1989, Núñez–Yépez *et al.* 1990). This easilly follows since, as the Gaussian curvature K of a two-surface $z = V(x, y)$ is defined as in equation (12), the sign of K obviously coincides with that of α and, hence, with the condition for real eigenvalues. There is thus a relationship between chaos producing instabilities and the sign of the Gaussian curvature of the PES of the system, which is what we have called CC (Akhiezer *et al.* 1991). Notice that the criterion only says that if the system is chaotic then its PES should have a negative curvature region. The just stated criterion cannot predict if a given two-dimensional Hamiltonian system is chaotic or not.

However, CC does ascertain that a system with no negative-curvature region in its PES has always to be regular, since this is just the converse of the CC statement. As we have explained in section 4 if the motions of a system are known to be chaotic, CC would then be able to predict if chaos can be made to dissapear on further increasing the energy. That is, CC can be used used to predict whether a chaotic system shall exhibit an order-chaos-order transition or not. For this to happen, it is only necessary that the negative curvature region of the system's PES be confined to a bounded region in configuration space; for, in such a case, at large enough energies the system would spend most of its time moving in a positive curvature region and thus the motion would become regular again (Bolotin *et al.* 1989, Arizmendi *et al.* 1995). This is basically what happens with the rolling elastic cylinder potential-energy function. Thence, as the system is chaotic for almost all c values, we can ascertain the existence of order-chaos-order transitions in all the chaotic cases.

6. Acknowledgements

This work was partially supported by CONACyT grant 4846-E9406. ALSB wants to thank Fundación Ricardo J. Zevada for partial support. The authors want to acknowledge N. Kuro, G. Abdul, K. Quiti, M. X'sac and Ch. Suri for useful comments on a first draft of the compuscript. The helpful advice of M. Ek and R. Tiki was invaluable in correcting the compuscript. This paper is dedicated to the memory of F. D. Micha who passed away on January 6, 1994. This paper is also dedicated to the EZLN (the Revolutionary Zapatista Army).

7. References

A. I. Akhiezer, V. I. Truten', N. F. Shul'ga 1991 *Phys. Reports*, **203** 289-343.

M. C. Arizmendi, R. Carretero-González, H. N. Núñez–Yépez, A. L. Salas–Brito 1995 to be published.

V. I. Arnold 1978 *Mathematical Methods of Classical Mechanics*, Springer Verlag, New York, USA.

Yu. L. Bolotin, V. Yu. Gonchar, V. N. Tarasov, N. A. Chekanov 1989 *Phys. Lett.* **135A** 29-31.

P. Brumer, W. Duff 1976 *J. Chem. Phys.* **65** 1665-1675.
R. Carretero–González 1992 Senior thesis, FC-UNAM, México D.F.
R. Carretero–González, H. N. Núñez–Yépez, A. L. Salas–Brito 1994 *Eur. J. Phys.* **15** 139-148.
A. J. Lichtenberg and M. A. Lieberman 1992 *Regular and chaotic motion*, Springer Verlag, New York, USA.
N. W. McLachlan 1947 *Theory and application of Mathieu functions*, Oxford University Press, Oxford U.K.
H. N. Núñez–Yépez 1990 M. Sc. thesis, FC-UNAM, México D. F.
H. N. Núñez–Yépez, A. L. Salas–Brito, C. A. Vargas, L. Vicente 1990 *Phys. Lett.* **145A** 101-105.
A. L. Salas-Brito 1984 *Am. J. Phys.* **52** 1012-1016.
M. Toda 1974 *Phys. Lett.* **48A** 335-336.
E. V. Wilms, H. Cohen 1990 ASME Journal of Applied Mechanics **57** 793-794.

HAMILTONIAN SYSTEM GOVERNED BY A FINITE-DIFFERENCE EQUATION: THE RELATIVISTIC OSCILLATOR IN AN EXTERNAL FIELD

NATIG M. ATAKISHIYEV*
IIMAS–Cuernavaca, UNAM
Apartado Postal 139–B, 62191 Cuernavaca, Morelos, México
E-mail: natig@ce.ifisicam.unam.mx

and

SHAKIR M. NAGIYEV
Institute of Physics, Azerbaijan Academy of Sciences, Baku 370143, Azerbaijan

and

KURT BERNARDO WOLF
IIMAS–Cuernavaca, UNAM
Apartado Postal 139–B, 62191 Cuernavaca, Morelos, México
E-mail: bwolf@ce.ifisicam.unam.mx

ABSTRACT

We consider a quantum Hamiltonian model of the relativistic harmonic oscillator in a homogeneous external field that is described by a finite-difference equation. This system exhibits both discrete and continuous energy spectra; its eigenfunctions are built explicity in configuration and momentum representations by Lie algebraic techniques.

1. Introduction

The quasipotential approach [1], [3] allows the development of a relativistic quantum mechanical description [4], [6], based on the relativistic three- dimensional configuration $\vec{r}$–space [4]. Although this approach is closely analogous to nonrelativistic quantum mechanics *à la Schrödinger*, its essential characteristic is that the relative motion wavefunction satisfies a differential-difference equation with step equal to the Compton wavelength of the particle, $\lambda = \hbar/mc$. For two scalar particles with equal masses m, local quasipotentials $V(\vec{r})$ lead to the equation [5]

$$[H_0(\vec{r}) + V(\vec{r})]\,\psi(\vec{r}) = E\,\psi(\vec{r})\,, \tag{1.1}$$

*Permanent Address: Institute of Physics, Academy of Sciences of Azerbaijan, Baku 370143, Azerbaijan. Visiting Scientist at IIMAS-UNAM/Cuernavaca with Cátedra Patrimonial CONACYT, México.

with the *free Hamiltonian*

$$H_0(\vec{r}) = mc^2 \left[\cos h \frac{i\hbar}{mc}\,\partial_r + \frac{i\hbar}{mcr} \sin h \frac{i\hbar}{mc}\partial_r + \frac{\vec{L}^2}{2(mcr)^2} \exp \frac{i\hbar}{mc}\partial_r \right] , \qquad (1.2)$$

where $\vec{L} = -i\hbar(\vec{r} \times \vec{\nabla})$ is the orbital angular momentum operator, $\partial_r = \partial/\partial r$, and we have the finite-difference operator action $\exp(a\partial_r)\, f(r) = f(r+a)$.

In framework of the quasipotential description, the transformation between configuration and its canonically conjugate momentum space is given by the relativistic plane wave [7]

$$\xi(\vec{p},\vec{r}) = \left(\frac{p_0 - \vec{p}\cdot\vec{n}}{mc} \right)^{-1-imcr/\hbar} ,$$

$$\vec{r} = r\vec{n}, \quad \vec{n}^2 = 1, \qquad 0 < r < \infty , \qquad (1.3)$$

rather than by the Fourier kernel $\exp(i\vec{p}\cdot\vec{r}/\hbar)$ in the nonrelativistic case. The momenta of positive mass particles belong to the upper sheet of the mass hyperboloid $p_0^2 - \vec{p}^{\,2} = m^2c^2$, and form a three-dimensional Lobachevsky space whose group of motion is the Lorentz group. The functions Eq. (1.3) are eigenfunctions of the Hamiltonian $H_0(\vec{r})$ in Eq. (1.2), and in the nonrelativistic limit (*i.e.*, when $r \gg \hbar/mc$ and $|\vec{p}| \ll mc$) they coincide with the Euclidean plane waves $\exp(i\vec{p}\cdot\vec{r}/\hbar)$.

Within this version of relativistic quantum mechanics, various exactly solvable phenomenological quasipotentials were considered in Refs. [4], [6], [8], [14]. In particular, a relativistic model of the linear oscillator was studied in detail in Refs. [8], [10]. This model was considered for the the case of a constant external force $F(x) = -g$, *i.e.*, in the presence of a homogeneous external field $V_g(x) = gx$ in Ref. [12]. The wavefunctions were found both in the relativistic configuration x and momentum p realizations; it was shown that, in contrast with the nonrelativistic case, here one has both a discrete energy spectrum and a continuous one, depending on the value of the force $|g|$ relative to $mc\omega$, where ω is the oscillator frequency (see below). The generalized coherent states for this model were built out of the bound states for this model in [13].

There are also algebraic approaches to the description of quantum systems by the theory of dynamical symmetries [15,16], potential groups [17], Casimir operators with mixed spectrum [18], and the 'Euclidean connection' [19]. In these approaches, the key role is played by the dynamical algebra whose generators connect states of different representations of the symmetry subalgebra. In particular, it has been used to find the energy spectrum, the wavefunctions, and transition probabilities. Among the generators of the dynamical algebra is the Hamiltonian, the symmetry commutant algebra of the latter, and raising and lowering operators which do not commute with it. As a consequence, the eigenfunctions of the Hamiltonian form a basis of an irreducible representation of the dynamical algebra.

In Section 2 we apply the algebraic approach to solve the finite-difference equations for the relativistic linear oscillator model in a homogeneous external field. The explicit form of the wavefunctions for the discrete and continuous spectra (both in the configuration and momentum realizations) are discussed in Sections 3 and 4, respectively.

2. The relativistic oscillator in a homogeneous external field

The wavefunction in one-dimensional configuration space x for the relativistic model of a linear oscillator in a homogeneous external field is defined as [9]

$$\psi(x) = \frac{1}{\sqrt{2\pi\hbar}} \int_{-\infty}^{\infty} d\Omega_p \, \xi(p, x) \, \psi(p) \,, \tag{2.1}$$

where $d\Omega_p = mc\,dp/p_0$ is the invariant volume element in the momentum realization, $p_0^2 = m^2c^2 + p^2$, and the plane wave $\xi(p, x)$ has the form [*cf.* the kernel (1.3)]

$$\xi(p, x) = \left(\frac{p_0 - p}{mc} \right)^{-imcx/\hbar} . \tag{2.2}$$

The free Hamiltonian $H_0(x)$ and momentum P_x operators are the finite difference operators

$$H_0(x) = mc^2 \cos h \frac{i\hbar}{mc} \partial_x, \tag{2.3a}$$

$$P_x = -mc \sin h \frac{i\hbar}{mc} \partial_x, \tag{2.3b}$$

$$H_0(x)^2 - c^2 P_x^2 = m^2 c^4, \tag{2.3c}$$

which commute with each other and have a mutual eigenfunction Eq. (2.2) with eigenvalues p_0 and p, respectively. They satisfy the following commutation relations with the relativistic coördinate x:

$$[x, P_x] = \frac{i\hbar}{mc^2} H_0(x) \,, \tag{2.4a}$$

$$[x, H_0(x)] = \frac{i\hbar}{m} P_x \,. \tag{2.4b}$$

The relativistic model of the one-dimensional harmonic oscillator studied in detail in Refs. [8,9] has the Hamiltonian

$$\mathcal{H}^{osc}(x) = H_0(x) + V(x)$$

$$= H_0(x) + \frac{m\omega^2}{2} x \left(x + \frac{i\hbar}{mc} \right) \exp \frac{i\hbar}{mc} \partial_x. \tag{2.5}$$

The "prolonged derivative" or *generalized momentum operator* [*cf.* Eq. (2.4*b*)] is

$$\mathcal{P}_x = \frac{im}{\hbar} [\mathcal{H}^{osc}(x), x] = P_x - \frac{1}{c} V(x), \tag{2.6}$$

and satisfies the following commutation relations with x and $\mathcal{H}^{osc}(x)$:

$$[x, \mathcal{P}_x] = \frac{i\hbar}{mc^2} \mathcal{H}^{osc}(x), \tag{2.7a}$$

$$[\mathcal{H}^{osc}(x), \mathcal{P}_x] = im\hbar\omega^2 x. \tag{2.7b}$$

Besides, for this particular potential Eq. (2.5) we have [*cf.* Eq. (2.3*c*)]

$$(\mathcal{H}^{osc}(x))^2 - c^2 \mathcal{P}_x^2 = m^2 c^4 \left(1 + \frac{\omega^2}{c^2} x^2 \right). \tag{2.8}$$

From Eq. (2.6) and Eq. (2.7) it follows that the triplet x, $\mathcal{P}_x$, and $\mathcal{H}^{osc}(x)$ closes into the Lie algebra of $Sp(2, \Re)$. Indeed, the operators

$$K_0 = \frac{1}{\hbar\omega} \mathcal{H}^{osc}(x), \quad K_1 = \frac{mc}{\hbar} x, \quad K_2 = -\frac{c}{\hbar\omega} \mathcal{P}_x, \tag{2.9a}$$

obey the standard commutation relations for the $Sp(2, \Re)$ generators,

$$[K_0, K_1] = iK_2, \quad [K_2, K_0] = iK_1, \quad [K_1, K_2] = -iK_0, \tag{2.9b}$$

and provide the Casimir invariant

$$K^2 = K_0^2 - K_1^2 - K_2^2 = s(s+1)\, 1. \tag{2.10a}$$

From Eq. (2.8) we see that K^2 is $(mc^2/\hbar\omega)^2\, 1$ and thus $s(s+1) = \nu(\nu - 1)$, with

$$\nu = \frac{1}{2} + \sqrt{\frac{1}{4} + \left(\frac{mc^2}{\hbar\omega} \right)^2} \geq 1. \tag{2.10b}$$

The eigenvalue $s = -\nu$ is not necessarily an integer and corresponds to the unitary irreducible representation of the *covering group* $\widetilde{\mathrm{Sp}}(2, \Re)$, denoted [20] $D^+(-\nu)$. The spectrum of the compact operator K_0 is bounded from below and has the values $-s + n = \nu + n$, $n = 0, 1, 2, \ldots$. Consequently, the energy levels of the model Hamiltonian Eq. (2.5) are

$$E_n = \hbar\omega(n + \nu), \qquad n = 0, 1, 2, \ldots. \tag{2.11}$$

Thus the introduction of the generalized momentum operator $\mathcal{P}_x$ leads directly to the dynamical algebra $\widetilde{\mathrm{Sp}}(2,\Re)$ for this model. We should underline that in Eq. (2.6)–(2.8) we provide companion operators to the Hamiltonian that close into an algebra, and will raise and lower the latter's eigenvalues Eq. (2.11). This is distinct from the more common (and more complicated) procedure of factorizing the Hamiltonian under consideration (*cf.* Ref. [9].)

The same line of reasoning is also valid for the relativistic model of the linear harmonic oscillator Eq. (2.5) in a homogeneous external field, governed by the difference equation

$$\mathcal{H}^g(x)\,\psi^g(x) = [\mathcal{H}^{osc}(x) + gx]\,\psi^g(x) = E\psi^g(x). \tag{2.12a}$$

Since $\mathcal{H}^{osc} = \hbar\omega K_0$ and $x = (\hbar/mc)K_1$, the equation Eq. (2.12*a*) can be written in terms of the $Sp(2,\Re)$ generators of the model Eq. (2.5) in the absence of external force, as

$$\left(\hbar\omega K_0 + \frac{g\hbar}{mc}K_1\right)\psi^g(x) = E\psi^g(x). \tag{2.12b}$$

From the commutation relations Eq. (2.9) for the $Sp(2,\Re)$ generators, it follows that

$$\exp^{-i\theta K_2} K_0 \exp^{i\theta K_2} = K_0\cosh\theta + K_1\sinh\theta, \tag{2.13a}$$

$$\exp^{-i\theta K_2} K_1 \exp^{i\theta K_2} = K_0\sinh\theta + K_1\cosh\theta. \tag{2.13b}$$

Therefore, by means of the unitary transformation

$$\psi^g(x) \mapsto \widetilde{\psi}^g(x) = \exp^{-i\theta K_2}\,\psi^g(x),$$

the eigenvalue equation Eq. (2.12*b*) can be written as

$$(aK_0 + bK_1)\widetilde{\psi}^g(x) = E\widetilde{\psi}^g(x), \tag{2.14}$$

where $a = \hbar\omega\cosh\theta + (g\hbar/mc)\sinh\theta$ and $b = \hbar\omega\sinh\theta + (g\hbar/mc)\cosh\theta$. Consequently, by choosing a suitable angle θ, one of the coefficients a or b can be annulled, according to whether $|g| < mc\omega$ or $|g| \geq mc\omega$, respectively. Then, the left-hand side of Eq. (2.14) will contain only one term, either in K_0 or in K_1 up to a constant factor [15]. Let us consider these two cases separately.

3. The states of the discrete spectrum

In the case $|g| < mc\omega$ we can choose $\tanh\theta_g = -g/mc\omega$, leading to $b = 0$ and $a = (\hbar/mc)\sqrt{(mc\omega)^2 - g^2} = \hbar\omega\sin\phi_g$, where the angle ϕ_g is defined through

$g = m c \omega \cos\phi_g$, $0 < \phi_g < \pi$. Equation (2.14) is then reduced to the eigenvalue equation for the compact generator,

$$\hbar\omega \sin\phi_g \, K_0 \widetilde{\psi}^g(x) = E \widetilde{\psi}^g(x), \tag{3.1}$$

having discrete spectrum $n + \nu$. Therefore, the energy spectrum for the bound states of the model Eq. (2.12) is [*cf.* Eq. (2.11)]

$$E_n = \hbar\omega(n+\nu)\sin\phi_g, \qquad n = 0, 1, 2 \ldots. \tag{3.2}$$

where ν is related to the Casimir operator eigenvalue $\nu(\nu - 1)$ in Eq. (2.10). Since Eq. (3.1) coincides with the equation for the relativistic linear oscillator in the absence of external force Eq. (2.5), the unitary transformation in the case when $|g| < m c \omega$

$$\psi^g(x) = S_g \widetilde{\psi}^g(x), \qquad S_g = \exp(i\theta K_2) \tag{3.3}$$

relates bound states for arbitrary g with bound states [13] for $g = 0$, *i.e.*,

$$\psi_n^g(x) = S_g \widetilde{\psi}_n^g(x) = S_g \psi_n^0(x). \tag{3.4}$$

We note that in the nonrelativistic limit, the operator S_g coincides with the unitary shift operator $\exp(x_0\, d/dx)$, $x_0 = g/m\omega^2$, and the relation (3.4) reflects the fact that in quantum mechanics the presence of the external field $V_g(x) = gx$ does not change the form of the harmonic oscillator potential, but only shifts its equilibrium position [21], namely $\psi_n^g(x) = \exp(x_0\, d/dx)\psi_n^0(x) = \psi_n^0(x + x_0)$.

To find an explicit form [12] for the wavefunctions ψ_n^g one can start with the momentum realization, in which the Hamiltonian Eq. (2.12*a*) is a second-order differential operator, and to define them in the x-realization afterwards by using the transformation Eq. (2.1). It is also possible to look directly for solutions of the difference equation Eq. (2.12) via expansion in generalized powers by using the method of undetermined coefficients (see, *e.g.*, [22]). In the next Section, however, it will be shown that this particular difference equation is simply solved with the aid of Gauss' recursion relation for the hypergeometric functions [23] $F(\alpha, \beta; \gamma; z)$, *i.e.*,

$$\alpha(1-z)\, F(\alpha+1, \beta; \gamma; z) + (\alpha - \gamma)\, F(\alpha - 1, \beta; \gamma; z)$$
$$+ [\alpha(2-z) + \gamma - \beta z]\, F(\alpha, \beta; \gamma; z) = 0. \tag{3.5}$$

The reason for leaving this derivation to the next Section is that the continuous case turns out to be more *general*, *i.e.*, it will contain also the discrete case.

The orthonormalized eigenfunctions of Eq. (2.12a) in the momentum realization are given [12] in terms of the generalized Laguerre polynomials [23] by

$$\psi_n^g(p) = c_n^g \eta^\nu \exp(i\eta e^{i\phi_g})\, L_n^{2\nu-1}(2\eta \sin\phi_g)\,,$$

$$\eta = \frac{c}{\hbar\omega}(p_0 + p), \qquad c_n^g = \sqrt{\frac{n!}{mc\Gamma(2\nu+n)}}\; i^n (2\sin\phi_g)^\nu. \tag{3.6}$$

The orthonormalized wavefunctions in the x-realization Eq. (3.4), related with Eq. (3.6) by the transformation Eq. (2.1), are expressed through the Meixner-Pollaczek polynomials $P_n^\nu(x;\varphi)$ and their weight function [24] as

$$\psi_n^g(x) = \tilde{c}_n^g P_n^\nu(\tilde{x};\phi_g)\Gamma(\nu + i\tilde{x})\left(\frac{\hbar\omega}{mc^2} e^{i(\pi/2-\phi_g)}\right)^{i\tilde{x}},$$

$$\tilde{x} = \frac{mc}{\hbar}x, \qquad \tilde{c}_n^g = \frac{mc\, i^\nu e^{-i(n+\nu)\phi_g}\, c_n^g}{\sqrt{2\pi\hbar}}. \tag{3.7}$$

4. The states of the continuous spectrum

In the case when $|g| \geq mc\omega$, we can choose $\tanh\theta_g = -mc\omega/g$; then $a = 0$ and $b = (\hbar/mc)\sqrt{g^2 - (mc\omega)^2}$, reducing Eq. (2.14) to the equation

$$\frac{\hbar}{mc}\sqrt{g^2 - (mc\omega)^2}\, K_1 \widetilde{\psi}^g(x) = E\widetilde{\psi}^g(x). \tag{4.1}$$

Since the noncompact $Sp(2,\Re)$ generator K_1 has the continuous real spectrum $\kappa \in \Re$, from Eq. (4.1) it follows that the corresponding wavefunction $\widetilde{\psi}^g(x)$ belongs to the continuous energy spectrum

$$E = (\kappa\hbar/mc)\sqrt{g^2 - (mc\omega)^2} = \kappa\hbar\omega\sinh\varphi_g, \tag{4.2}$$

where $|g| = mc\omega\cosh\varphi_g$ and $0 \leq \varphi_g < \infty$. Thus the eigenfunctions of the operator Eq. (2.12a) in this case satisfy the difference equation

$$\left([\tilde{x}(\tilde{x}+i) + \nu(\nu-1)]\, e^{i\partial_{\tilde{x}}} + \nu(\nu-1)\, e^{-i\partial_{\tilde{x}}} \pm 2\sqrt{\nu(\nu-1)}\,\tilde{x}\cosh\varphi_g\right)\psi^g(x)$$

$$= 2\kappa\sqrt{\nu(\nu-1)}\;\sinh\varphi_g\;\psi^g(x), \tag{4.3}$$

in which $\tilde{x} = mcx/\hbar$ is the dimensionless variable in configuration x-space, and the $\pm$ signs correspond to the positive and negative values of g, respectively. To symmetrize the coefficients of the difference operators $\exp(\pm i\partial_{\tilde{x}})$ in the equation Eq. (4.3) let us make the substitution

$$\psi^g(x) = [\nu(\nu-1)]^{-i\tilde{x}/2}\Gamma(\nu + i\tilde{x})\, f(x;g), \tag{4.4}$$

that converts Eq. (4.3) into

$$[(\nu - i\tilde{x})\, e^{i\partial_{\tilde{x}}} + (\nu + i\tilde{x})\, e^{-i\partial_{\tilde{x}}} \pm 2\tilde{x}\cosh\varphi_g]\, f(x;g) = 2\kappa \sinh\varphi_g\, f(x;g). \tag{4.5}$$

We note firstly that in the case of the discrete spectrum, this substitution separates out the factor that determines the asymptotic behavior of $\psi_n^g(x)$. Secondly, since the energy E does not depend on the sign of g [see Eq. (4.2)], from Eq. (4.5) it follows that $f(x,-g) = f(-x;g)$. Hence, the negative sign in Eq. (4.5) may be supressed without affecting the generality of its solution.

Now it remains only to compare Eq. (4.5) with Eq. (3.5) in order to find that

$$f(x;g) = e^{-\tilde{x}(\pi/2 + i\varphi_g)}\Gamma(\nu + i\tilde{x})\, F(\nu + i\tilde{x}, \nu - i\kappa; 2\nu; 1 - e^{-2\varphi_g}). \tag{4.6}$$

On combining the formulas Eq. (4.4) and Eq. (4.6), we get the explicit form of the wavefunctions for the continuous spectrum as

$$\psi^g(x) = \left(\frac{\hbar\omega}{mc^2} e^{i\pi/2 - \varphi_g}\right)^{i\tilde{x}} \Gamma(\nu + i\tilde{x})\, F(\nu + i\tilde{x}, \nu - i\kappa; 2\nu; 1 - e^{-2\varphi_g}). \tag{4.7}$$

Knowing this, it is easy to define the wavefunctions in the momentum realization by using the transformation inverse to Eq. (2.1), which can be written in terms of the dimensionless light-front variable η in Eq. (3.6), as

$$\psi^g(p) = \frac{1}{\sqrt{2\pi\hbar}} \int_{-\infty}^{\infty} dx \left(\frac{\hbar\omega}{mc^2}\,\eta\right)^{-i\tilde{x}} \psi^g(x). \tag{4.8}$$

Indeed, substituting Eq. (4.7) into Eq. (4.8) and integrating the hypergeometric series termwise with the aid of

$$\frac{1}{2\pi}\int_{-\infty}^{\infty} dt\, \Gamma(\mu + it)\, z^{-it} = z^{\mu} e^{-z}, \tag{4.9}$$

yields the explicit form of the wavefunctions in the momentum realization as

$$\psi^g(p) = \frac{\sqrt{2\pi\hbar}}{mc}(-i\eta e^{\varphi_g})^{\nu} \exp(i\eta e^{\varphi_g})\; {}_1F_1(\nu - i\kappa; 2\nu; -2i\eta\sinh\varphi_g), \tag{4.10}$$

where ${}_1F_1(\alpha;\beta;z)$ is the confluent hypergeometric function.

Before closing this section we emphasize that the convenience of using the parametrizations $g = \pm mc\omega \cosh \varphi_g$ and $g = mc\omega \cos \phi_g$ for the continuous and discrete cases of the previous and this Section, respectively, is that they can be transformed into each other by the formal relation $\varphi_g = i\phi_g$. Under this substitution, the formulas Eq. (4.7) and Eq. (4.10) yield, up to a common normalization constant, the explicit form of the wavefunctions for the discrete spectrum Eq. (3.7) and Eq. (3.6) respectively, after the corresponding replacement $\kappa = -i(n+\nu)$, $n = 0, 1, 2, \ldots$. To verify this one needs the explicit formulas for the generalized Laguerre and Meixner-Pollaczek polynomials [24]

$$L_n^{\mu-1}(z) = \frac{(\mu)_n}{n!}\ {}_1F_1(-n;\mu;z), \tag{4.11a}$$

$$P_n^{\lambda}(x;\phi) = \frac{(2\lambda)_n}{n!} e^{in\phi} F(-n, \lambda + ix; 2\lambda; 1 - e^{-2i\phi}). \tag{4.11b}$$

In the special case when $|g| = mc\omega$, the initial equation (2.12b) has the form $\hbar\omega(K_0 \pm K_1)\psi^g(x) = E\psi^g(x)$, and the energy spectrum is again continuous and positive, *i.e.*, $E > 0$ (*cf.* [25], [26]).

5. References

1. A.A.Logunov and A.N.Tavkhelidze, *Nuovo Cimento* **29** No.2 (1963) 308.
2. V.G.Kadyshevsky, *Nucl. Phys. B* **6** No.2 (1968) 125.
3. A.Klein and T.-S.H.Lee, *Phys. Rev. D* **10** No.12 (1974) 4308.
4. V.G.Kadyshevsky, R.M.Mir-Kasimov, and N.B.Skachkov, *Nuovo Cimento A* **55** No.2 (1968) 233.
5. M.Freeman, M.D.Mateev, and R.M.Mir-Kasimov, *Nucl. Phys. B* **12** No.1 (1969) 137.
6. V.G.Kadyshevsky, M.D.Mateev, and R.M.Mir-Kasimov, *Yadernaya Fizika* **11** No.3 (1969) 692.
7. I.S.Shapiro, *Sov. Phys. Doklady* **1** No.1 (1956) 91.
8. A.D.Donkov, V.G.Kadyshevsky, M.D.Mateev and R.M. Mir-Kasimov, *Teor. i Mat. Fizika* **8** No.1 (1971) 61.
9. N.M.Atakishiyev, R.M.Mir-Kasimov, and Sh.M.Nagiyev, 'Quasipotential models of the relativistic oscillator', JINR preprint E2–12367, Dubna, 1979 (32 pages); *Teor. i Mat. Fizika* **44** No.1 (1980) 47.
10. N.M.Atakishiyev, *Teor. i Mat. Fizika* **58** No.2 (1984) 254.
11. N.M.Atakishiyev, R.M.Mir-Kasimov, and Sh.M.Nagiyev *Ann. Phys.* (Leipzig) **42** No.1 (1985) 25.

12. R.M.Mir-Kasimov, Sh.M.Nagiyev and E.D.Kagramanov, A relativistic linear oscillator under constant external force and a bilinear generating function for the Pollaczek polynomials, Physics Institute preprint No. 214 Baku (1987) (12 pages).
13. N.M.Atakishiyev and K.B.Wolf, *Rep. Math. Phys.* **27** No.3 (1989) 305.
14. N.M.Atakishiyev and R.M.Mir-Kasimov, *J. Math. Phys.* **30** No.7 (1989) 1631.
15. A.O.Barut, *Dynamical Groups and Generalized Symmetries in Quantum Theory* (University of Canterbury, Christchurch, New Zealand, 1972).
16. I.A.Malkin and V.I.Man'ko, *The Dynamical Symmetries and Coherent States of Quantum Systems* (Nauka, Moscow, 1979).
17. Y.Alhassid, F.Gürsey, and F.Iachello, *Ann.Phys. (N.Y.)* **148** (1983) 346; *ib.* **167** (1986) 181.
18. A.Frank and K.B.Wolf, *Phys. Rev. Lett.* **52** No.20 (1984) 1737; *ibid.*, *J. Math. Phys.* **26** No.5 (1985) 973.
19. A.Frank, Y.Alhassid, and F.Iachello, *Phys. Rev. A* **34** No.1 (1986) 677.
20. V.Bargmann, *Ann. Math.* **48** No.3 (1947) 568.
21. L.Landau and E.Lifshitz, *Quantum Mechanics* (Addison-Wesley, Reading, MA, 1968).
22. N.M.Atakishiyev and S.K.Suslov, in: *Progress in Approximation Theory: an International Perspective*, A.A.Gonchar and E.B.Saff Eds., Springer Series in Computational Mathematics, Vol. **19** (Springer Verlag, New York, 1992), p. 1.
23. I.S.Gradshteyn and I.M.Ryzhik, *Table of Integrals, Series, and Products*, Fourth Edition (Academic Press, New York and London, 1965).
24. A.Erdélyi, W.Magnus, F.Oberhettinger, and F.G.Tricomi, *Higher Transcendental Functions*, Vol.2 (McGraw-Hill, New York, 1953).
25. G.Lindblad and B.Nagel, *Ann. Inst. Henri Poincaré, A* **13** No.1 (1970) 27.
26. D.Basu and K.B.Wolf, *J. Math. Phys.* **23** No.2 (1982) 189.
27. G.Szegő, *Orthogonal Polynomials*, Fourth Edition (American Mathematical Society, Providence, R.I., 1975).

CAPTURE ORBITS IN A NEWTONIAN N-BODY PROBLEM

Solon F. Blundell
College of General Studies, Boston University, 871 Commonwealth Avenue
Boston, MA 02215, USA
E-mail: fred@math.bu.edu

ABSTRACT

We exhibit capture phenomena in a conservative system governed by Newtonian gravitation, demonstrating that explanations for the aggregation of material in dense bounded structures need not rely on such dissipative phenomena as radiant cooling. We study the motion of two thin concentric spherical shells of small mass m centered about a central pointlike object of unit mass. The shells do not rotate, but expand or contract radially under the influence of their gravitational interactions with themselves, each other, and the central object. In the case of negative total energy, for each large integer n, there exists an orbit in which the spherical shells cross one another n times before one of them escapes to arbitrarily large radius as the other is captured. In the case of zero total energy, an open set of initial conditions lead to the escape of one sphere to infinite radius while the other sphere bounces rapidly within a small bounded radius. The analysis of the problem of two concentric spheres is made possible using a coordinate transformation developed by R. McGehee in connection with the collinear three body problem.

1. Introduction

The purpose of this paper is to state the results of a mathematical investigation of a problem closely related to the collinear three body problem, and to explain how they suggest a mechanism for capture phenomena in celestial mechanics. Most of the results are stated without proof, and the reader is refered to the dissertation[1] for details.

1.1. Statement of The Problem

We consider the motion of two irrotational concentric spherical shells of mass m centered about a fixed core of mass one. We assume that the only significant forces acting on the shells are their gravitational attraction to the central core, and their gravitational attraction to each other and themselves.

The simplest coordinates for the phase space are the radii and radial momenta of the spheres, which we will denote (r_1, r_2, p_1, p_2). We will refer to these as the Euclidean coordinates.

According to Newton's law of gravitation, each point on each sphere feels a net force towards the origin.

Points on the interior sphere feel no net attraction from the outer sphere; each point on the interior sphere feels only the attraction of the fixed core, and the attraction of the other points on the interior sphere. The net effect of the attraction of the other points on the interior sphere is a force directed towards the origin, equivalent in magnitude to the force that would be produced if there were a point mass having half the mass of the interior sphere located at the origin in addition to the fixed core.

Points on the outer sphere experience a net attraction from the inner sphere, as if the inner sphere were concentrated at the origin. In addition, each point on the outer sphere experiences an attraction to the other constituent points of the outer sphere, as if there were located at the origin a point mass equivalent to half the mass of the outer sphere. Finally, the outer sphere feels the attraction of the fixed core.

Thus, when $|r_2| > |r_1|$ the gravitational potential energy of the first sphere is given by

$$u1 = -\left(1+\frac{m}{2}\right)\left(\frac{m}{|r_1|}\right); \tag{1}$$

the gravitational potential energy of the second sphere is given by

$$u2 = -\left(1+\frac{3m}{2}\right)\left(\frac{m}{|r_2|}\right), \tag{2}$$

and when $|r_1| > |r_2|$, the situation is reversed so that

$$\begin{aligned} u1 &= -\left(1+\frac{3m}{2}\right)\left(\frac{m}{|r_1|}\right); \\ u2 &= -\left(1+\frac{m}{2}\right)\left(\frac{m}{|r_2|}\right). \end{aligned} \tag{3}$$

Note that the potentials are discontinuous along the submanifold $|r_2| = |r_1|$ where the spheres pass through each other and singular along the submanifolds $r_1 = 0$, and $r_2 = 0$ where the spheres collapse.

The Hamiltonian system described by these potentials is completely integrable in regions away from the singularities and discontinuities, however it is the passage of trajectories through these integrability barriers that produces interesting behavior. In particular, the singularity at $r_1 = 0 = r_2$ which we shall call the *double collapse* plays a dominant role in the global dynamics of this system.

The interesting behavior in this example can be understood intuitively as follows. Each time the shells pass through each other, so that the formerly outermost shell becomes the innermost, an exchange of potential energy occurs, while the individual kinetic energies are conserved.

Motion in the integrable regions of phase space can be analyzed by the Hamilton-Jacobi method. A rigorous treatment of the double collapse and other collisions requires a different approach. This approach is made possible by a change of coordinates developed by R. McGehee[11] in connection with the collinear three-body problem, concerning the motion of three gravitating point masses on a line.

The coordinate change introduced by McGehee is useful in the general n-body problem because it aids in the analysis of trajectories passing close to multiple collisions. Although our main problem does not involve point masses, the form of the gravitational potential for the two concentric spheres and central core is essentially similar to the potential that arises in the three-body problem: *both of these potentials are homogeneous of degree -2* in the appropriate "radial" coordinates. Consequently, McGehee's change of coordinates can also be used in our problem to analyze the double collapse.

We will briefly describe what it means to *regularize* a singularity in a mechanical system. This description is based on an account given by McGehee[12]. The essential problem is this: trajectories with arbitrarily similar initial conditions may, upon passing near a singularity, be thrown far apart, or may return to close proximity of one another after a finite period of time.

For example, if a satellite with negative energy passes near a fixed central gravitating body, it will return along a highly elliptical orbit to near its initial position with its momentum nearly reversed, and any nearby trajectory will do the same. Thus we may form the limit of a sequence of trajectories whose initial conditions converge, and predestine ever closer brushes with collision. In the limit, linear inbound and outbound trajectories are obtained and may be glued together so that a collision orbit with a bounce is defined and preserves the continuity of solutions with respect to initial conditions. The continuous extension of the phase portrait to contain such orbits while maintaining continuity with respect to initial conditions is called the *topological or block regularization* of the flow. This particular concept of regularization is attributed to Easton[8]. Such a procedure is not always possible.

R. Devaney[7] and R. McGehee[11] have studied a modification of the Kepler problem in which the Newtonian gravitational potential is replaced by $V(\vec{Q}) = (-1/k)|\vec{Q}|^{-k}$ with $k < 2$ with $k \neq 2 - 1/n$ for positive integers n. In this problem the orbits passing near collision rebound at plus or minus an angular deflection depending on k. The choice between the plus or minus branches depends discontinuously on initial conditions, the determining factor being whether the nearly linear inbound orbit is in fact slightly clockwise or counterclockwise about the center of mass. We see that in this modified Kepler problem, topological regularization is not usually possible. This situation contrasts with the usual Kepler problem in which the angular deflection is plus or minus 2π radians so that the existence of two branches induces no discontinuity with respect to initial conditions.

R. Devaney[7 3 4 5] has also studied the *anisotropic* Kepler Problem in which the usual Kepler problem is modified so that the satellite's inertial mass depends on the direction of the applied force, while the satellites's gravitational charge remains the same in all directions. In this case, as in the previously described modification of the Kepler problem, trajectories passing near collision with the fixed central mass may be ejected along radically different paths depending whether the inbound trajectory leads slightly to the right or to the left of the fixed central mass. So in the anisotropic Kepler problem, topological regularization of the collision is impossible.

But the anisotropic Kepler problem does admit a more general kind of extension, which we will refer to as the *augmentation* of the flow at collision. The natural phase space (minus the singularity) may be embedded in a larger phase space, its image being an open set. The embedding carries the original flow to a new flow in the larger phase space. The new flow may be *extended continuously to the boundary* of the embedded image of the natural phase space. Trajectories passing close to this boundary are the embedded images of trajectories passing close to the singularity in the natural phase space. The McGehee coordinate transformation accomplishes this embedding for singularities arising due to collisions in the n-body problem, or in other classical mechanical problems involving a homogeneous potential. In principle, other coordinate transformations might be found to augment flows at other types of singularities by a similar procedure of embedding into an extended phase space.

To make our definition explicit, throughout this report *augmentation at collision* will signify the embedding of the natural phase space (minus the singularity) into a region of a larger phase space such that the boundary of this region corresponds to the original singularity and such that the embedded flow admits a continuous extension to the boundary. Not only is this kind of augmentation more widely applicable, it is also more informative than treatments conducted entirely in the natural phase space. First, as we shall see in the details of several specific problems outlined below, the global topological effects of the singularity are revealed by our more general type of augmentation. Second, the numerical analysis of the flow is facilitated in the new coordinates introduced in this kind of augmentation.

Our problem may now be stated as follows: augment the flow at double collapse in the system of two concentric gravitating spheres of equal mass centered about a fixed central body of unit mass, and analyze the global behavior of trajectories in the phase space with attention to the following features: existence of fixed points, stable and unstable manifolds, existence and topology of periodic, escape, and other orbits, and bifurcations associated with changes in the masses of the spheres. Problems of this type have been studied by McGehee[12], Devaney[7], Simo[18], and Simo[17], and Moekel[13].

2. The Problem of Two Concentric Shells in McGehee Coordinates

Following the approach of the previously cited papers of McGehee, Devaney, and Simo, we introduce new coordinates (ρ, θ, v, w), which we shall refer to as the *McGehee coordinates*. We shall define these in terms of the natural or *Euclidean* coordinates (r_1, r_2, p_1, p_2) where r_i is the radius of sphere i, and $p_i = m dr_i/dt$. The McGehee coordinates are defined by the following system of equations:

$$\begin{aligned} r_1 &= \rho^{2/3}\cos\theta \\ r_2 &= \rho^{2/3}\sin\theta \\ p_1 &= m\rho^{-1/3}(v\cos\theta - w\sin\theta) \\ p_2 &= m\rho^{-1/3}(v\sin\theta + w\cos\theta). \end{aligned} \tag{4}$$

If we restrict our attention to the three dimensional submanifold of a fixed constant negative total energy, then the submanifold M defined by $\rho = 0$ becomes an invariant two dimensional surface of revolution whose equation is in the form $v^2 + w^2 = 2\tilde{U}(\theta)$. This surface M is called the *collision manifold.* In the problem of two concentric spheres, the McGehee coordinates in conjunction with a rescaling of time allow the Hamiltonian flow to be extended continuously to the collision manifold.

We obtain the following formulation of the equations of motion of the problem of two concentric gravitating spheres:

$$\begin{aligned} \frac{d\rho}{d\tau} &= (3/2)v\rho; \\ \frac{d\theta}{d\tau} &= w; \\ \frac{dv}{d\tau} &= \frac{v^2}{2} + w^2 - \frac{\tilde{U}(\theta)}{m}; \\ \frac{dw}{d\tau} &= -(1/2)vw + \left(\frac{1}{m}\right)\frac{d\tilde{U}(\theta)}{d\theta}; \\ \frac{dt}{d\tau} &= \rho. \end{aligned} \tag{5}$$

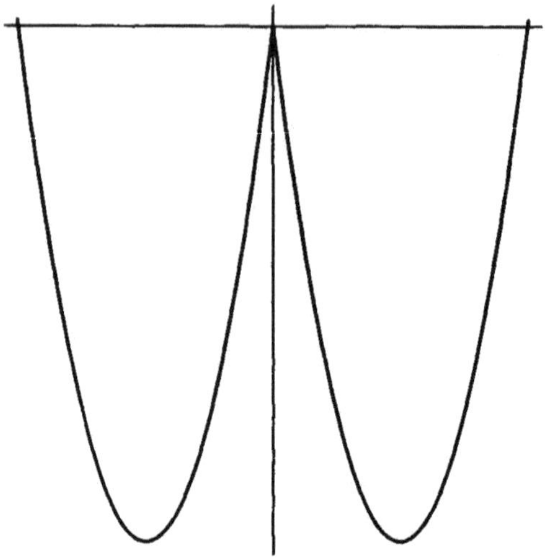

Fig. 1. Exaggerated View Of The Graph of $\tilde{U}$ Near The Central Cusp, In The Case $m = 0.1$.

We must define the rescaled potential $\tilde{U}$ appearing in equations 5. Let R^+ be the region of phase space for which $|r_1| < |r_2|$, i.e. $\theta > \pi/4$. Let R^- be the region for which $|r_1| > |r_2|$, i.e. $\theta < \pi/4$. Then $\tilde{U}$ is defined as follows:

$$\begin{aligned} for\ \theta > \pi/4,\ \tilde{U} &= m(1+m/2)\sec\theta + m(1+3m/2)\csc\theta; \\ for\ \theta < \pi/4,\ \tilde{U} &= m(1+m/2)\csc\theta + m(1+3m/2)\sec\theta. \end{aligned} \tag{6}$$

Note that $\tilde{U}$ is continuous, but ***not differentiable*** at $\theta = \pi/4$.

The energy relation is as follows:

$$\rho^{2/3}h = \frac{m}{2}(v^2 + w^2) - \tilde{U}(\theta). \tag{7}$$

An exaggerated representation of the graph of $\tilde{U}$ for $\theta \approx \pi/4$ is given in figure 1. An exaggerated representation of the region of the collision manifold M for $\theta \approx \pi/4$ is given in figure 2. The exaggerated scale is required because, in reality, the locally maximal radius which occurs at $\theta = \pi/4$ is only a small fraction greater than the globally minimal radius of the collision manifold. This differs qualitatively from the isosceles three body problem in that the collision manifold in the present case has a cusp at $\theta = \pi/4$, whereas the collision manifold of the isosceles three body problem is differentiable at $\theta = \pi/4$. See Devaney[6].

3. The Flow on the Collision Manifold

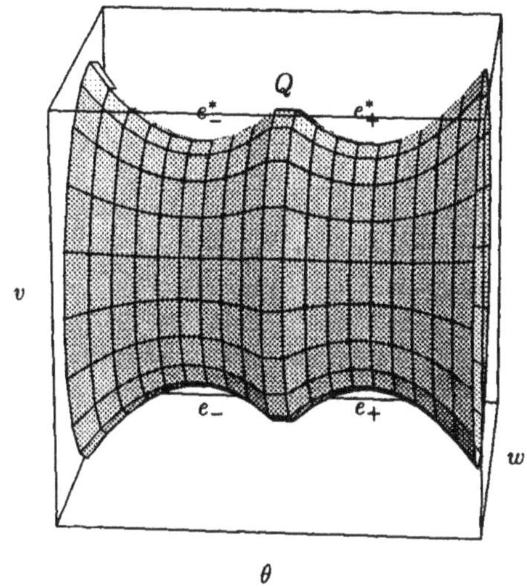

Fig. 2. Exaggerated View Of The Region of M Near The Central Cusp, In The Case $m = 0.1$. (The smooth appearance of the cusp at Q is an artifact of the graphing procedure.)

Considering equations 5, we note that the surface $\rho = 0$ is invariant under the flow defined by equations 5. We refer to this $\rho = 0$ surface as the *collision manifold* M. Recalling the energy relation (7), we see that the collision manifold is a surface of revolution given as follows:

$$v^2 + w^2 = (2/m)\tilde{U}(\theta), \tag{8}$$

where $\tilde{U}$ is defined as in equation 6. The surface M is illustrated in figure 2.

Using equation (8) we can greatly simplify the equations of motion (5) restricted to the surface M. The simplified equations are as follows:

$$\begin{aligned} \frac{d\rho}{d\tau} &= 0; \\ \frac{d\theta}{d\tau} &= w; \\ \frac{dv}{d\tau} &= \frac{w^2}{2} \geq 0; \\ \frac{dw}{d\tau} &= \frac{-vw}{2} + \frac{\tilde{U}'(\theta)}{m}. \end{aligned} \tag{9}$$

Note that v is a Lyapunov function for the flow on M. Also, there are rest points where $w = 0$ and $\tilde{U}'(\theta) = 0$. These points have been labeled e_-, e_+, e_-^*, and e_+^* in figure 2. In the case of the isosceles three body problem, the point labeled Q in figure 2 would be a fixed point and local analysis would be easy near Q, but in the present case, the vector field is not defined at Q, so that special attention is needed here.

A further change of variables would extend the flow continuously to the single collapses at $\theta = 0$, and $\theta = \pi/2$. However, the equations of motion become more complicated in this case, so we will leave them as they are.

3.1. Local Analysis Near The Rest Points On M

In the dissertation[1] we determine the stability type of the rest points e_-, e_+, e_-^*, and e_+^*, by computing the eigenvalues and eigenvectors of the flow restricted to the collision manifold M. Each of these rest points is a saddle point with stable and unstable eigenvectors denoted V^s and V^u respectively.

3.2. Global Analysis Of The Unstable Manifolds Of e_- And e_-^*

We will explore the global behavior of the unstable manifolds of e_- and e_-^* using a combination of analytic and numerical techniques.

By time reversal symmetry, the stable manifold of e_- is the image of the unstable manifold of e_-^* under sign reversal of the v and w coordinates. Similarly, the stable manifold of e_-^* is the image of the unstable manifold of e_- under sign reversal of the v and w coordinates. Thus we may obtain a complete analysis of the invariant manifolds of e_- and e_-^* from the unstable manifolds alone, without explicit consideration of the stable manifolds.

By the symmetry of the system under the exchange of the labeling of the two concentric spheres as “shell number one” and “shell number two”, we may eliminate the need for explicit consideration of the invariant manifolds of the points e_+ and e_+^*. The transformation corresponding to the exchange of names is reflection across the plane $\theta = \pi/4$, combined with sign reversal of the w coordinate. This transformation carries the invariant manifolds of e_- and e_-^* over to the invariant manifolds of e_+ and e_+^*. Thus, the rest points with $\theta > \pi/4$ need not be explicitly considered.

3.2.1. The Unstable Manifold Of e_-

Both branches of the unstable manifold of the rest point e_- of the flow restricted to the collision manifold were computed numerically. They are shown in figures 3 and 4. The trajectory along either of these manifolds bounces twice (once at each end of the collision manifold) before it spirals up one of the arms as v grows without bound. In the dissertation[1], it is proven that in fact this behavior occurs for all values of the shell mass m. The numerical computation of the unstable manifold was accomplished by noting that the first intersections of the two branches of the unstable manifold of e_- with the plane $\theta = \pi/4$ can be determined exactly by hand as in the dissertation[1]. The forward and reverse time orbits of these two intersection points then fill out the unstable manifold of e_- as illustrated in figures 3 and 4.

3.2.2. The Unstable Manifold Of e_-^*

Since v is a Lyapunov function for the flow restricted to the collision manifold M, we may immediately deduce some qualitative features of at least one branch of the unstable manifold of the rest point e_-^*. We consider the branch of the unstable manifold of e_-^* which emerges from e_-^* along the direction of the eigenvector with positive component in the θ direction. We will denote this branch of the unstable manifold of e_-^* by the symbol $W^{u+}(e_-^*)$. We claim the following:

Proposition 1 *The unstable branch $W^{u+}(e_-^*)$ must intersect the plane $\theta = \pi/4$ and must spiral towards the point Q where the v coordinate is locally maximal on the surface M. The unstable branch $W^{u+}(e_-^*)$ crosses the plane $P : \theta = \pi/4$ infinitely many times as it spirals.*

The proof that the spiral crosses the plane P infinitely many times depends on several lemmas concerning the amount $\Delta\tau_n$ of rescaled time that is required for each revolution of the spiraling orbit. These lemmas lead to the conclusion that the sum $\sum_{n=1}^{\infty} \Delta\tau_n = \infty$ so that the orbit never actually reaches the point Q. We note that because the vector field is discontinuous on the plane P (and more specifically, non-differentiable at the point Q) we cannot treat the point Q as an ordinary rest point to be studied by local linearization. It is the discontinuity at Q that necessitates the approach taken in the dissertation[1].

3.2.3. Intersections Of Invariant Manifolds In M With P

The global dynamics on the collision manifold M is determined by the relative positions of the stable and unstable manifolds of the fixed points. Recall that P is the plane $\theta = \pi/4$, and $C = M \cap P$. Finding where the various branches of the various stable and unstable manifolds intersect C completely determines the relative positions of these manifolds in M, and hence determines the global dynamics on M. These intersections comprise twenty-four points on C, but the symmetries under time reversal and under the exchange of names of the two spheres reduce the number of points needing independent consideration

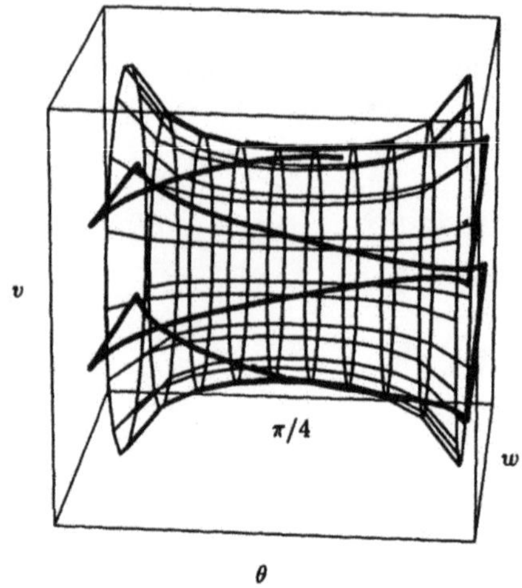

Fig. 3. Both Branches Of The Unstable Manifold Of The Rest Point e_- When $m = 0.1$, Shown With The Collision Manifold

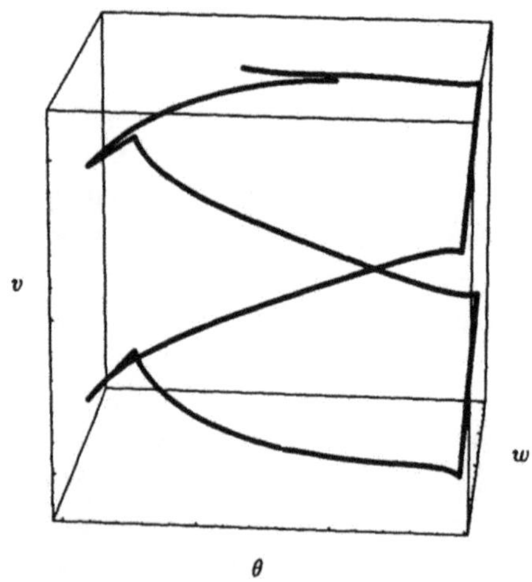

Fig. 4. Both Branches Of The Unstable Manifold Of The Rest Point e_- When $m = 0.1$, Shown Without The Collision Manifold

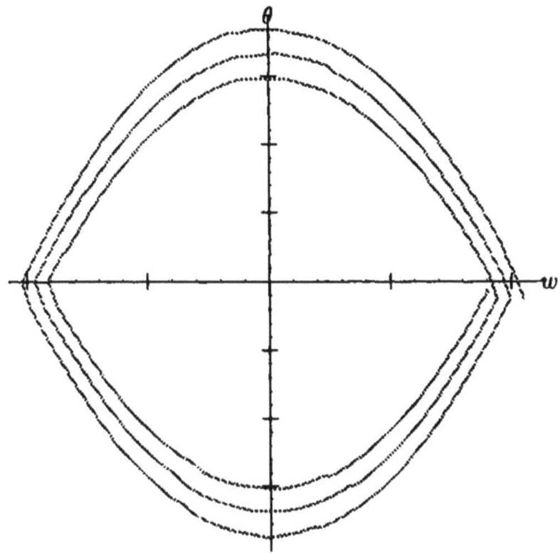

Fig. 5. Top View Of An Orbit Spiraling Towards Q

to six.

Notation There are sixteen branches of stable and unstable manifolds attached to the four rest points e_-, e_+, e_-^*, e_+^*. We denote these as follows: $W^{s-}(e_-)$, $W^{s+}(e_-)$, $W^{u-}(e_-)$, $W^{u+}(e_-)$, $W^{s-}(e_+)$, $W^{s+}(e_+)$, $W^{u-}(e_+)$, $W^{u+}(e_+)$, $W^{s-}(e_-^*)$, $W^{s+}(e_-^*)$, $W^{u-}(e_-^*)$, $W^{u+}(e_-^*)$, $W^{s-}(e_+^*)$, $W^{s+}(e_+^*)$, $W^{u-}(e_+^*)$, $W^{u+}(e_+^*)$. The subscripts $\pm$ indicate whether the rest point to which the denoted invariant branch is attached lies in the region $R_- : \theta < \pi/4$, or in $R_+ : \theta > \pi/4$. The superscript $*$ indicates that the rest point lies in the upper region, where $v > 0$, while the absence of the superscript $*$ indicates that the rest point has $v < 0$. The superscripts s or u indicate whether the denoted branch is stable or unstable. For the *unstable* branches, the superscript $+$ indicates that the denoted unstable branch departs from its rest point into the region $w > 0$, while the superscript $-$ indicates departure toward $w < 0$. In contrast, for the *stable* branches, the superscript $+$ indicates that the denoted stable branch arrives at its rest point from the region $w > 0$, while the superscript $-$ indicates arrival from $w < 0$.

Each branch of an invariant manifold may intersect the circle C several times. For the *unstable* branches, we will use notations in the form $(W^{u+}(e_-) \cap C)_n$ to indicate the nth intersection of the branch with C, counting outward from the fixed point, along the *forward*-time direction of the orbit along the branch. For the *stable* branches, we will use notations in the form $(W^{s+}(e_-) \cap C)_n$ to

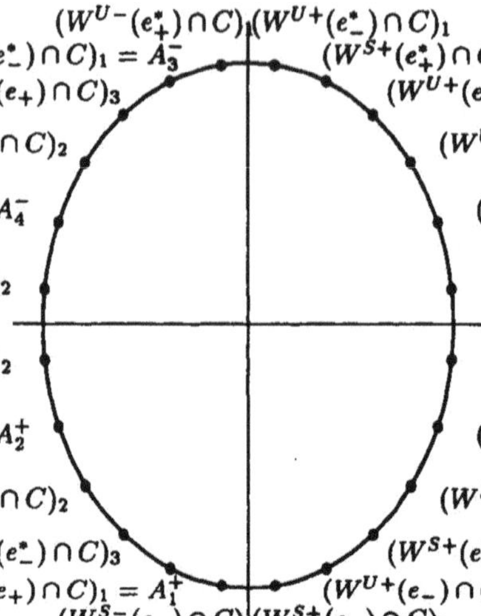

Fig. 6. Intersections Of Invariant Manifolds With The Circle C

indicate the nth intersection of the branch with C, counting outward from the fixed point, along the *backward*-time direction of the orbit along the branch. In the dissertation[1], the arrangement of these intersections around the circle C is shown to be as illustrated in figure 6.

The origins and fates of all orbits on the collision manifold M that pass through the circle C are determined by our knowledge of the orbits through the twenty-four points illustrated in figure 6.

The significance of the ordering illustrated in figure 6, is summarized in the following theorem.

Theorem 1 *The restriction of the flow to the collision manifold contains an open set of initial conditions leading to orbits that spiral into the point Q, and open sets of initial conditions leading to orbits that spiral up the arms at $\theta = 0$ and $\theta = \pi/2$. The separatrices between these regions are orbits asymptotic to e_-^* and e_+^*.*

The time reversal symmetry implies the existence of an open set of initial conditions leading to orbits that spiral into the point $-Q$ in reverse time, and open sets of initial conditions leading to orbits that spiral down the legs at $\theta =$ 0 and $\theta = \pi/2$ in reverse time. The separatrices between regions are orbits asymptotic to e_- and e_+ in reverse time.

Moreover, the orbits fit together on the collision manifold in such a way that the connections among the possible asymptotic past and future behaviors may be determined. Orbits from the left leg can go either up the right arm or into

Q, and a separatrix comes from the left leg to the point e_+^*. Orbits from the central cusp at $-Q$ can go up either arm, and the separatrices go into e_- and e_+. Orbits from the right leg can go either up the left arm or into Q, and a separatrix comes from the right leg to the point e_-^*.

Recall the equation we have used to define the surface M which we have, until now, called the collision manifold:

$$M = \{(\theta, v, w) : \frac{m}{2}(v^2 + w^2) - \tilde{U}(\theta) = 0\} \tag{10}$$

The surface M admits more than one physical interpretation.

Until now we have considered M to be the set $\rho = 0$. Indeed, in the case of negative total energy ($h < 0$), M admits no other interpretation. In this case ρ must be zero, because we have $0 = (m/2)(v^2 + w^2) - \tilde{U}(\theta) = \rho^{2/3}h$ and $h \neq 0$.

However, the zero-energy manifold M_0 satisfies

$$\begin{aligned} M_0 &= \{(\rho, \theta, v, w) : H(\rho, \theta, v, w) = 0\} \\ &= \{(\rho, \theta, v, w) : \rho \in \Re^+, (\theta, v, w) \in M\} \\ &= \Re^+ \times M. \end{aligned} \tag{11}$$

These equations show that in the zero energy case, M should no longer be interpreted as the $\rho = 0$ surface, but rather as the image of the zero-energy manifold M_0 under projection onto the variables (w, θ, v).

Suppose we want to know the behavior of a particular trajectory in the zero-energy manifold M_0. Recall that the equations of motion 5 decouple so that the (w, θ, v) components of the vector field do not depend on ρ. Since the evolution of the variables (w, θ, v) is independent of ρ, we can study the *projection* of the trajectory from M_0 onto M. But the flow on M has already been discussed in section . Having determined the (w, θ, v) components of the flow as in section , we can study the equation $d\rho/d\tau = (3/2)v\rho$ to understand $\rho(\tau)$.

Note that in the case of nonzero energy, when we studied the flow away from M in section , we never needed to consider the equation $d\rho/d\tau = (3/2)v\rho$, since the energy relation 14 let us immediately compute ρ from w, θ, and v. However, when $h = 0$ we can longer do this. In the zero energy case, to obtain the actual flow in M_0 away from M, we must integrate the equation $d\rho/d\tau = (3/2)v\rho$. We can perform this integration because, knowing the projection of the flow from M_0 onto M, we know $v(\tau)$.

In the case of nonzero energy, the trajectories in M, are in a sense fictitious, because M exists only as a boundary attached to the original phase space of the problem. The original flow has been continuously, but nevertheless artificially, extended to this boundary. In the case of zero energy, however, the trajectories in M. correspond to actual orbits of the original system. The orbits in M that spiral into the point Q as described in are especially interesting. Their interpretation is contained in theorem 2. Theorem 2 simply reinterprets theorem 1 in the zero energy case.

Theorem 2 *In the case of zero total energy, there exists a simply connected open set of initial conditions leading to orbits in which both spheres escape to infinite radius, crossing one another infinitely many times, with the ratio of their radii approaching one, and their outward velocities dr/dt asymptotic to $2^{5/4}\sqrt{\frac{1+m}{r}}$, where each shell has mass m. Other open sets of trajectories lead to the escape of one or the other of the spheres to infinite radius while the other sphere bounces inside. The separatrices between these behaviors are orbits asymptotic to homothetic expansion or collapse.*

Proof Of Theorem 2 There is nothing to prove here, except to interpret the results of section in the case of zero total energy. The discussion of section shows that all of the orbits in the central cusp region cross $\theta = \pi/4$ infinitely many times as they spiral towards the point Q. Since $\theta(Q) = \pi/4$, the radii are asymptotically of ratio one. Since $v^2(Q) = 4\sqrt{2}(1+m)$, and from equations 5, $\frac{d\rho}{dt} = (3/2)v$, we have

$$\frac{d\rho}{dt} \sim 3(2^{1/4})\sqrt{1+m},$$

but $r = \rho^{2/3}$ so,

$$\frac{dr}{dt} \sim 2(2^{1/4})r^{-1/2}\sqrt{1+m}.$$

Consequently,

$$r \sim (9\sqrt{2}(1+m)t^2)^{1/3}. \tag{12}$$

The orbits discussed in section that spiral up one of the arms at $\theta = 0$ or $\theta = \pi/2$ are interpreted as escape orbits in the case of zero energy, and the orbits along the stable and unstable manifolds of the saddle points are asymptotic to homothetic (constant θ) orbits as stated.

□

The other types of solutions found in section are also of interest in the zero-energy case. In particular, the solutions along the unstable manifolds of the lower saddle points that spiral up the arms of the collision manifold acquire physical meaning in the zero energy case: first one shell collapses to zero radius and rebounds, then the other shell follows suit, and finally, one sphere is trapped within a bounded radius, bouncing rapidly as the other escapes to infinite radius.

We mention one last interesting immediate consequence of the ordering of points indicated in figure 6:

Theorem 3 *In the zero energy case, every orbit in which both spherical shells arrive from infinite radius departs in such a way as to leave one shell trapped within a bounded radius as the other escapes to infinity.*

Proof of Theorem 3 The orbits on the zero energy surface that spiral out of $-Q$ either must remain between the unstable branches $W^{u+}(e_-)$ and $W^{u+}(e_+)$ or (in the analogous case under the name-exchange symmetry) must remain between the unstable branches $W^{u-}(e_-)$ and $W^{u-}(e_+)$. If an orbit remains between $W^{u+}(e_-)$ and $W^{u+}(e_+)$ it must escape up the arm at $\theta = \pi/2$, since using the notation of section the entire strip bounded by these two branches eventually arrives at the circle $C : \theta = \pi/4$ above $(W^{s-}(e_+^*) \cap C)_2$ and below $(W^{s+}(e_+^*) \cap C)_1$. For similar reasons, if an orbit remains between $W^{u-}(e_-)$ and $W^{u-}(e_+)$ it must escape up the arm at $\theta = 0$. (The exceptions are the orbits that spiral out of $-Q$ along the stable manifolds of e_- and e_+.)

□

4. The Flow with Negative Energy

We now consider the flow in the case of negative finite total energy. This is the region of phase space containing the orbits that do not involve the simultaneous collapse of both spherical shells in finite time. This region is the set

$$I = \{(\rho, \theta, v, w) : \rho > 0, H(\rho, \theta, v, w) = h < 0\}.$$

We begin by recalling the equations of motion 5 in the McGehee coordinates:

$$\begin{aligned} \frac{d\rho}{d\tau} &= (3/2)v\rho; \\ \frac{d\theta}{d\tau} &= w; \\ \frac{dv}{d\tau} &= \frac{v^2}{2} + w^2 - \frac{\tilde{U}(\theta)}{m}; \\ \frac{dw}{d\tau} &= \frac{-vw}{2} + \frac{1}{m}\frac{d\tilde{U}(\theta)}{d\theta}; \\ \frac{dt}{d\tau} &= \rho. \end{aligned} \tag{13}$$

Since the equations of motion for the coordinates w, θ, and v, are independent of ρ, we may study these three equations alone, and afterwards (provided the energy h is non-zero) use the energy relation 7, to determine ρ. Recall that the energy relation is as follows:

$$\rho^{2/3}h = \frac{m}{2}(v^2 + w^2) - \tilde{U}(\theta). \tag{14}$$

So, in the case of non-zero total energy h, we will study the flow in the three dimensional region $\tilde{I}$ obtained by projecting I onto the three variables (w, θ, v). In the case of negative total energy $h < 0$, the region $\tilde{I}$ is the interior of the solid cylinder bounded by the collision manifold. In the case of positive total energy, the region $\tilde{I}$ would be the exterior of this cylinder. Our interest in this section will be in the case of negative total energy.

4.1. Invariant Manifolds

Our understanding of the flow in $\tilde{I}$ will benefit from a study of invariant manifolds in this region. Let the dimension of the phase space be reduced to three by fixing the total energy H at some finite negative value h. Let $\tilde{I}_< = \{(w,\theta,v) \in \tilde{I} : \theta < \pi/4\}$, and let $\tilde{I}_> = \{(w,\theta,v) \in \tilde{I} : \theta > \pi/4\}$. One family of invariant manifolds has already been mentioned: In each of the regions $\tilde{I}_<$ and $\tilde{I}_>$, the surfaces $S_{h_1} = \{(w,\theta,v) : H_1(w,\theta,v) = h_1\}$ are invariant. We mention two additional invariant manifolds: Let $W^u(e_-^*)$ denote the two dimensional unstable manifold of e_-^*, and let $W^u(e_-) \subset M$ denote the one dimensional unstable manifold of e_-. These invariant manifolds will also reveal much about the flow throughout $\tilde{I}$.

As discussed in section , the line $L = \{(w,\theta,v) : w = 0,\ \theta = \theta(e_-^*)\}$ is invariant under the flow. Trajectories along this line flow downward, from e_-^* to e_-. Consequently, we expect that $W^u(e_-^*)$ will connect nicely with $W^u(e_-)$. This expectation is made precise in the following lemma.

Lemma 1 $W^u(e_-) \subset \overline{W^u(e_-^*)}$.

4.2. The Central Region

The following theorem is a consequence of lemma 1.

Theorem 4 *In the case of negative total energy, for each large integer n, there exists an orbit in which the spherical shells cross one another n times before one of them escapes to arbitrarily large radius as the other is captured.*

Proof of Theorem 4 Using the composition of the name-exchange and time reversal symmetries, we see that the stable manifold $W^s(e_+)$ spirals into the vertical line through Q and $-Q$, winding in the opposite direction from the spiral of the unstable manifold $W^u(e_-^*)$, as illustrated in figure 7. Consider the plane $\Pi : v = v(Q) - \epsilon$. Let O be the point $\theta = 0 = w$ in Π. Then $W^s(e_+) \cap \Pi$, and $W^u(e_-^*) \cap \Pi$ are oppositely directed spirals around O in Π as illustrated in figure 8. So $W^s(e_+) \cap W^u(e_-^*) \cap \Pi$ contains infinitely many points. Each of these points lies on an orbit connecting the saddle point e_-^* to the saddle point e_+.

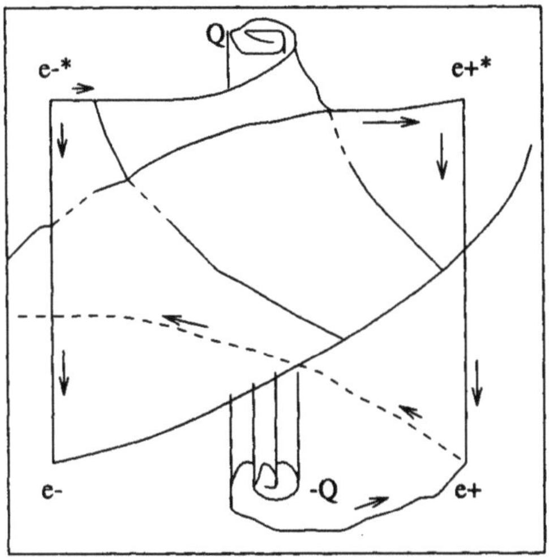

Fig. 7. $W^s(e_+)$ And $W^u(e_-^*)$

Let X be one of the points in $W^s(e_+) \cap W^u(e_-^*) \cap \Pi$. The complete orbit of X $(-\infty < \tau < \infty)$ spirals around the line from Q to $-Q$ some finite number of times before limiting onto e_+. Let N be the number of revolutions of the complete orbit of X about the line from Q to $-Q$. We claim that for every $n > N$ there exists a point $X_n \in (W^s(e_+) \cap W^u(e_-^*) \cap \Pi)$ whose forward orbit revolves exactly n times before limiting onto e_+. This will follow immediately by induction if we can show the existence of a point X_{N+1} that revolves $N+1$ times.

Looking down on the plane Π from the point Q so that (w, θ) appears as a right handed coordinate system on Π, follow $W^u(e_-^*) \cap \Pi)$ counterclockwise from X_N to its next intersection Y with $W^s(e_+)$. We note that Y revolves $N+1$ times, so that the induction is completed by letting $X_{N+1} = Y$. Having these saddle connections that revolve n times, now yields the desired orbits leading to the escape of one sphere, since in some small neighborhood of each saddle connection, there must be orbits that pass near the saddle point e_+ and then follow near the unstable branch $W^{u-}(e_+)$ which escapes up the arm at $\theta = 0$.

□

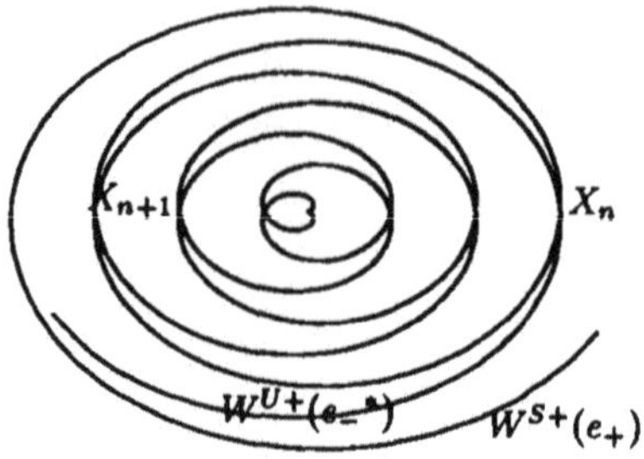

Fig. 8. $W^s(e_+) \cap W^u(e_-^*) \cap \Pi$

5. Acknowledgements

The author wishes to thank his thesis advisor, Professor Glen R. Hall, for his inspiration and patient instruction. The author is also deeply indebted to Professors Robert Devaney and Paul Blanchard for their clear and careful lectures and seminars in dynamics at Boston University.

6. References

1. Solon F. Blundell. *A Spherically Symmetric Model In Celestial Mechanics.* PhD thesis, Boston University, 1993.
2. Gregory Buck, *Mass Distributions in Central Configurations.* PhD thesis, Boston University, 1988.
3. Robert L. Devaney, Collision orbits in the anisotropic kepler problem. *Invent. Math.*, 45(3):221–251, 1978.
4. Robert L. Devaney, Nonregularizability of the anisotropic kepler problem. *Journal of Differential Equations*, 29(2):252–268, 1978.
5. Robert L. Devaney, Transverse heteroclinic orbits in the anisotropic kepler problem. In *The Structure of Attractors in Dynamical Systems.* Springer-Verlag, Berlin, 1978. This book is Volume 668 of the Lecture

Notes in Mathematics series.

6. Robert L. Devaney, Triple collision in the planar isosceles three body problem. *Invent. Math.*, 60(3):249–267, 1980.
7. Robert L. Devaney, Singularities in classical mechanical systems. In *Ergodic Theory and Dynamical Systems I*, pages 211–233. Birkhauser, Boston, 1981.
8. Robert Easton, Regularization of vector fields by surgery. *Journal of Differential Equations*, 10:92–99, 1971.
9. Michel R. Herman, *Sur les Courbes Invariantes par les Diffeomorphismes de l'Anneau.* Société Mathématique de France, 1983.
10. Kevin Hockett, Chaotic numerics from an integrable hamiltonian system. *Proceedings of the American Mathematical Society*, 108(1):271–281, 1990.
11. R. McGehee, Triple collision in the collinear three-body problem. *Invent. Math.*, 27:191–227, 1974.
12. R. McGehee, Triple collision in newtonian gravitational systems. In J. Moser, editor, *Dynamical Systems Theory and Applications.* Springer-Verlag, Berlin, 1975. This book is Volume 38 of the Lecture Notes in Physics series.
13. R. Moekel, Orbits of the three-body problem which pass infinitely close to triple collision. *American Journal of Mathematics*, 103(6):1323–1341, 1981.
14. Jurgen Moser, On invariant curves of area-preserving mappings on an annulus. *Nachr. Aka. Wiss., Gottingen Math. Phys. K1 II*, pages 1–20, 1962.
15. H. Russmann, Uber invariante kurven differenzierbarer abbildungen eines kreisrings. *Nachr. Aka. Wiss., Gottingen II Math. Phys. K1 II*, pages 67–105, 1970.
16. C. Siegel and J. Moser, *Lectures on Celestial Mechanics.* Springer-Verlag, 1971.
17. C. Simo, Masses for which triple collision is regularizable. *Celestial Mechanics*, 21:25–36, 1980.
18. C. Simo, Analysis of triple collision in the isosceles problem. In *Clasical Mechanics and Dynamical Systems*, pages 203–224. Marcel Dekker, New York, 1981.
19. K. Sundman, Memoire sur le probleme de trois corps. *Acta Mathematica*, 36:105–179, 1912.

SUBHARMONIC SOLUTIONS OF HAMILTONIAN SYSTEMS IN THE PLANE

HILDEBERTO EULALIO CABRAL
Departamento de Matemática - UFPE
Recife, PE - Brasil, 50740-540
E-mail: hild@pdamt.ufpe.br

ABSTRACT

We present a geometrical proof of the Melnikov theorem for subharmonic solutions of planar Hamiltonian systems

1. Introduction

We consider a Hamiltonian of the form $H(x, y, I, \theta; \epsilon)$, 2π-periodic in θ. Suppose that by fixing its value, h, we can solve for I to get

$$I = K(x, y, \theta, ; h, \epsilon) = K_0(x, y, \theta; h) + \epsilon K_1(x, y, \theta; h) + \cdots$$

thereby defining a Hamiltonian system in terms of the "time" θ,

$$\frac{dx}{d\theta} = \frac{\partial K_0}{\partial y} + \epsilon\frac{\partial K_1}{\partial y} + \cdots, \qquad \frac{dy}{d\theta} = -\frac{\partial K_0}{\partial x} - \epsilon\frac{\partial K_1}{\partial x} - \cdots.$$

This system is a 2π-periodic perturbation of the Hamiltonian system defined in the plane by K_0.

Assume that the unperturbed system has a hyperbolic equilibrium of saddle type at the origin $x = 0$, $y = 0$ and a homoclinic loop to $(0,0)$ which encloses a region with periodic orbits emanating from an elliptic equilibrium. We assume that the periods $T(c)$ of these orbits, where c is the value that K_0 takes on the orbit, increase in a non degenarate way in the sense that $\frac{dT}{dc} > 0$.

By using the Melnikov method for subharmonic solutions a local analysis can be made around the elliptic equilibrium enclosed by the homoclinic loop to investigate the existence of periodic solutions of the unperturbed system with periods that are rational multiples of 2π. This local analysis will appear elsewhere and in this note we present a proof of the Melnikov theorem that, we believe, has a more geometrical flavor than the one in [GH].

2. The Subharmonic Melnikov Method

According to [GH], we will consider the following setting. We take a time-dependent system in the plane

$$\dot{x} = f(x) + \epsilon g(x, t, \epsilon),$$

where $f(x)$ and $g(x,t,\epsilon)$ are of class C^2 and bounded on bounded sets, and g is T-periodic in t. Assume that the unperturbed system ($\epsilon = 0$) has the following properties:

(1) It is Hamiltonian with a hyperbolic equilibrium at x_0.

(2) There is a homoclinic orbit to x_0 enclosing a region filled with periodic solutions of the unperturbed system, growing from an elliptic equilibrium inside the region up to the homoclinic loop.

(3) The period of this family of periodic solutions is a smooth function of the area of the bounded region defined by the solution. In fact, if c is the value of f on the periodic solution of $\dot{x} = f(x)$, and $T(c)$ is the period, we require that $\frac{dT}{dc} > 0$.

Let $x^\beta(t)$, $-1 < \beta < 0$ be the family of periodic solutions of the unperturbed system within the homoclinic loop with initial conditions $x^\beta(0)$ on a smooth arc joining the equilibrium inside the homoclinic loop, $\lim_{\beta\to -1} x^\beta(0)$, to a point on the loop, $\lim_{\beta\to 0} x^\beta(0)$.

Denote by T_β the period of the solution $x^\beta(t)$. Let $x^\alpha(t)$ be a periodic solution of the unperturbed system with period $T_\alpha = \frac{m}{n}T$ and consider, for each β, the integral

$$M^\beta(t_0) = \int_0^{nT_\beta} f(x^\beta(t+t_0)) \wedge g(x^\beta(t+t_0), t, 0)dt.$$

$M^\alpha(t_0)$ is known as the Poincar-Melnikov integral of the resonant solution $x^\alpha(t)$.

The Melnikov theorem that guarantees the existence of solutions of the perturbed system is the following.

Theorem. *Let $x^\alpha(t)$ be a solution of the unperturbed system with period $T_\alpha = \frac{m}{n}T$. If $M^\alpha(t_0)$ has a simple zero at t^*, then there exists a positive number $\epsilon(m,n)$ such that for $0 < |\epsilon| < \epsilon(m,n)$, the perturbed system (2) has a periodic solution with period mT and initial condition close to $x^\alpha(t_0)$.*

For the proof of the theorem we first establish a lemma.

Let $x^\beta_\epsilon(t,t_0)$ be the solution of the perturbed system with the same initial condition as $x^\beta(t+t_0)$, i.e., $x^\beta_\epsilon(0,t_0) = x^\beta(t_0)$. Then, by a standard result in asymptotic expansions (Cf. A.N.Thikonov et al., [T], Ch. VII), there is an $\epsilon(m,n) > 0$ such that for all $|\epsilon| < \epsilon(m,n)$ the approximation

$$x^\beta(t,t_0) = x^\beta(t+t_0) + \epsilon x^\beta_1(t,t_0) + o(\epsilon^2),$$

is valid uniformly for t in the interval $[0, nT_\beta]$, with $x_1^\beta(t, t_0)$ given by the variational equation

$$\dot{x}_1^\beta(t, t_0) = Df(x^\beta(t + t_0) \cdot x_1^\beta(t, t_0) + g(x^\beta(t + t_0), t, \epsilon), \qquad x_1^\beta(0, t_0) = 0.$$

We now can state the lemma.

Lemma. *With notation as above, we have*

(1) $f(x^\beta(t_0)) \wedge [x_\epsilon^\beta(nT_\beta, t_0) - x_\epsilon^\beta(0, t_0)] = \epsilon M^\beta(t_0) + o(\epsilon^2)$.

(2) $f(x^\beta(t_0)) \wedge [x_\epsilon^\beta(nT_\alpha, t_0) - x_\epsilon^\beta(0, t_0)] = \epsilon M^\beta(t_0) - \epsilon\eta f(x^\beta(t_0)) \wedge \dot{x}_1^\beta(nT_\beta, t_0) + o(\eta^2) + o(\epsilon^2)$,

where $\eta = nT_\beta - nT_\alpha$ *is an increasing smooth function of* β.

Proof. We have

$$f(x^\beta(t_0)) \wedge [x_\epsilon^\beta(nT_\beta, t_0) - x_\epsilon^\beta(0, t_0)] = \int_0^{nT_\beta} \frac{d}{dt}[f(x^\beta(t + t_0)) \wedge x_\epsilon^\beta(t, t_0)]dt.$$

In view of the above expression for $x_\epsilon^\beta(t, t_0)$ and the equation obeyed by $x_1^\beta(t, t_0)$, the integrand becomes

$$Df(x^\beta(t + t_0)) \cdot f(x^\beta(t + t_0)) \wedge x^\beta(t + t_0) + \epsilon[Df(x^\beta(t + t_0)) \wedge x_1^\beta(t, t_0) + f(x^\beta(t+t_0)) \wedge Df(x^\beta(t+t_0)) \cdot x_1^\beta(t, t_0)] + \epsilon f(x^\beta(t+t_0)) \wedge g(x^\beta(t+t_0), t, \epsilon) + o(\epsilon^2).$$

The first summand is equal to $\frac{d}{dt}[f(x^\beta(t + t_0) \wedge x^\beta(t + t_0)]$ while the second is equal to

$$\epsilon[\text{trace } Df(x^\beta(t + t_0))] f(x^\beta(t + t_0)) \wedge x_1^\beta(t, t_0).$$

Therefore, (1) follows because nT_β is a period of $x^\beta(t + t_0)$, and the trace is zero in case the unperturbed system is Hamiltonian.

Now, (2) follows from (1) and the computations

$$x_\epsilon^\beta(nT_\beta - \eta, t_0) - x_\epsilon^\beta(0, t_0) = x_\epsilon^\beta(nT_\beta, t_0) - x_\epsilon^\beta(0, t_0) - \eta \dot{x}_\epsilon^\beta(nT_\beta, t_0) + o(\eta^2),$$

and

$$\dot{x}_\epsilon^\beta(nT_\beta, t_0) = f(x^\beta(t_0)) + \epsilon \dot{x}_1^\beta(nT_\beta, t_0) + o(\epsilon^2).$$

Proof of the Theorem. Let $V(\beta, t_0, \epsilon) = x_\epsilon^\beta(nT_\alpha, t_0) - x_\epsilon^\beta(0, t_0)$, and consider the zero-level sets $\Sigma^{\|}$ and $\Sigma^{\perp}$ of the smooth functions

$$\Phi(\beta, t_0, \epsilon) = f(x^\beta(t_0)) \wedge V(\beta, t_0, \epsilon)$$

and

$$\Psi(\beta, t_0, \epsilon) = f(x^\beta(t_0)) \cdot V(\beta, t_0, \epsilon).$$

Then, $V(\beta, t_0, \epsilon)$ vanishes exactly on the intersection of $\Sigma^{\|}$ and $\Sigma^{\perp}$. Notice tha the zero set V consists precisely of the points (β, t_0, ϵ) for which $x^\beta_\epsilon(0, t_0)$ is the initial condition of an mT-periodic solution of the ϵ-perturbed system.

For any t_0, we have $\Psi(\alpha, t_0, 0) = 0$ and so the line $(\alpha, t_0, 0)$ is contained in the set $\Sigma^{\perp}$. Now, in action-angle coordinates inside the homoclinic loop,

$$I = I(\beta, t_0), \qquad \theta = \theta(\beta, t_0),$$

we have

$$f(x^\beta(0)) = (0, f_2(\beta)), \qquad x^\beta(nT_\alpha) - x^\beta(0) = (0, \omega(I))$$

with I an increasing function of β and $\omega(I)$ a decreasing function of I. Since

$$\frac{\partial \Psi}{\partial \beta}(\alpha, \bar{t}_0, 0) = f(x^\alpha) \cdot \frac{\partial}{\partial \beta}(x^\beta(nT) - x^\beta(0))\bigg|_{\beta=\alpha} = f_2(\alpha)\omega'(I)I'(\alpha) \neq 0,$$

the implicit function theorem shows that, in a neighborhood of $t_0 = \bar{t}_0$, $\epsilon = 0$, $\Sigma^{\perp}$ is the graph of a smooth function, $\beta = \beta(t_0, \epsilon)$.

Letting $\eta = \lambda\epsilon$ and $\tilde{\Phi}(\lambda, t_0, \epsilon) = \frac{1}{\epsilon}\Phi(\beta(\lambda\epsilon), t_0, \epsilon)$, the above lemma shows that $\tilde{\Phi}(\lambda, t_0, \epsilon) = M^\beta(t_0) + o(\epsilon)$. Now, $\tilde{\Phi}$ is a smooth function function of (λ, t_0, ϵ) and $\tilde{\Phi}(\lambda, t_0, \epsilon) = M^\alpha(t_0)$, so that if t_0^* is a simple zero $M^\alpha(t_0)$, the implicit function theorem again shows that the set S: $\tilde{\Phi}(\lambda, t_0, \epsilon) = 0$ defines a smooth surface, graph of a function $t_0 = t_0(\lambda, \epsilon)$ in a neighborhood of $\lambda = 0$, $\epsilon = 0$, $t_0^* = t_0(0, 0)$.

Finally, notice that the smooth map

$$G(\lambda, t_0, \epsilon) = (\beta(\lambda\epsilon), t_0, \epsilon)$$

has a smooth inverse $G^{-1}(\beta, t_0, \epsilon) = (\frac{1}{\epsilon}\eta(\beta), t_0, \epsilon)$, defined for $\epsilon \neq 0$. By the way G is defined and the position of the surface $\Sigma^{\perp}$, the surface $G^{-1}(\Sigma^{\perp} - (\alpha, t_0, 0))$ intersects the surface S along a smooth curve and therefore in a sufficiently small neighborhood of $(\alpha, t_0^*, 0)$, the set $G(S) = \Sigma^{\|}$ intersects $\Sigma^{\perp}$ along a smooth curve $\beta = \beta(\epsilon)$, $t_0 = t_0(\epsilon)$, each point of which is an initial condition for an mT-periodic solution of the perturbed system (on applying the implicit function theorem we may have to decrease the number $\epsilon(m, n)$).

3. An Expression for $M^{m,n}(t_0)$

We consider a solution $(x_0(t), y_0(t))$ of the unperturbed system with period $\frac{m}{n}T$. We have then a simple expression for the Melnikov function $M^{m,n}(t_0)$, associated to this periodic solution, which is given by the following theorem.

Theorem. *We have* $M^{m,n}(t_0) = F'(t_0)$, *where*

$$F(t_0) = \int_0^{mT} K_1(x_0(t), y_0(t), t + t_0)dt.$$

Proof. We compute

$$\frac{d}{dt}K_1(x_0(t), y_0(t), t + t_0) = \frac{\partial K_1}{\partial t_0}(x_0(t), y_0(t), t + t_0)+$$

$$\left[\frac{\partial K_1}{\partial x}(x_0(t), y_0(t), t + t_0)\frac{dx_0}{dt} + \frac{\partial K_1}{\partial y}(x_0(t), y_0(t), t + t_0)\frac{dy_0}{dt}\right]$$

and notice that since $(x_0(t),\ y_0(t))$ is a solution of the unperturbed system defined by K_0, the expression inside the brackets is given by

$$\frac{\partial K_0}{\partial y}\frac{\partial K_1}{\partial x} - \frac{\partial K_0}{\partial x}\frac{\partial K_1}{\partial y},$$

evaluated at $(x_0(t), y_0(t), t + t_0)$. But this is the negative of the integrand of the Melnikov integral for the system. Since the integral of the left-hand side in the above equality, between 0 and mT is zero, it follows that $M^{m,n}(t_0)$ is given by

$$M^{m,n}(t_0) = \int_0^{mT} \frac{\partial K_1}{\partial t_0}\left(x_0(t), y_0(t), t + t_0\right)\ dt,$$

and the theorem is established.

Remark. $\frac{1}{mT}F(t_0)$ is the mean-value of the function K_1 along the solution $(x_0(t), y_0(t))$ defined in the interval $0 \leq t \leq mT$. Therefore, this theorem shows that the Melnikov method is related to the method of Poincaré described in Section 42 of the chapter on periodic solutions in his book "Méthodes Nouvelles de la Mécanique Céleste".

4. References

[GH] Guckenheimer, J. & Holmes, P., Nonlinear Oscillations, Dynamical Systems, and Bifurcation of Vector Fields, Springer-Verlag, N.Y., 1983.

[P] Poincaré, H., Les Méthodes Nouvelles de la Mécanique Céleste, Vol. 1, Dover Publ., Inc., N.Y., 1957.

[T] Tikhonov, A. N. et al., Differential Equations, Springer-Verlag, N.Y., 1985.

BIFURCATIONS OF LINKS OF PERIODIC ORBITS ON INTEGRABLE HAMILTONIAN SISTEMS ON S^3
*

B. CAMPOS
Dpt. Matemàtiques, Universitat Jaume I, Campus de Penyeta Roja
Castelló, 12071, Spain
E-mail: campos@vents.uji.es

J. MARTINEZ ALFARO
Dpt. Matemàtica Aplicada, Universitat de València
Burjassot, 42100, Spain
E-mail: martinej@vm.ci.uv.es

and

P. VINDEL
Dpt. Matemàtiques, Universitat Jaume I, Campus de Penyeta Roja
Castelló, 12071, Spain
E-mail: vindel@vents.uji.es

ABSTRACT

We consider an integrable Hamiltonian flow on S^3 depending on a parameter. A characterization of the set of periodic orbits of Bott integrable Hamiltonian systems which lie on energy levels diffeomorphic to 3-spheres has been recently developed. Relating local bifurcations with this characterization, we obtain topological restrictions on the realization of a given codimension one bifurcation. We also get that if the NMS periodic orbits that are not iterated toral knots of two links differ, they are not related by a sequence of generic codimension one bifurcations.

1. Introduction

The periodic orbits of a vector field on a three-dimensional manifold can be considered as knots, in the topological sense of this term. In the qualitative study of a system, the topological characterization of a trajectory, an invariant subset, etc. can produce restrictions to the rest of the flow. This paper tries to continue a series of works on the relation between periodic orbits and knots, that began with J. Birman and R.F. Williams ([1]), ([2]), which have now an extensive literature.

In particular, the characterization of the set of periodic orbits of a non-singular Morse-Smale flow on the three-dimensional sphere has been studied by A.T. Fomenko and T.Z. Nguyen ([6]), K. Sasano ([8]), M. Wada ([10]) and K.

*Partially supported by A-42-MA (Convenio Bancaixa-Universitat Jaume I).

Yano ([11]). Casasayas et al ([4]) give a similar description of the knots and links formed by the periodic orbits of Bott integrable Hamiltonian systems which lie on energy levels diffeomorphic to 3-spheres.

The aim of this work is to study the topological restrictions a link requires for suffering a given generic codimension one bifurcation, and, to obtain the final links in terms of Wada's operations. If the link after the bifurcation is not an admissible link in a integrable Hamiltonian system on S^3, as the system after the bifurcation is still integrable, there must be other bifurcations simultaneously. In this paper we study the cases where this link turns out to be an admissible link.

1.1. Wada's theorem

Wada's theorem ([10]) characterizes the set of indexed links which arises as the closed orbits of a non-singular Morse-Smale flow on S^3 in terms of a generator, the Hopf link h with indices 0 and 2 attached to the components, and six operations.

"*Every indexed link which consists of all the closed orbits of a non-singular Morse-Smale flow on S^3* is obtained from Hopf links by applying the following six operations. Conversely, every indexed link obtained from Hopf links by applying the operations is the set of all the closed orbits of some non-singular Morse-Smale flow on S^3".

OPERATIONS: For given indexed links l_1 and l_2, the six operations are defined as follows. Let $l_1 \cdot l_2$ denote the split sum of l_1 and l_2 and $N(k, M)$ the regular neighborhood of k in M.

$\mathrm{I}(l_1, l_2) = l_1 \cdot l_2 \cdot u$, where u is an unknot with index 1.

$II(l_1, l_2) = l_1 \cdot (l_2 - k_2) \cdot u$, where k_2 is a component of l_2 of index 0 or 2.

$III(l_1, l_2) = (l_1 - k_1) \cdot (l_2 - k_2) \cdot u$, where k_1 is a component of l_1 of index 0 and k_2 is a component of l_2 of index 2.

$IV(l_1, l_2) = (l_1 \# l_2) \cup m$. The connected sum $(l_1 \# l_2)$ is obtained by composing a component k_1 of l_1 and a component k_2 of l_2, each of which has index 0 or 2. The index of the composed component $k_1 \# k_2$ is equal to either $i(k_1)$ or $i(k_2)$. Finally, m is a meridian of $k_1 \# k_2$ with $i = 1$.

$V(l_1)$: Choose a component k_1 of l_1 of index 0 or 2, and replace $N(k_1, S^3)$ by $D^2 \times S^1$ *with three indexed circles in it;* $\{0\} \times S^1$, k_2 *and* k_3. *Here,* k_2 *and* k_3 *are parallel (p,q)-cables on* $\partial N(\{0\} \times S^1, D^2 \times S^1)$, *where* p *is the number of longitudinal turns and* q *the number of the transverse ones. The indices of* $\{0\} \times S^1$ *and* k_2 *are either 0 or 2, and one of them is equal to* $i(k_1)$. *The index of* k_3 *is 1.*

$VI(l_1)$: *Choose a component* k_1 *of* l_1 *of index 0 or 2. Replace* $N(k_1, S^3$ *by*

$D^2 \times S^1$ with two indexed circles in it; $\{0\}\times S^1$ and the (2,q)-cable k_2 of $\{0\}\times S^1$. The index of $\{0\}\times S^1$ is 1, and $i(k_2)=i(k_1)$.

1.2. Operations in Hamiltonian Systems

We assume that the Hamiltonian systems are Liouville integrable by means of a Bott integral. This is an integral whose critical points may be degenerate but are of necessity organized as non degenerate smooth manifolds.

A smooth Bott integral f cannot have isolated critical points on a non singular compact constant-energy surface Q, they fill either isolated smooth critical circles or smooth two-dimensional tori, or Klein bottles.

The non singular level surfaces of the integral f and Q are compact and by the Liouville theorem are unions of tori. Let S^1 be a critical circle on a level set of the integral f on Q , Fomenko ([5],[lemmas 2.1.1 and 2.1.2]) ascribes an index to it, depending on the index of the function f restricted to a two dimensional disk normal to the circle. The circles of index 0 are local minima of the integral f, those of index 2 are maxima, and those of index 1 are saddle circles.

As Casasayas et al. show in ([4]), Wada's operations used in Hamiltonian systems are IV, V and VI. They give a description of the knots and links formed by the periodic orbits of Bott integrable Hamiltonian systems which lie on energy levels that are 3-spheres.

Given a Hamiltonian H integrable on a non-singular level Q by means of a Bott integral f, and a periodic orbit p of H, in ([4]) it is defined a NMS type periodic orbit as an orbit p such that neither $f(p)$ is a regular value of f nor p lies on a critical torus of f. The set of NMS type orbits of H in Q is denoted by P.

Theorem [4][**th. 17**]: Let $Q \cong S^3$ be a non-singular energy level of a Hamiltonian H integrable on Q by means of a Bott integral f. The indexed link formed by the orbits of P may be obtained from the (0,2)-Hopf link by applying the operations IV, V and VI of Wada.

Conversely, given an indexed link l obtained from the (0,2)-Hopf link by applying the operations IV, V and VI of Wada, there exists a Hamiltonian and a Bott integral on $Q \cong S^3$ such that the associated set P forms the link l.

So, all the links obtained for a Hamiltonian system are connected links, i.e. they have no split component.

2. Generic Bifurcations in Hamiltonian Systems

Generic bifurcations are produced when the characteristic multipliers of the periodic orbits are kth roots of the unity. See ([7]).

If the link after the bifurcation is not an admissible link for a Hamiltonian system on S^3, as the system after the bifurcation is still Hamiltonian, there must be other bifurcations simultaneously. In this paper we study the cases where this link turns out to be a new type NMS link and changes are only produced on the orbit that bifurcates.

2.1. Extremal orbit

Consider the case when the characteristic multipliers are equal to $+1$. *For the critical value of the parameter, we have an extremal orbit, and after the bifurcation two new periodic orbits appear: one of them is elliptic (in a Hamiltonian system they are associated with knots with index* 0 *or* 2 *and we denote them as d) and the other is hyperbolic (it is associated with a knot with index* 1*).*

We denote $ext_1[l]$ *a link obtained from* l *adding two periodic orbits of the same knot type, one elliptic and the other one hyperbolic, that are linked similarly to a d-component of the link, and* $ext_2[l]$ *the link obtained when the two periodic orbits are linked similarly to a component of the link with index* 1.

Proposition 1 $\forall\, l$ *link on* S^3*:*

1) $ext_1[l]$ *is always an admissible link on* S^3.

2) $ext_2[l]$ *is an admissible link on* S^3 *when* $l = VI(l_o)$, l_o *is an admissible link on* S^3.

Proof: When the bifurcation occurs, the two new orbits appear linked to a component k of the original link l because the result must be a connected link.

When the two new orbits that appear are parallel and they are not linked between each other, they can be considered as meridians of this component k , then:

1) if k has index 0 *or* 2, $ext_1[l]$ *is equivalent to the link obtained applying operation IV of Wada on a link l and a Hopf link h,making the connected sum on k. So:*

$$ext_1[l] = IV(l, h).$$

2) if k has index 1, *we can not apply directly Wada operations on it. We apply operation IV on a component with index* 0 *or* 2 *and then, it is necessary operation* VI *to change the index of the component, but operation* VI *generates a* $(2, q)$*-cable of k that cannot be eliminated, so it will be necessary that this* $(2, q)$*-cable were present in the link l, this means that l is the result of applying operation* VI *on a link* l_o*, then we obtain the following relation:*

$$ext_2[VI(l_o)] = VI[IV(l_o, h)]$$

When the new two orbits that appear are linked between each other, they can be represented by (p,q)-toral knots. Then, they appear as (p,q)-cables of a component k that can not be thrown out since in Hamiltonian Systems the first three operations of Wada are not used, then

1) if k has index 0 or 2, $ext_1[l]$ is equivalent to the link obtained applying operation V of Wada on the component k of l :

$$ext_1[l] = V(l).$$

2) if k has index 1, as we have seen before, we must use operation VI after operation V for changing the index of k, then we obtain the same restriction for l, so:

$$ext_2[VI(l_o)] = VI[V(l_o)].$$

In the particular case that $p = q = 1$ the new orbits are trivial knots linked between each other.

2.2. Period doubling

Consider the case when the characteristic multipliers are equal to -1. Before the bifurcation there exists a hyperbolic (elliptic) periodic orbit that becomes elliptic (hyperbolic) after the bifurcation and a doubled period hyperbolic (elliptic) orbit appears, i.e. the stability type of the orbits changes from hyperbolic to elliptic, or vice versa. At the transitional point a new period two orbit appears on one side of μ_o; this is called period doubling *in the literature.*

Let k be a component with index 0 or 2 (or 1) of a link l. We denote $flip_1[l]$ the link obtained changing the index of k from 0 or 2 to 1 and adding a doubled period orbit with index 0 or 2, and $flip_2[l]$ the link obtained changing the index of k from 1 to 0 or 2 and adding a doubled period orbit with index 1.

Proposition 2 *$\forall$ l link on S^3*

1) $flip_1[l]$ is always a link on S^3.

2) $flip_2[l]$ is a link on S^3 if $l = VI(l_o)$, where l_o is a link on S^3 and a $(2,1)$-cable is added .

Proof: In the first case, the link obtained after the bifurcation is equivalent to the result of applying operation VI of Wada on the previous one since it consists in changing the index of a component k with index 0 (or 2) to 1 and adding a $(2,q)$-cable with index 0 or 2, i.e., a double period orbit for k, so:

$$\forall l \quad \text{link on } S^3 \qquad flip_1[l] = VI(l).$$

In the second case, the doubled-period orbit that appears after the bifurcation is a $(2,q)$*-cable with index* 1 *and the only way to obtain this type of cables is using operation* V. *But, when operation* V *is applied, two parallel* $(2,q)$*-cables appear. As the bifurcation consists in the appearance of only one cable, the other one have to be in the previous link, so the link before the bifurcation must have an orbit with index* 1 *with a* $(2,q)$*-cable of index* 0 *or* 2, *that is, it comes from operation* VI *of Wada. Then, we start from a link* l_o *and choose a component* k *of it . If we make* $VI(l_o)$ *we have a link with a component* k *of index* 1 *which has a* $(2,q)$*-cable with index* 0 *or* 2. *When we make* $V(l_o)$ *we have a link where the component* k *has index* 0 *or* 2 *and with two* $(2,q)$*-cables, one of them with index* 0 *or* 2 *and the other one with index* 1. *In the case* $q=1$ *the cable with index* 1 *can be carried continuously onto the base component. Therefore, we can see that* $V(l_o)$, *when a* $(2,1)$*-cable is used, is the link corresponding to the bifurcation of* $VI(l_o)$. *So:*

$$flip_2[VI(l_o)] = V(l_o)$$

2.3. k-Bifurcation orbits

In this case the characteristic multipliers of bifurcation point are k*th roots of the unity, with* $k \geq 3$. *Two different cases of bifurcations are obtained depending on the value of* k. *If* $k=3$ *there is a 3-bifurcation point for* μ_o. *Then there is a hyperbolic periodic orbit of period three which exits for values of* μ *near both sides of* μ_o *and tends to the 3-bifurcation point as* $\mu \to \mu_o$ *from either side.*

If there is a 4-bifurcation point, then:

a) there is a hyperbolic periodic orbit of period 4 *that exists for values of* μ *near both sides of* μ_o *and tends to the 3-bifurcation as* $\mu \to \mu_o$ *from either side.*

b) there exists and elliptic and a hyperbolic periodic orbit of period 4 *when* $\mu < \mu_o$ *(or* $\mu > \mu_o$*) and no periodic orbit of period* 4 *when* $\mu > \mu_o$ *(respectively* $\mu < \mu_o$*). As* $\mu \to \mu_o$ *from the appropriate side, both the elliptic and hyperbolic orbits tend to the 4-bifurcation point.*

If $k \geq 5$ *there exits an elliptic and also a hyperbolic period orbit of period* k *for* $\mu < \mu_o$ *(respectively* $\mu > \mu_o$*) and no periodic orbit of period* k *when* $\mu > \mu_o$ *(respectively* $\mu < \mu_o$*). As* $\mu \to \mu_o$ *from the appropriate side, both the elliptic and hyperbolic orbits tend to the k-bifurcation point.*

We see that the case for $k=4$ *is sometimes as the case for* $k=3$ *and sometimes as the case for* $k \geq 5$ *depending on the relative size of the twist term and the resonance term.*

In the two first cases, the number and the type of the orbits before and after the bifurcation point are the same, so we are not able to distinguish them from a topological point of view. But, in the two last cases, after the bifurcation point, two new periodic orbits of period k ($k \geq 4$) appear, one hyperbolic ($i = 1$) and one elliptic ($i = 0,2$), then, we can identify this type of bifurcation with operation V of Wada. Both types of bifurcations can be considered, depending on the knot type obtained when we apply this operation .

We denote $bif_i[l]$ a link obtained from l adding two periodic orbits of period $k \geq 4$, one elliptic and the other one hyperbolic, that are, of the same type of knot linked similarly to the rest of the link, $bif_1[l]$ will be the link obtained when the new orbits are linked to a k-component with index 0 or 2, and $bif_2[l]$ when they are linked to a component with index 1.

Proposition 3 *$\forall\, l$ link on S^3 :*

1) $bif_1[l]$ is always an admissible link on S^3.

2) $bif_2[l]$ is an admissible link on S^3 when $l = VI(l_o)$, where l_o is an admissible link on S^3.

Proof: As we have seen in proposition 1, every link l on S^3 can suffer this type of bifurcation because if two parallel orbits of period $k \geq 4$ are added to a link l, they correspond to the cables of a component k (when the index of k is 0 or 2) of l, so:

$$bif_1[l] = V(l)$$

and the new two orbits can be represented by (p,q)-toral knots with $p \geq 4$. So, this bifurcation can occur on every link of S^3, and the two new orbits must appear around one component k of l.

When this component has index 1 it is necessary to apply operation VI for changing the index of k, and as we have seen in the proof of proposition 1 the existence of a $(2,q)$-cable implies that the link that suffers the bifurcation must come from operation VI, so:

$$bif_2[VI(l_o)] = VI[V(l_o)].$$

3. Conclusions

Theorem 1 *A link l can suffer a period doubling bifurcation when the doubled period orbit that appears has index 1, only when this orbit appears parallel to another one with index 0 or 2.*

Proof: As we have seen in proposition 2 the doubled-period orbit that appears after the bifurcation is a $(2,1)$-cable with $i = 1$ and the only way to get it is *using operation V, but it is not possible to throw out the other cable that appears when this operation is applied, so it has to be in the previous link and the new orbit that appears must do it parallel to this cable..*

From the expressions of the links obtained in the propositions we deduce that when generic codimension one bifurcations occur,only iterated toral knots are involved.

Theorem 2 *When two new orbits appear linked between each other for a generic codimension one bifurcation, they have to be represented as cables of a component k of the previous link.*

Proof: As we have seen in propositions 1 and 3, this type of orbits are associated with (p,q)-cables generated by operation V of Wada, and the base component can not be thrown out because in Hamiltonian systems we can only use the last three Wada operations, so the new orbits always appear around a component k as cables of it.

We can also associate some Wada operations with some type of generic codimension one bifurcations to the effect that the link obtained is equivalent to the result of applying these operations over the previous one.

- *In proposition 1 operation IV on a Hopf link is associated with that bifurcation when the new orbits that appear after the bifurcation are trivial knots linked to a component k of the link but not linked between each other.*
- *Operation VI can be associated with a period doubling bifurcation (see proposition 2), in the case that the orbit that bifurcates has index* 0 *or* 2.
- *In propositions 1 and 3 we see that operation V can be associated with those bifurcations when the new orbits are linked between them.*

From this associations we deduce that:

Theorem 3 *Given two Bott integrable Hamiltonian system whose links of NMS periodic orbits on Q consist of iterated toral knots, one can be obtained from the other by a sequence of generic codimension one bifurcations. If two links differ in periodic orbits that are not iterated toral knots,they can not be related by a sequence of of generic codimension one bifurcations.*

Proof: If we have two set of periodic orbits in a integrable Hamiltonian system which can be represented as links composed of iterated toral knots we can rewrite the different components in terms of Wada operations which have been associated with certain bifurcations, so we can obtain the sequence of generic codimension one bifurcations that carries one link onto the other. The second statement follows from the fact that generic codimension one bifurcations only produce iterated toral knots.

4. References

1. *Birman, J. and R.F. Williams.* Knotted periodic orbits in dynamical systems I: Lorenz's equations. *Topology* **22**, *(1983), 47-82.*
2. *Birman, J. and R.F. Williams.* Knotted periodic orbits in dynamical systems II: knot holders for fibered knots. *Contemporary Mathematics* **20** *(1983), 1-60.*
3. *Campos, B., Martínez Alfaro, J. and Vindel, P.* Bifurcations of links of periodic orbits in Non-Singular Morse-Smale Systems on S^3. *To be published*
4. *Casasayas, J., Martínez Alfaro, J. and Nunes A.* Knots and Links in Integrable Hamiltonian Systems. *Warwick Prepints:* **48** *(1992)*
5. *Fomenko, A.T.* Integrability and Non integrability in Geometry and Mechanics. *Kluwe Academic Publishers, 1988*
6. *Fomenko, A.T., T.Z. Nguyen.* Topological Classification of Integrable Non degenerate Hamiltonians on the Isoenergy Three-dimensional Sphere. *Advances in Soviet Math.* **6**, *(1991), 267-297*
7. *Meyer, K. R. and Hall, G. R.* Introduction to Hamiltonian Dynamical Systems and the N-Body Problem. *Springer-Verlag.1992.*
8. *Sasano, K.* Links of closed orbits of non-singular Morse-Smale flows. *Proc. A.M.S.* **88**, *(1983), 727-734.*
9. *Uezu, T. and Y. Aizawa.* Three-Dimensional Ordinary Differential Equation System. *Progress on Theor. Phys.*, **68**, *no. 6, (1982)*
10. *Wada, M.* Closed orbits of non-singular Morse-Smale flows on S^3. *J. Math. Soc. Japan,* **41**, *n°3, (1989), 405-413.*
11. *Yano, K.* A note on non-singular Morse-Smale flows on S^3. *Proc. Japan Acad.*, **58**, *(1982), 447-450.*

POINCARE-MELNIKOV METHOD FOR PARABOLIC POINTS

JOSEFINA CASASAYAS
Departament de Matemàtica Aplicada i Anàlisi, Universitat de Barcelona, Gran Via 585, 08071 Barcelona, Spain

ERNEST FONTICH
Departament de Matemàtica Aplicada i Anàlisi, Universitat de Barcelona, Gran Via 585, 08071 Barcelona, Spain

and

ANA NUNES
Departamento de Física, Universidade de Lisboa, Campo Grande, C1, 1700 Lisboa, Portugal

1. Introduction

The subject of this paper is the application of the Poincaré-Melnikov method to detect the existence of transversal intersections of invariant manifolds of parabolic points or orbits. We give a set of sufficient conditions for the applicability of the method and we also provide examples where the method would give erroneous results.

The study of this question was motivated by the fact that in several problems in Celestial Mechanics we obtain, after regularization of the collision or infinity singularities, a one parameter family of vector fields possessing a parabolic periodic orbit and, for a certain value of the parameter, a cylinder of orbits homoclinic to the parabolic orbit (see [CFN2], [Mc], [MP], [X]). In some of these problems, such as the one considered in [Mc], the linear part of the Poincaré map associated to the parabolic orbit is the identity. In some others, such as the one studied in [CFN2], the linear part of the Poincaré map associated to the orbit is the identity plus a nilpotent term.

The present paper is in a way a continuation of [CFN2], but instead of restricting ourselves to the highly degenerate model considered in that previous work, which corresponds to a perturbed central force problem with a $1/r^2$ potential, we shall present a more general discussion of the applicability of Poincaré-Melnikov method in the case of parabolic orbits with nilpotent linear part.

Perhaps the most natural example that gives rise to a problem of this sort is that of a time dependent perturbation of a Hamiltonian of the form

$$H(x,p) = \frac{p^2}{2} + V(x),$$

where $V(x) = \mathcal{O}(x^3)$. So, to put ourselves in the simplest possible setting, we shall consider a family of systems with one and a half degrees of freedom having the following properties:

(1) The system is a Hamiltonian vector field plus a (not necessarily Hamiltonian) perturbation.

(2) For every value of the parameter there exists a periodic orbit and the Poincaré map around this orbit has a parabolic fixed point whose linear part is the identity plus a nilpotent term.

In Section 2 we state the results that we have obtained for these families of systems (the proofs can be found in [CFN3]). These results can be readily extended to other cases, such as, for instance, that of autonomous perturbations of 2-degrees of freedom Hamiltonians, or the case when the unperturbed vector field is not Hamiltonian. In Section 3 we show by an example that the formal application of the Poincaré-Melnikov method may give erroneous results. However, there are cases not covered by the results of Section 2 where the Poincaré-Melnikov method still holds. This is illustrated by two numerical examples.

2. Statement of the results

In this section we shall state without proof the main results of [F] and [CFN3].

Consider the system

$$\dot{z} = f(z) + \varepsilon g(z, t, \varepsilon), \tag{1}$$

with $z = (x, y) \in U \subset \mathbb{R}^2$, $f = JD_zH$ where H is a Hamiltonian, g is $T(\varepsilon)$-periodic in t, and f and g are analytic.

We shall also assume that system (1) possesses a fixed point at $z = (0, 0)$, that is $f(0) = g(0, t, \varepsilon) = 0$, and that this fixed point is parabolic with linear part of the form

$$\begin{pmatrix} 0 & \nu(t,\varepsilon) \\ 0 & 0 \end{pmatrix}$$

and $\int_0^{T(0)} \nu(t, 0)\, dt \neq 0$. Furthermore, we assume the existence of a homoclinic orbit σ associated to $z = 0$ for $\varepsilon = 0$.

Proposition 1 ([CFN3]). *There exists a polynomial change of variables $z = (x, y) \to (\xi, \eta)$ such that the time T-Poincaré map of system (1) up to third* *order terms is*

$$F_\varepsilon \begin{pmatrix} \xi \\ \eta \end{pmatrix} = \begin{pmatrix} \xi + \eta \\ \eta + g_0(\varepsilon)\xi^2 + g_1(\varepsilon)\xi\eta + g_2(\varepsilon)\xi^3 + g_3(\varepsilon)\xi^2\eta \end{pmatrix} + O_4(\xi, \eta)$$

See [CFN3] for the explicit expression of the coefficients $g_i(\epsilon)$, $i = 0, 1, 2, 3$, in terms of the functions f and g of (1).

Theorem 2 ([F]).
(a) If $g_0(0) > 0$, there exists a local stable invariant manifold $W^s_{\delta,\varepsilon}(0)$ =graf ϕ_ε, with $\phi_\varepsilon(x) = \phi(x, \varepsilon)$ and $\phi : (0, \delta) \times (\varepsilon_1, \varepsilon_1) \to \mathbb{R}$ depends analytically on x and ε. Furthermore, we have the asymptotic expression

$$\phi_\varepsilon(x) \approx -\sqrt{\frac{2}{3} g_0(\varepsilon)} x^{3/2}.$$

(b) If $g_0(\varepsilon) \equiv 0$,and $g_2(0) > 0$ or and $g_1(0) < 0$, statement (a) holds with

$$\phi_\varepsilon(x) \approx \frac{g_1(\varepsilon) - \sqrt{(g_1(\varepsilon))^2 + 8g_2(\varepsilon)}}{4} x^2.$$

Applying Theorem 2 to the inverse map, similar results hold for the unstable manifolds. For higher order degeneracies we must compute the normal form up to higher order terms and use the corresponding invariant manifold results, see [F].

Consider again system (1) and assume for simplicity that the period T of the perturbation does not depend on ϵ (this assumption can be easily removed, see [CFN3]). If we look at this system as an autonomous system defined for $(z, \theta) \in U \times \mathbb{R}/T\mathbb{Z}$,

$$\begin{aligned} \dot{z} &= f(z) + \varepsilon g(z, \theta, \varepsilon), \\ \dot{\theta} &= 1, \end{aligned} \tag{2}$$

the time-T map given by Proposition 1 is the Poincaré map of system (2) taking the plane $\theta = \theta_0$ as a surface of section. We shall also assume that this Poincaré map falls into one of the cases of Theorem 2, so that equations (2) have stable and unstable manifolds $\widetilde{W^{s,u}_\varepsilon}$ associated to the periodic orbit at $z = (0, 0)$ for all values of ε in $(-\varepsilon_0, \varepsilon_0)$. In fact, the invariant manifolds $W^{s,u}_\varepsilon$ given by Theorem 2 are the intersection of $\widetilde{W^{s,u}_\varepsilon}$ with the plane $\theta = \theta_0$. Also, under the assumption of the existence of the homoclinic orbit σ, two branches of $\widetilde{W^u_0}$ and $\widetilde{W^s_0}$ coincide. By the analytic dependence of $\widetilde{W^{s,u}_\varepsilon}$ in ε, the invariant manifolds for the perturbed system will be close to $\widetilde{W_0^{s,u}}$ if ε is small.

Taking arbitrary θ_0 and t_0 in $\mathbb{R}/T\mathbb{Z}$ and $\mathbb{R}$ respectively, consider the plane $\theta = \theta_0$ and a point $p = \sigma(-t_0)$ on the homoclinic orbit of the unperturbed system. Denote by L_p the straight line through p in $\theta = \theta_0$ orthogonal to σ at p and by $z^u(t_0, \theta_0, \varepsilon)$ (resp. $z^s(t_0, \theta_0, \varepsilon)$) the intersection $L_p \cap \widetilde{W_\varepsilon^u}$ (resp. $L_p \cap \widetilde{W_\varepsilon^s}$). By the analytic dependence of $W_\varepsilon^{s,u}$ on ε, these intersections exist for ε small enough.

We define the signed distance between $\widetilde{W_\varepsilon^u}$ and $\widetilde{W_\varepsilon^s}$ at the point p, denoted by $d(t_0, \theta_0, \varepsilon)$, to be

$$\begin{aligned} d(t_0, \theta_0, \varepsilon) &= (z^u(t_0, \theta_0, \varepsilon) - z^s(t_0, \theta_0, \varepsilon)) \cdot \frac{D_z H(p)}{\|D_z H(p)\|} \\ &= \pm\,(z^u(t_0, \theta_0, \varepsilon) - z^s(t_0, \theta_0, \varepsilon)), \end{aligned} \tag{3}$$

where the last equality comes from the fact that $D_z H(p)$ is orthogonal to $f(p)$.

The Melnikov function is obtained by considering the Taylor expansion of $d(t_0, \theta_0, \varepsilon)$ in ε about $\varepsilon = 0$, which gives

$$d(t_0, \theta_0, \varepsilon) = d(t_0, \theta_0, 0) + \varepsilon \frac{\partial d}{\partial \varepsilon}(t_0, \theta_0, 0) + O(\varepsilon^2),$$

where $d(t_0, \theta_0, 0) = 0$. The Melnikov function is defined to be

$$\begin{aligned} M(t_0, \theta_0) =& \frac{\partial d}{\partial \varepsilon}(t_0, \theta_0, 0)\|D_z H(\sigma(-t_0))\| = \\ =& D_z H(\sigma(-t_0)) \cdot \left(\frac{\partial z^u}{\partial \varepsilon}(t_0, \theta_0, 0) - \frac{\partial z^s}{\partial \varepsilon}(t_0, \theta_0, 0) \right), \end{aligned} \tag{4}$$

where the last equality is obtained from (3).

To have a computable expression of $M(t_0, \theta_0)$ without needing to know the solutions of the perturbed system, one starts to define the time-dependent Melnikov function as follows

$$M(t; t_0, \theta_0) = D_z H\left(\sigma(t - t_0)\right) \cdot \left(\frac{\partial z^u}{\partial \varepsilon}(t; t_0, \theta_0, 0) - \frac{\partial z^s}{\partial \varepsilon}(t; t_0, \theta_0, 0) \right),$$

and using that $\partial z^{u,s}/\partial \varepsilon$ satisfy the first order variational equation, and the fact that the unperturbed system is Hamiltonian, one arrives at

$$\begin{aligned} M(t_0, \theta_0) = M(0; t_0, \theta_0) = & \int_{-\tau}^{\tau} D_z H\left(\sigma(t - t_0)\right) \cdot g\left(\sigma(t - t_0), t + \theta_0, 0\right) dt \\ & + (\Delta^u(-\tau) - \Delta^s(\tau)), \end{aligned}$$

where

$$\Delta^{u,s}(t) = D_z H\left(\sigma(t - t_0)\right) \cdot \frac{\partial z^{u,s}}{\partial \varepsilon}(t; t_0, \theta_0, 0).$$

Our main result is the following

Theorem 3. *Consider system (1) and assume that for ϵ small enough its time T Poincaré map falls into one of the four cases considered in Theorem 2. Then,*

$$M(t_0,\theta_0) = \int_{-\infty}^{+\infty} D_zH\left(\sigma(t-t_0)\right)\cdot g\left(\sigma(t-t_0),t+\theta_0,0\right)dt.$$

Moreover, suppose we have a point $(t_0,\theta_0)=(\widetilde{t_0},\widetilde{\theta_0})$ such that $M(\widetilde{t_0},\widetilde{\theta_0})=0$ and $\frac{\partial M}{\partial \widetilde{t_0}}(\widetilde{t_0},\widetilde{\theta_0})\neq 0$. Then W^s_ε and W^u_ε intersect transversally at $(\sigma(-\widetilde{t_0})+O(\varepsilon),\widetilde{\theta_0})$.

Sketch of the Proof: To prove the formula we have to show that $\Delta^u(-\tau)$ and $\Delta^s(\tau)$ go to 0 as $\tau\to\infty$. Since $D_zH(\sigma(t-t_0))$ goes to zero as $t\to\pm\infty$, it is enough to prove that $\frac{\partial z^s}{\partial\varepsilon}(t;t_0,\theta_0,0)$ is bounded for positive t and $\frac{\partial z^u}{\partial\varepsilon}(t;t_0,\theta_0,0))$ is bounded for negative t. For that it is enouh that

$$\|z^s(t;t_0,\theta_0,\varepsilon)-z^s(t;t_0,\theta_0,0)\|\leq c\varepsilon$$

for $t\geq 0$, and the analogous bound for z^u for $t\leq 0$. We first check that the bound is true for z^s for values of time $t=mT$, $m\geq 0$, by considering the Poincaré map of the equation. For that we change variables to put the map in normal form and then we prove that there exist $c_1>0$, $c_2>0$ such that if $z_1\in W^s_0$ and $z_2\in W^s_\varepsilon$ with $\|z_2-z_1\|<c_2\varepsilon$ then

$$\|F^m_\varepsilon(z_2)-F^m_0(z_1)\|<c_1\varepsilon,\quad \forall m\in N.$$

by defining $\Delta_m=\|F^m_\varepsilon(z_2)-F^m_0(z_1)\|$ and getting the recurrent relation

$$\Delta_{m+1}\leq\varepsilon(1+3c_3)(\pi_1F^m_\varepsilon(z_2))^\beta+\Delta_m,$$

which gives the bound using the analytical dependence on the invariant manifolds with respect to ε and the estimates on the convergence of the points on the invariant manifolds to the fixed point, which is such that

$$\sum_{j=0}^{\infty}\left(\pi_1F^j_\varepsilon(z_2)\right)^\beta$$

is convergent.

For $t\geq 0$, $t\neq mT$, the result easily follows from the continuous dependence of the solution on initial conditions and parameters. The analogous bounds hold for z^u.

The claim about the transversal intersection is proved using the implicit function theorem as in the hyperbolic case.

3. Three examples where Theorem 3 does not apply

If for ϵ small enough the Poincaré map of system (1) does not fulfill neither (a) nor (b) of Theorem 2 it is easy to find examples where the perturbed system no longer possesses stable and unstable invariant manifolds associated to the parabolic orbit. However, the case when one of the coefficients $g_i(\epsilon)$, $i = 0, 1$, or 2 becomes zero at $\epsilon = 0$ is open since Theorem 3 does not apply. In this section we shall consider three examples where this situation occurs. In all three of them, the Melnikov integral converges and gives, up to first order in the parameter, the distance between the invariant manifolds, provided that these manifolds exist.

Consider first a simple example of a Hamiltonian with a parabolic fixed point and a heteroclinic solution where we can compute explicitely the distance between the invariant manifolds. We take a perturbed system whose perturbation has smaller order in x than the nonlinear part of the unperturbed system at the parabolic point. The perturbation has been chosen so that the separation between the invariant manifolds can be computed explicitely. Let

$$H_\epsilon(x,y) = \frac{1}{2}y^2 - \frac{1}{n+3}x^{n+1}(x-2)^2 - \epsilon\left(\frac{x^4}{4} - 2\frac{x^3}{3}\right).$$

The equations of motion are

$$\begin{aligned} \dot{x} &= y \\ \dot{y} &= x^n\left(x - \frac{2(n+1)}{n+3}\right)(x-2) + \epsilon x^2(x-2). \end{aligned}$$

The unperturbed system has three fixed points, $(0,0)$, $(\alpha, 0)$ and $(2,0)$, with $\alpha = 2(n+1)/(n+3) \in (0,2)$. The first one is parabolic (if $n > 1$), the second is elliptic and the third is hyperbolic. The fixed points $(0,0)$ and $(2,0)$ are connected by two heteroclinic orbits, given by the graphs of the functions

$$\phi^\pm(x) = \pm\sqrt{\frac{2}{n+3}}x^{(n+1)/2}(2-x)$$

for $0 < x < 2$. The perturbed system also has the fixed points $(0,0)$ and $(2,0)$ with the same nature. The elliptic point is perturbed and remains elliptic if ϵ is small.

The relation $H_\epsilon(x,y) = H_\epsilon(0,0)$ gives the invariant manifolds of $(0,0)$ as graphs of

$$y = \phi^{\pm}(x,\epsilon) = \pm\left[\frac{2}{n+3}x^{n+1}(x-2)^2 + 2\epsilon x^3\left(\frac{x}{4}-\frac{2}{3}\right)\right]^{1/2},$$

and $H_\epsilon(x,y) = H_\epsilon(2,0)$ gives the ones of $(2,0)$

$$y = \psi^{\pm}(x,\epsilon) = \pm\left[\frac{8\epsilon}{3} + \frac{2}{n+3}x^{n+1}(x-2)^2 + 2\epsilon x^3\left(\frac{x}{4}-\frac{2}{3}\right)\right]^{1/2}.$$

Notice that $\phi^+(x,0) = \psi^+(x,0)$ correspond to a heteroclinic connection of the unperturbed Hamiltonian.

Let $r \in (0,2)$. The distance between the perturbed invariant manifolds measured along the line $x = r$ is given by

$$\overline{\Delta}(r,\epsilon) = \phi^+(r,\epsilon) - \psi^+(r,\epsilon) = \frac{1}{2}\xi^{-1/2}\left(-\frac{8}{3}\epsilon\right),$$

with

$$0 < \xi - \frac{2}{n+3}r^{n+1}(r-2)^2 + 2\epsilon r^3\left(\frac{r}{4}-\frac{2}{3}\right) < \frac{8}{3}\epsilon,$$

so that $\xi = \frac{2}{n+3}r^{n+1}(r-2)^2 + O(\epsilon)$ and hence

$$\overline{\Delta}(r,\epsilon) = -\frac{4}{3\sqrt{\frac{2}{n+3}}r^{(n+1)/2}(2-r)}\epsilon + O(\epsilon^2) = -\frac{4}{3\phi^+(r)}\epsilon + O(\epsilon^2).$$

To have the distance between the perturbed manifolds measured along a line orthogonal to the unperturbed heteroclinic orbit to first order in ϵ we have to multiply $\overline{\Delta}$ by the cosinus of the angle between the vectors $(0,1)$ and $J\nabla H_0(x,y)^\perp = (\frac{\partial H_0}{\partial y}, -\frac{\partial H_0}{\partial x})^\perp = (-x^n(x-\frac{2n+2}{n+3})(x-2), y)$ at $(x,y) = (r, \phi^+(r))$. This gives

$$\Delta(r,\epsilon) = \overline{\Delta}(r,\epsilon).\frac{\phi^+(r)}{||J\nabla H_0(x,y)||} + O(\epsilon) = -\frac{4}{3}\frac{1}{||\nabla H_0(x,y)||}\epsilon + O(\epsilon^2).$$

On the other hand, a formal Poincaré-Melnikov theory predicts that the distance between the perturbed invariant manifolds, measured as above, is

$$\Delta(r,\epsilon) = \frac{\epsilon}{N}\int_{-\infty}^{+\infty}[yx^2(x-2)\circ\sigma]\,dt + O(\epsilon^2),$$

where $N = ||J\nabla H_0(x,y)||$ is the modulus of the unperturbed vector field at $(x,y) = (r, \phi^+(r))$ and $\sigma(t)$ is the heteroclinic solution of the unperturbed system. The integral in the above expression is easily computed:

$$\begin{aligned}\int_{-\infty}^{+\infty} [yx^2(x-2)\circ\sigma]\,dt &= \int_{-\infty}^{+\infty} \sigma_1^2(t)\,(\sigma_1(t)-2)\,\sigma_2(t)\,dt \\ &= \int_{-\infty}^{+\infty} \sigma_1^2(t)\,(\sigma_1(t)-2)\,\dot\sigma_1(t)\,dt \\ &= [\frac{\sigma_1^4(t)}{4} - \frac{2\sigma_1^3(t)}{3}]_{-\infty}^{+\infty} = \frac{2^4}{4} - \frac{2^4}{3} = -\frac{4}{3}.\end{aligned}$$

Therefore, both calculations agree. Notice however that the invariant manifolds of $(0,0)$ exist for the perturbed system only when $\epsilon > 0$. For $\epsilon < 0$ the computation of the Melnikov integral has no meaning.

Consider now the system associated to the Hamiltonian

$$H(x,y,t) = \frac{1}{2}y^2 - x^4 + x^5 - \epsilon x^3(\sin t + 1),$$

whose equations are

$$\begin{aligned}\dot x &= y \\ \dot y &= 4x^3 - 5x^4 + 3\epsilon x^2(1+\sin t).\end{aligned}$$

The origin is a fixed point with linear part

$$\begin{pmatrix} 0 & 1 \\ 0 & 0 \end{pmatrix}$$

and the unperturbed system has a homoclinic orbit to the parabolic point $(0,0)$ given by $y^2 = 2x^4 - 2x^5$.

The time-2π Poincaré map from $t = 0$ to $t = 2\pi$ around the orbit at $(x,y) = (0,0)$ is

$$\begin{aligned}F_\epsilon \begin{pmatrix} x \\ y \end{pmatrix} &= \begin{pmatrix} x + 2\pi y \\ y \end{pmatrix} \\ &+ 2\pi\epsilon \begin{pmatrix} 3(1+\pi)x^2 & 4\pi^2 xy & 2(\pi^3-9)y^2 \\ 3x^2 & 6(\pi-1)xy & 2\pi(2\pi-3)y^2 \end{pmatrix} \\ &+ \begin{pmatrix} \cdots \\ [8\pi + 6\pi\epsilon^2(4\pi^2-9)]x^3 + \ldots \end{pmatrix},\end{aligned}$$

or, in the normal form given by Proposition 1,

$$F_\epsilon \begin{pmatrix} \xi \\ \eta \end{pmatrix} = \begin{pmatrix} \xi + \eta \\ \eta + 12\pi^2\epsilon\xi^2 + 24\pi^2\epsilon\xi\eta + [16\pi^2 + 12\pi^2\epsilon^2(2\pi^2 - 3)]\xi^3 \end{pmatrix} + \cdots.$$

In terms of the notation of Proposition 1 and Theorem 2 we have

$$g_0(\epsilon) = 12\pi^2\epsilon,\ g_1(\epsilon) = 24\pi^2\epsilon,\ g_2(\epsilon) = 16\pi^2 + 12\pi^2\epsilon^2(2\pi^2 - 3).$$

The Melnikov integral is

$$\begin{aligned} M(t_0, \theta_0) &= \int_{-\infty}^{\infty} 3\dot{x}(t - t_0)x^2(t - t_0)(1 + \sin(t + \theta_0))\,dt \\ &= \int_{-\infty}^{\infty} 3\dot{x}(t)x^2(t)\sin(t + t_0 + \theta_0)\,dt \\ &= \cos(t_0 + \theta_0)\int_{-\infty}^{\infty} 3\dot{x}(t)x^2(t)\sin t\,dt \end{aligned}$$

where $x(t - t_0)$ is the solution along the homoclinic orbit of zero energy of the unperturbed system with initial condition $x(0) = 1$, and we have used that $x(t)$ is an even function. Taking $t_0 = \theta_0 = 0$ we may write

$$M(0,0) = 2\int_0^{+\infty} 3\dot{x}(t)x^2(t)\sin(t)\,dt.$$

When $t \in (0, +\infty)$ we have $dx/dt = -\sqrt{2x^4 - 2x^5}$. Hence, performing a change of variables $t = T(x)$, with

$$T(x) = \int_x^1 \frac{dx}{\sqrt{2x^4 - 2x^5}} = \frac{1}{\sqrt{2}}\left(\frac{\sqrt{1-x}}{x} + \tanh^{-1}(\sqrt{1-x})\right),$$

we obtain

$$M(0,0) = -6\int_0^1 x^2 \sin\left(\frac{\sqrt{1-x}}{\sqrt{2}x} + \frac{\tanh^{-1}(\sqrt{1-x})}{\sqrt{2}}\right)dx = -1.22004,$$

which is in good agreement with the distance between the invariant manifolds computed numerically by extending the local manifolds (see Table 1).

ϵ	Δ	ϵ	Δ
1.000×10^{-3}	1.2168×10^{-3}	5.000×10^{-4}	6.0921×10^{-4}
9.000×10^{-4}	1.0954×10^{-3}	4.000×10^{-4}	4.8750×10^{-4}
8.000×10^{-4}	9.7397×10^{-4}	3.000×10^{-4}	3.6572×10^{-4}
7.000×10^{-4}	8.5245×10^{-4}	2.000×10^{-4}	2.4387×10^{-4}
6.000×10^{-4}	7.3086×10^{-4}	1.000×10^{-4}	1.2197×10^{-4}

Table 1. The distance Δ between the invariant manifolds for several values of ϵ

Again, however, these computations are meaningful only when $\epsilon > 0$.

The last example that we shall consider is the system associated to the Hamiltonian

$$H(x,y,t) = \frac{1}{2}y^2 + \frac{1}{3}y^3 + \frac{x^4 y}{4} - \varepsilon\frac{x^4}{4}(\sin t + 1),$$

whose equations are

$$\begin{aligned}\dot{x} &= y + y^2 + \frac{x^4}{4}\\ \dot{y} &= -x^3 y + \epsilon x^3(1+\sin t).\end{aligned}$$

The origin is a fixed point with linear part

$$\begin{pmatrix} 0 & 1 \\ 0 & 0 \end{pmatrix}$$

and the unperturbed system has a homoclinic orbit to the parabolic point $(0,0)$ given by $x^4 + 2y + 4y^2/3 = 0$.

The time-2π Poincaré map from $t = 0$ to $t = 2\pi$ around the orbit at $(x,y) = (0,0)$ is

$$\begin{aligned}F_\epsilon\begin{pmatrix} x \\ y \end{pmatrix} &= \begin{pmatrix} x + 2\pi y \\ y \end{pmatrix}\\ &+ \begin{pmatrix} 2\pi y^2 + a_1(\epsilon)x^3 + a_2(\epsilon)x^2 y + a_3(\epsilon)xy^2 + a_4(\epsilon)y^3 \\ b_1(\epsilon)x^3 + b_2(\epsilon)x^2 y + b_3(\epsilon)xy^2 + b_4(\epsilon)y^3 \end{pmatrix} + O_4(x,y),\end{aligned}$$

where $a_i(\epsilon) = O(\epsilon)$, $b_i(\epsilon) = O(\epsilon)$, $i = 1, ..., 4$, or, in the normal form given by Proposition 1,

$$F_\epsilon\begin{pmatrix} \xi \\ \eta \end{pmatrix} = \begin{pmatrix} \xi + \eta \\ \eta + 4\pi^2\epsilon\xi^3 \end{pmatrix} + O_3(\xi,\eta).$$

ϵ	Δ	ϵ	Δ
1.000×10^{-3}	-4.5261×10^{-4}	5.000×10^{-4}	-2.2653×10^{-4}
9.000×10^{-4}	-4.0743×10^{-4}	4.000×10^{-4}	-1.8125×10^{-4}
8.000×10^{-4}	-3.6223×10^{-4}	3.000×10^{-4}	-1.3597×10^{-4}
7.000×10^{-4}	-3.1701×10^{-4}	2.000×10^{-4}	-9.0665×10^{-5}
6.000×10^{-4}	-2.7178×10^{-4}	1.000×10^{-4}	-4.5341×10^{-5}

Table 2. The distance Δ between the invariant manifolds for several values of ϵ

In terms of the notation of Proposition 1 and Theorem 2 we have

$$g_0(\epsilon) \equiv 0, \ g_1(\epsilon) \equiv 0, \ g_2(\epsilon) = 4\pi^2\epsilon,$$

and the coefficient $g_2(\epsilon)$ has a simple zero at $\epsilon = 0$. When $\epsilon = 0$, the first non-zero coefficient for the Poincaré map in normal form is the coefficient of the term in $\xi^3\eta$. Computing the Melnikov function for this system in the same way as before we get $M(0,0) = -0.340$. Here, the unperturbed homoclinic orbit at $t = 0$ is at the point $(0, -3/2)$, where the tangent vector to the orbit is $(1,0)$. The modulus of the unperturbed vector field at $(0, -3/2)$ is $3/4$. Hence, the distance between the invariant manifolds measured along the line $x = 0$ is

$$-0.453\epsilon + O(\epsilon^2).$$

This is again in good agreement with the numerical results obtained as in the previous example (see Table 2).

4. Acknowledgements

This work has been partially supported by *Acciones integradas Hispano-Portuguesas 1993* and JNICT grant n. PB1082/92.

5. References

[CFN1] J. Casasayas, E. Fontich and A. Nunes, *Invariant manifolds for a class of parabolic points* (1992) Nonlinearity **5**, p. 1193-1210.

[CFN2] J. Casasayas, E. Fontich, and A. Nunes, *Transversal Homoclinic Orbits for a Class of Hamiltonians* (1993), Hamiltonian Systems and Celestial

Mechanics, Advanced Series in Nonlinear Dynamics, World Scientific **4**, p. 35-44.

[CFN3] J. Casasayas, E. Fontich and A. Nunes, *Homoclinic orbits to parabolic points* (1994), preprint.

[F] E. Fontich, (1994), preprint.

[Mc] R. McGehee, *A stable manifold theorem for degenerate fixed points with applications to Celestial Mechanics* (1973), Journal of Differential Equations **14**, p. 70-88.

[MP] R. Martinez and C. Pinyol, *Parabolic Orbits in the Elliptic Restricted Three Body Problem* (1994), Journal of Differential Equations **111**, p. 299-339.

[X] Z. Xia, *Melnikov Method and Transversal Homoclinic Orbits in the Restricted Three-body Problem* (1992), Journal of Differential Equations **96**, p. 170-184.

INTEGRABLE HAMILTONIANS ON SIMPLE, IRREDUCIBLE, 3-MANIFOLDS

JOSEFINA CASASAYAS
Departament de Matemàtica Aplicada i Anàlisi, Universitat de Barcelona, Gran via 585, 08071 Barcelona, Spain

and

ANA NUNES
Departamento de Física, Universidade de Lisboa, Campo Grande, C1, 1700 Lisboa, Portugal

ABSTRACT

We give a characterization of the simple, irreducible, closed, orientable 3-manifolds Q which are non-singular energy levels of Hamiltonian systems H Bott integrable on Q, and a topological description of the periodic orbits of H in Q.

1. Introduction

In this paper we consider a two-degrees of freedom Hamiltonian H which is Bott integrable on some non-singular 3-dimensional energy level Q.

In the case when Q is S^3, a completly characterization of the links formed by the singular periodic orbits of H was given in [FN] and independently in [CMN1,2]. The proof in [FN] strongly relies on Fomenko's previous work on the topological characterization of integrable Hamiltonian systems, see [F] and [FZ]. The method used in [CMN1,2] is based on the work of Morgan [M] and Wada [W] on non-singular Morse-Smale (NMS) flows, after reducing the problem to that classifying the links formed by the closed orbits of a class of NMS flows on S^3.

Here we deal with the case when Q is a simple, irreducible, closed, orientable 3-manifold, and we use the same approach as in [CMN1,2] to obtain topological restrictions on Q (Theorem 5) as well as a topological characterization of the periodic orbits of H in Q (Theorem 6).

For all the definitions concerning Hamiltonian systems (Bott integrability, non-singular energy level, etc) we refer to [F], for those concerning non-singular Morse-Smale flows and round handle decompositions see [M], and for the standard terms in 3-dimensional topology and knot theory we refer to [H], [J], and [R].

2. Preliminaries

Let M be a compact, smooth, symplectic, 4-dimensional manifold, $H : M \to \mathbb{R}$ a smooth function and Q a non-singular, orientable, energy level of the Hamiltonian system associated to H. Suppose that H is integrable on Q by means of a Bott integral $f : U \to \mathbb{R}$, where U is a neighbourhood of Q in M. We shall call *critical orbits of H in Q* the periodic orbits of H which are critical circles of f.

The following lemma may be found in [CMN2] but we shall prove it for the sake of completeness.

Lemma 1. *There exists a non-singular Morse-Smale flow on Q whose closed orbits are in one-to-one correspondence with the critical orbits of H in Q.*

Proof: The idea is to construct an NMS flow by a suitable perturbation of the vector field defined on Q by $grad\, f$.

First, notice that the critical submanifolds of f on Q are either diffeomorphic to circles or tori, because they must be invariant under the flow of H which has no equilibrium points on Q and, therefore, must be either compact (without boundary) 1-dimensional submanifolds or compact (without boundary) 2-dimensional submanifolds with zero Euler's characteristic.

Consider then the vector field $grad\, f$ defined on Q. The critical circles of f are normally hyperbolic circles of equilibrium points for the flow of $grad\, f$. By perturbing $grad\, f$ in a neighbourhood of these circles, we can change them into stable, unstable or hyperbolic closed orbits. The critical tori of f are also normally hyperbolic and, in a similar way, we can perturb the vector field in a neighbourhood of these tori to obtain an NMS flow on the thick tori with two closed orbits, one stable and one hyperbolic. Clearly, the construction of this modified vector field can be made so as to avoid saddle connections. Hence, the flow associated to the modified vector field is non-singular Morse-Smale.

By [M, Proposition], if we have a NMS flow on Q then Q has a round handle decomposition whose core circles are the closed orbits of this NMS flow. Each thick torus around a critical torus of f is formed by a round 0-handle or 2-handle and a round 1-handle. Clearly, if Q has this round handle decomposition it also has a round handle decomposition compressing the thick tori corresponding to these pairs (we exclude the trivial cases $T^2 \times I$ and $T^2 \times S^1$). Finally, by [A, Theorem N], to every round handle decomposition corresponds an NMS flow whose closed orbits are the cores of the round handles, and the result follows. //

Denote by $L(S^3)$ the set of indexed links obtained from $(0,2)$-Hopf links by performing operations of one of the following three types.

Operations. Let l_1 and l_2 be indexed links and denote by $N(k, M)$ a regular neighbourhood of k in M.

I. To make $(l_1 \sharp l_2) \cup m$. The connected sum $l_1 \sharp l_2$ is obtained by composing a component k_1 of l_1 and a component k_2 of l_2 each of which has index 0 or 2. The index of the composed component $k_1 \sharp k_2$ is equal to either $ind(k_1)$ or $ind(k_2)$. Finally, m is a meridian of $k_1 \sharp k_2$, and is of index 1.

II. Choose a component k_1 of l_1 of index 0 or 2, and replace $N(k_1, S^3)$ by $D^2 \times S^1$ with three indexed circles in it: $\{0\} \times S^1$, k_2 and k_3. Here, k_2 and k_3 are parallel (p,q)-cables on $\partial N(\{0\} \times S^1, D^2 \times S^1)$. The indices of $\{0\} \times S^1$ and k_2 are either 0 or 2, and one of them is equal to $ind(k_1)$. The index of k_3 is 1.

III. Choose a component k_1 of l_1 of index 0 or 2. Replace $N(k_1, S^3)$ by $D^2 \times S^1$ with two indexed circles in it: $\{0\} \times S^1$, and the $(2,q)$-cable k_2 of $\{0\} \times S^1$. The index of $\{0\} \times S^1$ is 1, and $ind(k_2) = ind(k_1)$.

The next theorem is the main result of [CMN2] and [FN].

Theorem 2. *Let $Q \cong S^3$ be a non-singular energy level of a Hamiltonian H integrable on Q by means of a Bott integral. Then, the indexed link L formed by the critical orbits of H in Q belongs to $L(S^3)$. Conversely, if $L \in L(S^3)$ then L may be obtained as the link formed by the critical orbits of some Hamiltonian H, Bott integrable and non-singular on S^3.*

Denote by $L(\tau)$ the set of indexed links obtained from $L(S^3)$ by removing from each $L \in L(S^3)$ a trivial knot of index 0 or 2.

Corollary 3. *Let Q be a non-singular energy level of a Hamiltonian H Bott integrable on Q, and let τ be a solid torus invariant under the flow of H. The indexed link L formed by the critical orbits of H in τ belongs to $L(\tau)$.*

3. Statement and proof of the results

Let Q be a simple, irreducible, closed, orientable 3-manifold such that there exists a Hamiltonian H Bott integrable and non-singular on Q. By Lemma 1, there exists a NMS flow ϕ_t on Q whose closed orbits are the critical orbits of H in Q. Let ψ_t be the primitive flow associated to ϕ_t, as defined in [K].

Lemma 4. *There exists on Q a Bott integrable, non-singular, Hamiltonian H_ψ whose critical orbits are the closed orbits of ψ_t.*

Proof: By [M, Proposition], to every NMS flow on Q corresponds a round handle decomposition (RHD) of Q whose core circles are the closed orbits of the flow. Moreover, it is easy to show (see [CMN2]) that the round 1-handles of the RHD associated to ϕ_t must be attached to the boundary tori along essential circles.

By construction of the primitive flow (see [K]), the RHD associated to ψ_t contains no round 1-handles which do not belong to the RHD associated to ϕ_t. On the other hand, in [BF] and [CMN2] it was shown that to every RHD such that the round 1-handles are attached along esential circles to the boundary tori, corresponds a non-singular Bott integrable Hamiltonian whose critical orbits are the cores of the round handles, and hence the lemma follows. //

We shall call the flow of H_ψ given by Lemma 4, the *primitive Hamiltonian flow of H in Q*, and the link formed by the critical orbits of H_ψ in Q, the *primitive link of H in Q.*

We note that by definition and [K, Lemma 3], the flow of H_ψ is such that the associated RHD,

$$\phi = M_0 \subset M_1 \subset \cdots \subset M_s = Q\,,$$

has the following property: each component of ∂M_i, i=1,$\cdots$,$s-1$, bounds a solid torus which contains only one critical orbit of H_ψ, and it is a stable orbit.

Theorem 5. *Let Q be a simple, irreducible, closed, orientable 3-manifold such that there exists a Hamiltonian H Bott integrable and non-singular on Q, and let L be the primitive link of H in Q. If Q is not a lens space (respectively, a lens space which is not $\mathbb{R}P^3$; $\mathbb{R}P^3$) then Q and L satisfy one of (b), (c), (d), (e) (respectively, one of (a), (b), (c); (a)) of the following.*

(a) *Q is a union of two solid tori V_1, V_2 where $V_1 \cap V_2 = \partial V_1 = \partial V_2$, and $L = l_1 \cup l_2$ where l_i (i=1,2) is a core of V_i.*

(b) *Q admits a Seifert fibration F with orbit manifold a projective plane with no exceptional fibers, and $L = l_1 \cup l_2 \cup l_3$ where each l_i is a regular fiber of F and l_3 is an untwisted hyperbolic closed orbit.*

(c) *Q admits a Seifert fibration F with orbit manifold a projective plane with one exceptional fiber, and $L = l_1 \cup l_2 \cup l_3$ where l_i (i=1,2) is a regular fiber of F, l_3 is the exceptional fiber of F, and l_2 is an untwisted hyperbolic closed orbit.*

(d) *Q admits a Seifert fibration F with orbit manifold a 2-sphere with three exceptional fibers, and $L = l_1 \cup l_2 \cup l_3 \cup l_4$ where l_i (i=1,2,3) is an exceptional fiber of F and is attracting or repelling, l_4 is a regular fiber of F and is an untwisted hyperbolic closed orbit.*

(e) *Q admits a Seifert fibration F with orbit manifold a 2-sphere with three exceptional fibers, and $L = l_1 \cup l_2 \cup l_3$ where each l_i is an exceptional fiber of F, l_3 is a twisted hyperbolic closed orbit, and the index of l_3 as an exceptional fiber is 2.*

Proof: In the case when Q is not the projective space $\mathbb{R}P^3$, the theorem follows from Lemma 4 and [K, Theorem 1].

In the case when Q is diffeomorphic to $\mathbb{R}P^3$, Kobayashi also proved in [K] that if a primitive flow ψ_t has one hyperbolic orbit (i.e., saddle type orbit) then the associated round 1-handle must be attached to a boundary torus along a trivial circle. As mentioned in the proof of Lemma 4, it was shown in [CMN2] that round 1-handles in a RHD corresponding to a NMS flow obtained from a Hamiltonian vector field must be attached along essential circles to the boundary tori. Therefore, by Lemma 4, ψ_t can have only orbits of indexes 0 or 2 and so L must satisfy (a). //

Our last result uses the preceding theorem and Corollary 3 to give a completely characterization of the indexed link formed by the critical orbits of integrable Hamiltonians on simple, irreducible, closed, orientable 3-manifolds.

Theorem 6. *Let Q be a simple, irreducible, closed, orientable 3-manifold and H a Hamiltonian non-singular and Bott integrable on Q. The foliation of Q given by the flow of H in Q is obtained from the foliation of the primitive Hamiltonian flow of H in Q by changing it only in tubular neighbourhoods N of the elements of the primitive link of indexes 0 or 2. Moreover, the link formed by the critical orbits of H in each N belongs to $L(\tau)$.*

Proof: The case when $Q \cong S^3$ is the result of Theorem 2, because the primitive link associated to any Bott integrable Hamiltonian flow on S^3 is the $(0,2)$-Hopf link.

The case when Q is a simple, irreducible, closed, orientable 3-manifold which is not diffeomorphic neither S^3 nor $\mathbb{R}P^3$, the theorem follows from [K, Theorem 3], Theorem 5 and Corollary 3. If $Q \cong \mathbb{R}P^3$ the result follows from [K, Theorem 4], Theorem 5 and Corollary 3. //

Remark. As pointed out in [K, Corollary 1], if Q is a lens space which is not of type $(4n, 2n-1)$, then every primitive link satisfies (a) of Theorem 5, as it does in the case $\mathbb{R}P^3$. This means that for these lens spaces and $\mathbb{R}P^3$, there exists a link with two elements which plays the same role as the Hopf link for S^3 in the topological description of the critical orbits.

4. Acknowledgements

The authors would like to thank Alain Chenciner for his hospitality at the Universit Paris VII when this manuscript was written.

5. References

[A] Asimov, D., *Round handles and non-singular Morse-Smale flows* (1975) Ann. Math. **102**, p. 41-54.

[BF] Brailov, A.V. and Fomenko, A.T., *The topology of integrable manifolds of completely integrable Hamiltonian systems* (1989), Math. USSR Sbornik **62**, p. 373-383.

[CMN1] Casasayas, J., Martinez-Alfaro, J., and Nunes, A., *Knotted periodic orbits and integrability* (1993), Hamiltonian Systems and Celestial Mechanics, Advanced Series in Nonlinear Dynamics **4**, p. 45-54.

[CMN2] Casasayas, J., Martinez-Alfaro, J., and Nunes, A., *Knots and links in integrable Hamiltonian systems* (1992), preprint.

[F] Fomenko, A.T., *Integrability and Nonintegrability in Geometry and Mechanics*, Klumer Academic Publishers (1988).

[FN] Fomenko, A.T., Nguyen, T.Z., *Topological classification of integrable nondegenerate Hamiltonians on the isoenergy three-dimensional sphere* (1991), Advances in Soviet Mathematics **6**, p. 267-296.

[FZ] Fomenko, A.T. and Zieschang, H., *On typical topological properties of integrable Hamiltonian systems* (1989), Math. USSR Izvestiya **32**, p. 385-412.

[H] Hempel, J., *3-manifolds*, Ann. of Math. Studies 86, Princeton University Press, Princeton, NJ (1976).

[J] Jaco, W., *Lectures on 3-manifold topology*, CBMS Regional Conference Series in Maths. 43, Amer. Math. Soc. (1980).

[K] Kobayashi, T., *Primitive links of non-singular Morse-Smale flows on the special Seifert fibered manifolds* (1985), Topology and its Applications **20**, p. 67-78.

[M] Morgan, J.W., *Non-singular Morse-Smale flows on 3-dimensional manifolds* (1978), Topology **18**, p. 41-53.

[R] Rolfsen, D., *Knots and Links*, Publish or Perish (1976).
[W] Wada, M., *Closed orbits of non-singular Morse-Smale flows on* S^3 (1989), J. Math. Soc. Japan **41:3**, p. 405-413.

GEOMETRIC PROPERTIES OF SOME SETS OF STABLE POLYNOMIALS AND ENTIRE FUNCTIONS

HERNAN CENDRA
Departamento de Matemática, Instituto de Matemática (CONICET),
Universidad Nacional del Sur, Av. Alem 1253
Bahía Blanca (8000), Buenos Aires, ARGENTINA
E-mail: uscendra@criba.edu.ar

and

ANA M. L. TORRESI
Departamento de Matemática, Instituto de Matemática (CONICET),
Universidad Nacional del Sur, Av. Alem 1253
Bahía Blanca (8000), Buenos Aires, ARGENTINA

ABSTRACT

This work is a selection of results on geometric properties of the set of stable polynomials $\mathbf{H}^n$. They have been proved by the authors and other people in several previous papers. Convexity properties of $\mathbf{H}^n$, the distance from a given stable polynomial to $\partial\mathbf{H}^n$ as a measure of the robustness, and a simplicial aproximation of $\partial\mathbf{H}_1^4$ are studied. Also a few generalizations in the context of entire functions are mentioned.

1. Introduction

A given monic polynomial of degree n, $a(s) = a_0 + a_1 s + \ldots + a_{n-1}s^{n-1} + s^n$ with real coefficients $a_i \in \mathbb{R}$ is *stable* or *Hurwitz* by definition, if all its roots have negatitve real part. It follows inmediately that if $a(s)$ is stable, then $a_i > 0, i = 0, \ldots, n-1$. We will identify a given polynomial $a(s) = a_0 + a_1 s + \ldots + s^n$ as before with the point $a = (a_0, a_1, \ldots, a_{n-1}) \in \mathbb{R}^n$.

Definition 1 $\mathbf{H}^n \subseteq \mathbb{R}^n$ *is the set of Hurwitz polynomials of degree* n.

The previous definition can be generalized in an obvious way in order to include the case of not necessarily monic polynomials. In fact, a given polynomial $a(s) = a_0 + a_1 s + \ldots + a_{n-1}s^{n-1} + a_n s^n$, $a_n > 0$ is *stable* or *Hurwitz*, if all its roots have negative real part. It is clear that both, a given polynomial of degree n and the monic polynomial obtained by dividing this polynomial by a_n, have the same roots. Although working with not necessarily monic polynomials is more convenient for certain problems, many questions of interest can be reduced to equivalent ones in terms of monic polynomials, by simply dividing by a_n. In the present work, we will deal only with the monic case, which has interest by itself.

Example 1 If $n = 1$ then

$$\mathbf{H}^1 = \{a_0 > 0\}.$$

We can also verify very easily that

$$\mathbf{H}^2 = \{(a_0, a_1) \in \mathbb{R}^2 : a_0 > 0, a_1 > 0\}.$$

However giving a description of $\mathbf{H}^n$, is not easy in general.

One of the important results in the field is the following old theorem of Hermite-Bieler. Before we can state it, we need some notation.

Let (a, b) be a pair of real polynomials, say

$$\begin{array}{ll} a(s) = a_0 + a_1 s + \ldots + a_{n-1} s^{n-1} + a_n s^n, & a_n \neq 0 \\ b(s) = b_0 + b_1 s + \ldots + b_{m-1} s^{m-1} + b_m s^m, & b_m \neq 0 \end{array}$$

where $m = n$ or $m = n - 1$.

Definition 2 *(a, b) is a positive pair, if the following conditions i) and ii) are satisfied:*
i) The roots λ_i, $i = 1, \ldots, n$ of $a(s)$ and μ_i, $i = 1, \ldots, m$ of $b(s)$ are all real and negative and $a_n > 0, b_n > 0$. ii) The following inequalities hold

$$0 > \lambda_1 > \mu_1 > \ldots > \lambda_n > \mu_n \quad if \quad m = n$$

or

$$0 > \lambda_1 > \mu_1 > \ldots > \mu_{n-1} > \lambda_n \quad if \quad m = n - 1.$$

Given any polynomial $a(s) = a_0 + a_1 s + \ldots + a_{n-1} s^{n-1} + a_n s^n$, $a_n \neq 0$ of degree n, we define, if $n = 2l$

$$\begin{array}{l} a^e(s) = a_0 + a_2 s + \ldots + a_{2l} s^l \\ a^o(s) = a_1 + a_3 s + \ldots + a_{2l-1} s^{l-1} \end{array}$$

and if $n = 2l + 1$

$$\begin{array}{l} a^e(s) = a_0 + a_2 s + \ldots + a_{2l} s^l \\ a^o(s) = a_1 + a_3 s + \ldots + a_{2l+1} s^l. \end{array}$$

Let us fix the following notation: λ_i^e are the roots of a^e and λ_i^o are the roots of a^o.

Theorem 1 (Hermite-Bieler) *A given polyomial $a(s) = a_0 + a_1 s + \ldots + a_{n-1}s^{n-1} + s^n$ of degree n is stable if and only if (a^e, a^o) is a positive pair. The later means that, if $n = 2l$, then*

$$0 > \lambda_1^e > \lambda_1^o > \ldots > \lambda_{l-1}^o > \lambda_l^e \quad \text{and } a_{2l-1} > 0$$

and if $n = 2l + 1$, then

$$0 > \lambda_1^e > \lambda_1^o > \ldots > \lambda_l^e > \lambda_l^o \quad \text{and } a_{2l} > 0.$$

For a proof see for instance [11].

Using the previous theorem, we can describe $\mathbf{H}^3$, as follows. Let $a(s) = a_0 + a_1 s + a_2 s^2 + s^3$, then

$$a^e(s) = a_0 + a_2 s$$
$$a^o(s) = a_1 + s$$

Then $\lambda_1^e = -(a_0/a_2)$, $\lambda_1^o = -a_1$. The condition $0 > \lambda_1^e > \lambda_1^o$ is equivalent to

$$0 > -(a_0/a_2) > -a_1.$$

Since $a_2 > 0$ it follows easily that

$$0 < a_0 < a_1 a_2, \quad a_1 > 0, a_2 > 0.$$

Thus, $\mathbf{H}^3$ is the subset of $\mathbb{R}^3$ defined by

$$\mathbf{H}^3 = \{(a_0, a_1, a_2) : a_0 > 0, a_1 > 0, a_2 > 0, a_0 < a_1 a_2\}.$$

Let us remark here that the boundary $\partial\mathbf{H}^3$, of $\mathbf{H}^3$, can be described as follows

$$\partial\mathbf{H}^3 = \partial\mathbf{H}_0^3 \cup \partial\mathbf{H}_1^3$$

where

$$\partial\mathbf{H}_0^3 = \{(a_0, a_1 a_2) : a_0 = 0, a_1 \geq 0, a_2 \geq 0\}$$

$$\partial\mathbf{H}_1^3 = \{(a_0, a_1 a_2) : a_0 = a_1 a_2, a_1 > 0, a_2 > 0\}.$$

Another important result which, however, we will not need in the present work, is a very neat description of $\mathbf{H}^n$ as a semialgebraic subset of $\mathbb{R}^n$, given by the following inequalities (see [11] or [17])

$$\triangle_r > 0, \; for \; r = 1, 2, \ldots$$

where the *Hurwitz determinants* $\triangle_r$ are defined as the succesive principal minors of the determinant

$$\mathbf{H} = \begin{vmatrix} a_{n-1} & a_{n-3} & \cdots & \cdots & \cdots \\ a_n & a_{n-2} & \cdots & \cdots & \cdots \\ 0 & a_{n-1} & a_{n-3} & \cdots & \cdots \\ 0 & a_n & a_{n-2} & \cdots & \cdots \\ \cdots & \cdots & \cdots & \cdots & \cdots \\ \cdots & \cdots & \cdots & \cdots & \cdots \end{vmatrix}.$$

The original source of interest for studying stable polynomials comes from one of the stability criteria for equilibrium points of ordinary differential equation. Without loss of generality, we let 0 be an equilibrium of the system

$$\dot{x} = Ax + g(x).$$

According to this criterion, 0 is a stable equilibrium if and only if the characteristic polynomial of the linear part of the system, given by the linear map A, is stable (or Hurwitz) in the sense defined above. In control theory, A depends on a control parameter μ. Thus the corresponding characteristic polynomial $a_\mu(s)$ also depends on μ. A basic problem in the field, is to establish ways of varying the control μ such that a_μ remains within $\mathbf{H}^n$. Moreover, a basic question in robust control theory, consists in making sure that a_μ remains *far*, in some prescribed sense, from the boundary $\partial\mathbf{H}^n$.

All this questions and others that arise naturally can be approached from a geometrical standpoint i.e. by reducing them to stablishing corresponding geometric properties of $\mathbf{H}^n$. In the present work, we are interested mainly in two kinds of questions. First, we will give some formulas to calculate the distance from a given point $a \in \mathbf{H}^n$ to the boundary $\partial\mathbf{H}^n$, once a convenient norm has been fixed in $\mathbb{R}^n$. Second, we will describe certain convexity properties of $\mathbf{H}^n$.

Another type of questions is concerned with finding the intersection of the subset of $\mathbb{R}^n$ described by an $\mathbb{R}^n$-valued function a_μ of real parameter $\mu \in \mathbb{R}^m$, with the boundary $\partial\mathbf{H}^n$. This is related to the question of finding the bifurcation set in m-parameter Hopf bifurcation problems [12]. We also give some results on this questions below.

Up to now we have considered the case of *real* polynomials $a(s)$. However the same definition of stability (negative real part of all roots) applies to any given complex polynomial $a(s)$, which is inspired by the stability criterion of equilibrium of complex ordinary differential equation. The set of stable complex polynomials of degree n will be denoted $\mathbf{H}^n_{\mathbb{C}}$. Although we will be mainly concerned with studying the set $\mathbf{H}^n$ in this work, we will also describe some properties of $\mathbf{H}^n_{\mathbb{C}}$.

Another important stability notion is that of *stability of a fixed point of a map.* Let $\phi : \mathbb{R}^n \to \mathbb{R}^n$ be a C^1 map such that $\phi(0) = 0$. Then the fixed point 0 is *stable* (in a sense which we will not study in this work in detail) if the eigenvalues of $D\phi(0)$ have all modulus less than 1. This leads immediately to the following question: to describe the set of all polynomials $a(s)$ such that the modulus of each of its roots is less than 1. This is referred to as the *discrete case* (because it arises naturally in the context of *difference* equations) as opposite to the *continuous case* described above (which arises in the context of *differential* equations). Of course, also the discrete case presents both, the real and the complex version.

More generally, it is important for both, theoretical purposes and applications, to consider the whole question of root location from a more general point of view. Namely, the natural generalization that occurs when instead of requiring that the roots of a given polynomial remain on the negative real complex half–plane, or inside the unit circle, we require that the roots remain in a prescribed open subset $D \subseteq \mathbf{C}$ which is a disjoint union of a finite number of pathwise connected components, each of them bounded by a piecewise C^1 curve. This leads naturally to define the set $\mathbf{H}^n(D)$ of all those polynomials whose roots belong to D. In [20] a formula to calculate the distance from a given $\overline{a} \in \mathbf{H}^n(D)$ to the boundary $\partial\mathbf{H}^n(D)$ has been devised.

Finally, we should mention the important fact that many of the previous results can be generalized in the context of certain classes of entire functions rather than polynomials. In particular, generalizations of the Hermite-Bieler theorem in the context of entire functions have been studied by many authors. See for instance [9], [18] and [27]. For instance, this kind of generalization has been carried out for the class of the so called *quasi-polynomials*, which are entire functions that can be written as follows:

$$\sum_{i=1}^{n} P_i(z)e^{\alpha_i z},$$

where $\alpha_i \in \mathbb{C}$, and P_i are real or complex polynomials. From the point of view of applications, this is related to the study of time–delay systems (see [24] and references therein).

In the present work, we briefly describe, without proof, some of the results (both published and still not published) obtained by some people from the Departments of Mathematics and of Electrical Engineering of the Universidad Nacional del Sur, at Bahía Blanca, Argentina, during the last few years. We give a list of some of the relevant references, related to the problems we considered herein only.

2. Distance to the boundary of the stability set

The boundary $\partial\mathbf{H}^n$ of $\mathbf{H}^n$ has a decomposition into a disjoint union

$$\partial\mathbf{H}^n = \partial\mathbf{H}_0^n \cup \partial\mathbf{H}_1^n$$

as we will see in a moment. First, it can be shown without difficulty that $a \in \partial\mathbf{H}^n$ if and only if $a(s)$ can be factorized in at least one of the following ways:

$$a(s) = b(s)(s^2+\lambda), \quad \lambda > 0, \quad b \in \overline{\mathbf{H}}^{n-2}$$

or

$$a(s) = c(s)s, \quad c \in \overline{\mathbf{H}}^{n-1}$$

(here, $\overline{A}$ is the closure of A for any given $A \subseteq \mathbf{R}^n$).

We set

$$\partial\mathbf{H}_0^n = \{a \in \overline{\mathbf{H}}^n : a(s) = c(s)s \text{ for some }, c \in \overline{\mathbf{H}}^{n-1}\}$$

and

$$\partial\mathbf{H}_1^n = \{a \in \overline{\mathbf{H}}^n : a(s) = b(s)(s^2+\lambda) \text{ for some } \lambda > 0, b \in \overline{\mathbf{H}}^{n-2} \\ \text{such that } b(0) \neq 0\}.$$

For instance, the reader can check using this that

$$\partial\mathbf{H}_0^3 = \{(a_0, a_1, a_2) : \ a_0 = 0, \ a_1 \geq 0, \ a_2 \geq 0\}$$

and

$$\partial\mathbf{H}_1^3 = \{(a_0, a_1, a_2) : \ a_0 = a_1 a_2, \ a_1 > 0, \ a_2 > 0\}.$$

Observe that $\partial\mathbf{H}_0^n$ can be naturally identified with $\overline{\mathbf{H}}^{n-1}$, in fact we have that
$(a_0, a_1, \ldots, a_{n-1}) \in \partial\mathbf{H}_0^n$ if and only if $a_0 = 0$ and $(a_1, a_2, \ldots, a_{n-1}) \in \bar{H}^{n-1}$. Likewise, it follows immediately from the definitions that $a \in \partial\mathbf{H}_1^n$ if and only if there exist $b \in \overline{\mathbf{H}}^{n-2}$ and $\lambda > 0$ such that the following is satisfied

$$\begin{aligned} a_0 &= b_0\lambda, \quad b_0 \neq 0 \\ a_1 &= b_1\lambda \\ a_2 &= b_2\lambda + b_0 \\ a_3 &= b_3\lambda + b_1 \end{aligned}$$

$$\begin{aligned}
&\vdots \\
a_{n-3} &= b_{n-3}\lambda + b_{n-5} \\
a_{n-2} &= \lambda + b_{n-4} \\
a_{n-1} &= b_{n-3}. \qquad (1)
\end{aligned}$$

For each $\lambda \in \mathbb{R}$, the system (2) defines a $(n-2)$-plane $\Pi(\lambda)$ parametrized by $b = (b_0, \ldots, b_{n-3}) \in \mathbb{R}^{n-2}$. As $b \in \overline{\mathbf{H}}^{n-2}$, $b_0 \neq 0$ and $\lambda > 0$, the point a describes exactly $\partial \mathbf{H}_1^n$.

By eliminating $b_0, \ldots, b_{n-3}$ in (2), we arrive at the following implicit description of $\Pi(\lambda)$: If $n = 2l$,

$$\Pi(\lambda) \left\{ \begin{array}{lcl} a_0 - a_2\lambda + a_4\lambda^2 - \ldots + (-1)^l\lambda^l & = & 0 \\ a_1 - a_3\lambda + a_5\lambda^2 - \ldots + (-1)^{l-1}a_{2l-1}\lambda^l & = & 0 \end{array} \right.$$

and if $n = 2l + 1$,

$$\Pi(\lambda) \left\{ \begin{array}{lcl} a_0 - a_2\lambda + a_4\lambda^2 - \ldots + (-1)^l a_{2l}\lambda^l & = & 0 \\ a_1 - a_3\lambda + a_5\lambda^2 - \ldots + (-1)^l\lambda^l & = & 0. \end{array} \right.$$

For a given $a = (a_0, \ldots, a_{n-1}) \in \mathbb{R}^n$ an for $0 < p < \infty$ let us define the *p-norm* in $\mathbb{R}^n$ with weights w_i, $i = 0, \ldots, n-1$ as usual:

$$|a|_p = \left[\sum_{i=0}^{n-1} w_i^p |a_i|^p \right]^{\frac{1}{p}}.$$

Then the distance $d_p(\overline{a})$ from a given $\overline{a} \in \mathbf{H}^n$ to the boundary $\partial\mathbf{H}^n$ is given by the following formula (see [6])

$$d_p^p(\overline{a}) = \min_{\lambda > 0} d_p^p(\overline{a}, \lambda)$$

with

$$d_p^p(\overline{a}, \lambda) = \frac{|\overline{a}^e(-\lambda)|^p}{(Q_p^e(\lambda))^{p-1}} + \frac{|\overline{a}^o(-\lambda)|^p}{(Q_p^o(\lambda))^{p-1}}$$

where, for $n = 2m$

$$Q_p^e(\lambda) = w_0' + w_2'\lambda^{\frac{p}{p-1}} + \ldots + w_{n-2}'\lambda^{\frac{(m-1)p}{p-1}}$$

$$Q_p^o(\lambda) = w_1' + w_3'\lambda^{\frac{p}{p-1}} + \ldots + w_{n-1}'\lambda^{\frac{(m-1)p}{p-1}}$$

and for $n = 2m + 1$

$$Q^e(\lambda) = w_0' + w_2'\lambda^{\frac{p}{p-1}} + \ldots + w_{n-1}'\lambda^{\frac{mp}{p-1}}$$

$$Q^{o}(\lambda) = w_0^{'} + w_3^{'}\lambda^{\frac{p}{p-1}} + \ldots + w_{n-2}^{'}\lambda^{\frac{(m-1)p}{p-1}}$$

with

$$w_i^{'} = \frac{1}{w_i^{\frac{p}{p-1}}}.$$

In [6] the cases $p = 1, 2, \infty$ are studied in detail. We would like to mention here that as a corollary to our study of the case $p = \infty$, *an entirely new proof of Kharitonov Theorem* (see Theorem 2) can be obtained. Let us give an idea of the proof. Choose weights $\omega_i > 0$, $i = 0, 1, ..., n-1$, and let $B_a^{\infty}(r)$ be the ball of center a and radius $r > 0$ with respect to the ∞-norm corresponding to those weights. Choose $a \in \mathbf{H}^n$, then $B_a^{\infty}(r) \subseteq \mathbf{H}^n$ for small $r > 0$. The ∞-distance from a to the boundary $\partial\mathbf{H}^n$ is

$$d_{\infty}(a) = sup\{r \in R : B_a(r) \subseteq \mathbf{H}^n\}.$$

In [6] we have also shown that the set $\partial B_a(d_{\infty}(a)) \cap \partial\mathbf{H}^n$ always contains one of the four Kharitonov vertices of $B_a(d_{\infty}(a))$. In other words, we have that $B_a(r) \subseteq \mathbf{H}^n$ for $r < d_{\infty}(a)$, and as r increases, $B_a(r) \subseteq \mathbf{H}^n$ *unless one of its Kharitonov vertices loses stability.* This leads to a proof of Kharitonov Theorem.

Some bounds are also found in [6] for the value of λ that realizes the minimum in the previous formulas, which helps with the problem of numerical determination of $d_p(\overline{a})$. Some more references are [13], [22], [23].

Similar formulas are devised for the distance from $\overline{a} \in \mathbf{H}^n(D)$ to $\partial\mathbf{H}^n(D)$ in [20], and the important particular cases $p = 1, \infty$ are considered too. However we will not write the general expressions here.

Now let us describe the complex case. For a given $0 < p < \infty$ and weights Γ_k, $k = 0, \ldots, n-1$, define the following norm in $\mathbf{C}^n$:

$$|a|_p = \left[\sum_{k=0}^{n-1} \left|\frac{a_k}{\Gamma_k}\right|^p\right]^{\frac{1}{p}}.$$

The distance $d_p(\overline{a})$ from $\overline{a} \in \mathbf{H}^n_{\mathbf{C}}$ to the boundary $\partial\mathbf{H}^n_{\mathbf{C}}$ is given by (see [10])

$$d_p(\overline{a}) = \min_{w \in \mathbf{R}} \frac{|\overline{a}(iw)|}{\left[\sum_{k=0}^{n-1}(\Gamma_k |w|^k)^{\frac{p}{p-1}}\right]^{\frac{p-1}{p}}}.$$

Let us consider the particular cases $p = 1, \infty$. We obtain, after some work, the following

$$d_{\infty}(\overline{a}) = \min_{w \in \mathbf{R}} \frac{|\overline{a}(iw)|}{\sum_{k=0}^{n-1} \Gamma_k |w|^k}$$

and

$$d_1(\overline{a}) = \min_{0 \le k \le n-1} d_{1k},$$

where

$$d_{1k}(\overline{a}) = \min_{w \in \Omega_k} \frac{|\overline{a}(iw)|}{\Gamma_k |w|^k}$$

with

$$\Omega_k = \{w : |w|^k \Gamma_k > |w|^i \Gamma_i, \ 0 < i < n-1, \ i \neq k\}.$$

See [10] for a proof of all this.

3. Simplicial approximation of $\partial \mathbf{H}^n$

As we have indicated in section **1**, in the determination of m-parameter Hopf bifurcation points, we are led to find the intersection of $\partial \mathbf{H}^n$ with an m-submanifoled of $\mathbb{R}^n$ described by a map, $a = a_\mu$ with $\mu = (\mu_1, \mu_2, \ldots, \mu_m) \in \mathbb{R}^m$. We are mainly interested in the case $m = 2$ in the present work. As far as we know, there is no systematic procedure in the literature to solve this problem, although it is obvious that there is a need for it. (See for instance [12], [14] and [19]). In those examples an *ad hoc* numerical solution is performed to find the intersection curve in the parameter space by the method of (numerical) *continuation* of such a curve, once a starting point $\overline{\mu}$ such that $a_{\overline{\mu}} \in \partial \mathbf{H}^n$ has been found. Finding such starting points is a problem in itself. The continuation also requires a good knowledge of at least local defining equations for $\partial \mathbf{H}^n$ which does not seem easy in general. ($\partial \mathbf{H}^n$ is a semialgebraic set and has singularities).

One way to solve the problem in a systematic manner, should be to find an arbitrarily precise simplicial approximation of $\partial \mathbf{H}^n$, and also of the surface a_μ. Then the problem is now reduced, in an aproximate sense, to finding the intersection of a $(n-1)$-dimensional simplicial complex with a 2-dimensional complex. This in turn also presents some delicate numerical problems, in order to get an efficient algorithm.

We have applied the previous ideas to the case of $\partial \mathbf{H}^4$ (the case $\partial \mathbf{H}^3$, is in some sense much easier, because a simple equation of $\partial \mathbf{H}^3_1$ as we have seen before is known. It is also very easy to find a triangulation of $\partial \mathbf{H}^3$. This case is also interesting in applications(See [19] and references therein). We find a *stable* (i.e. contained in $\overline{\mathbf{H}}^4$) arbitrarily precise simplicial approximation of $\partial \mathbf{H}^4_1$ (see [8]) which behaves nicely in the troublesome singular points of $\partial \mathbf{H}^4_1$.

Let us give a more precise idea of how to construct such a simplicial aproximation of $\overline{\partial H^4_1}$. According to (2) (see *Introduction*) we have that a point $a = (a_0, a_1, a_2, a_3) \in \partial \mathbf{H}^4_1$ if and only if there are $\lambda > 0, b_0 > 0,$ *and* $b_1 > 0$ such that

$$\begin{aligned} a_0 &= b_0\lambda \\ a_1 &= b_1\lambda \\ a_2 &= b_0+\lambda \\ a_3 &= b_1. \end{aligned} \tag{2}$$

Let us write $a = f(\lambda, b_0, b_1)$, where $a = (a_0, a_1, a_2, a_3)$, and f is the map given by (3). For given $a = f(\lambda, b_0, b_1)$, $\Delta\lambda$, Δb_0 *and* Δb_1, let

$$\begin{aligned} P_\lambda &= f(\lambda+\Delta\lambda, b_0, b_1) \\ P_0 &= f(\lambda, b_0+\Delta b_0, b_1) \\ P_1 &= f(\lambda, b_0, b_1+\Delta b_1) \\ P_{\lambda 1} &= f(\lambda+\Delta\lambda, b_0, b_1+\Delta b_1) \\ P_0 &= f(\lambda, b_0+\Delta b_0, b_1+\Delta b_1). \end{aligned}$$

It is easy to see that the four points a, P_0, P_1 *and* P_{01} lie on the 2-plane, parametrically defined by the map

$$f(\lambda, b_0+t, b_1+s),\ (t,s) \in \mathbb{R}^2.$$

For any given $A \subseteq \mathbb{R}^n$ let $C(A)$ be the convex hull of A. Then we have that $C(a, P_0, P_1, P_{01}, P_\lambda)$ is a pyramid which can be decomposed as the union of two 3-simplices. For instance:

$$\begin{aligned} S_1 &= C(a, P_0, P_1, P_\lambda) \\ S_2 &= C(P_0, P_1, P_{01}, P_\lambda), \end{aligned}$$

which have the 2-simplex $C(P_\lambda, P_0, P_1)$ in common. Of course, $S_3 = C(P_\lambda, P_1, P_{01}, P_{\lambda 1})$ is a 3-simplex having the 2-simplex $C(P_\lambda, P_1, P_{01})$ in common with $C(a, P_\lambda, P_0, P_1, P_{01})$. Assume that P_λ, P_0, P_1, P_{01} and $P_{\lambda 1}$ belong to $\partial\mathbf{H}^4$. Then it can be shown (after some work! see [8]) that the edges $C(P_\lambda, P_{\lambda 1})$, $C(P_0, P_{01})$, $C(a, P_\lambda)$, $C(a, P_1)$, $C(P_1, P_{\lambda 1})$, $C(P_1, P_{01})$ are contained in $\partial\mathbf{H}^4$.

Besides it can be shown that $C(P_\lambda, P_0)$ and $C(P_{\lambda 1}, P_{01})$ are contained in $\overline{\mathbf{H}}^4$ provided that $|\Delta b_0| = |\Delta b_1|$ and some of the following conditions

$$\begin{aligned} &i)\ sg\Delta\lambda = sg\Delta b_0 = sg\Delta b_1 > 0, \text{and } \lambda \geq b_0 \\ &ii)\ sg\Delta\lambda = sg\Delta b_0 > 0,\ sg\Delta b_1 < 0, \text{and } \lambda+\Delta\lambda \leq b_0 \\ &iii)\ sg\Delta\lambda = sg\Delta b_0 < 0,\ sg\Delta b_1 > 0, \text{and } \lambda \leq b_0 \\ &iv)\ sg\Delta\lambda = sg\Delta b_0 = sg\Delta b_1 < 0, \text{and } \lambda+\Delta\lambda \geq b_0 \end{aligned}$$

are satisfied.

Thus in any of those cases, we can conclude, using theorem 3 that all 3-simplices, S_1, S_2 and S_3 are contained in $\overline{\mathbf{H}}^4$.

Next let us choose $\Delta > 0$ and consider the set of points

$$L = \{(l\Delta, p\Delta, q\Delta) \ : \ l,\ p \text{ and } q \text{ are arbitrary non-negative integers}\}.$$

It can be shown that, for given $a = (l\Delta, p\Delta, q\Delta) \in L$ such that $q > 0$, there are choices of $\Delta\lambda = \pm\Delta$, $\Delta b_0 = \pm\Delta$ and $\Delta b_1 = \pm\Delta$, such that the corresponding simplices S_1, S_2 and S_3 are contained in $\overline{\mathbf{H}}^4$ and, moreover the union of all such simplices form a simplicial complex which approximate $\partial\mathbf{H}_1^4$. A little care should be excersiced if $a \in L$ is chosen such that $q = 0$. We have that $f(\lambda, b_0, 0) = f(b_0, \lambda, 0)$ thus some of the simplices become glued together, in this case. Observe that for $\lambda \geq 0$ and $b_0 \geq 0$ we have that $f(\lambda, b_0, 0) = (a_0, 0, a_2, 0)$ with $a_0 \leq \frac{1}{4}a_2^2$ and that for $\lambda = b_0$, we have $a_0 = \frac{1}{4}a_2^2$. Thus the exceptional 2-simplices of type $C(f(t\Delta, t\Delta, 0), f((t+1)\Delta, t\Delta, 0), f(t\Delta, (t+1)\Delta, 0))$ which are some of the faces of some S_i, $i = 1, 2, 3$ are actually *not* contained in $\overline{\mathbf{H}}^4$. However if Δ is choosen to be small, they fall within a distance $\mathcal{O}(\Delta^2)$ from $\overline{\mathbf{H}}^4$. Moreover as explained in [8], we can modify a little bit the previous construction about the troublesome line $b_0 = \lambda$ to get, finally, a stable arbitrarily precise simplicial approximation of $\partial\mathbf{H}_1^4$.

We have also devised an efficient algorithm and a computer program to find (numerically) the intersection of the simplicial approximation of $\partial\mathbf{H}_1^4$ with a given triangulated surface a_μ. In order to reduce the amount of computer time required to do this (the complete algorithm has order $\mathcal{O}(n^8)$), acceleration methods were used in the program. Only a small number of intersections of individual simplices from the two simplicial approximations have to be really performed thanks to a preselection of those hypercubes in the tetrahedrons space which do not have any chance to intersect each other. To achieve this, increasingly smaller areas of the space are tested following an "Octree searching scheme". Furthermore, intersections are quickly inspected using bounding volumes concepts.

We will not reproduce the source code of the program here because of its length.

4. Convexity properties

It is not difficult to prove that $\mathbf{H}^n$ is convex if and only if $n \leq 2$. In this section, we will describe some convexity properties of $\mathbf{H}^n$, for $n \geq 1$. For $A \subseteq \mathbb{R}^n$ let $C(A)$ denote the convex hull of A.

We start with a relatively new theorem which, nevertheless has become a classic in the field of linerized stability and robustness. First we need some notation. Let $v^e = (v_0, 0, v_2, 0, \ldots) \in \mathbb{R}^n$ such that the signs of v_{2i} alternate, i.e. $Sg(v_2) = -Sg(v_0)$, $Sg(v_4) = -Sg(v_2)$, ... Let $v^o = (v_1, 0, v_3, 0, \ldots) \in \mathbb{R}^n$ such that the signs of v_{2i+1} alternate, i.e. $Sg(v_3) = -Sg(v_1)$, $Sg(v_5) = -Sg(v_3)$, ... Then, given $P \in \mathbb{R}^n$, we can built four points as follows $P_0 \equiv P$, $P_1 = P_0 + v^o$, $P_2 = P_0 + v^e$, $P_1 = P_0 + v^o + v^e$. In this situation, P_0, P_1, P_2, P_3 are called *Kharitonov vertices*, (a non standard notation which we think is justified because of the theorem below). It is very easy to check that given four Kharitonov vertices P_0, P_1, P_2, P_3 as before, there is one and only one parallelepiped $K(\{P_0, P_1, P_2, P_3\})$ having all its edges parallel to the coordinate axis and such that P_0, P_1, P_2, P_3 are vertices of it.

Theorem 2 (Kharitonov) *If P_0, P_1, P_2, $P_3 \in \mathbf{H}^n$ are Kharitonov vertices, then $K(\{P_0, P_1, P_2, P_3\}) \subseteq \mathbf{H}^n$.*

For a proof see for instance [16]. This famous theorem has been generalized for the complex case, where eight Kharitonov vertices instead of four, have to be considered in this case [5].

It is *not* generally true that if P is a convex polyhedron whose vertices belong to $\mathbf{H}^n$ then $P \subseteq \mathbf{H}^n$.However we have the following interesting result (see [3]).

Theorem 3 *Let A be the union of the edges of a given convex polyhedron $P \subseteq \mathbb{R}^n$, thus in particular $P = \mathcal{C}(A)$. Then, $\mathcal{C}(A) \subseteq \mathbf{H}^n$, if $A \subseteq \mathbf{H}^n$.*

Theorems 2 and 3 have had a wide diffusion in the literature and many generalizations and applications have been found.

Let us now describe some more results about convexity properties which contribute to increase the power of theorems 2 and 3. First we need some more notation:

$$\mathbb{R}^n_e = \{a \in \mathbb{R}^n : a_1 = 0, a_3 = 0, \ldots\},$$

$$\mathbb{R}^n_o = \{a \in \mathbb{R}^n : a_0 = 0, a_2 = 0, \ldots\}.$$

Clearly $\mathbb{R} = \mathbb{R}^n_e \oplus \mathbb{R}^n_o$.

Theorem 4 *Let $a \in \mathbf{H}^n$ then $(a + \mathbb{R}^n_e) \cap \mathbf{H}^n$ and $(a + \mathbb{R}^n_o) \cap \mathbf{H}^n$ are convex polyhedra. Moreover they are (possibly unbounded) pyramids.*

The first part of theorem 4 is proved in [15]. The exact description of the convex polyhedra showing in particular that they are pyramids is given in [7].

Another result related to convexity type properties is given by the following theorem (see [21])

Theorem 5 *Let* $v = (v_0, v_1, \ldots, v_{n-1}) \neq 0$ *with* $v_i \geq 0$ *be given. Let*

$$l^v = \{\lambda v \ \lambda > 0\},$$
$$\lambda_o = \inf\{\lambda : \lambda v \in \mathbf{H}^n\},$$
$$l_o^v = \{\lambda v : \lambda > \lambda_o\}.$$

Then the following property is satisfied

$$l_o^v = l^v \cap \mathbf{H}^n.$$

Let us show how theorem 5 helps to increase the power of theorems 2 and 3. For given $B \subseteq \mathbb{R}^n$, let $\mathcal{T}(B) = \{\lambda b : \lambda \geq 1, b \in A\}$.Then the following property holds.

Corollary 1 *If* $B \subseteq \mathbf{H}^n$ *then* $\mathcal{T}(B) \subseteq \mathbf{H}^n$.

The proof of this corollary follows from theorem 5 immediately . Therefore under the hypothesis of theorem 2 we obtain the following:

$$\mathcal{T}K(\{P_o, P_1, P_2, P_3\}) \subseteq \mathbf{H}^n,$$

and under the hypothesis of theorem 3 we obtain the following:

$$\mathcal{T}\mathcal{C}(A) \subseteq \mathbf{H}^n.$$

In other words, via Corollary 1, we get *stable* subsets $\mathcal{T}K(\{P_0, P_1, P_2, P_3\})$ and $\mathcal{T}\mathcal{C}(A)$ which are *much bigger* than $K(\{P_0, P_1, P_2, P_3\})$ and $\mathcal{C}(A)$, respectively.

Now let us explain how we can get a better version of theorem 3. Given a convex polyhedra $P = \mathcal{C}(A)$ as before, such that $0 \notin P$, let us call a point in ∂P *visible* if $\mathcal{C}(\{0, x\}) \cap \partial P = \{x\}$ and *invisible* otherwise. Let $\mathcal{V}(A) \subseteq A$ be the subset of visible points of A. It not difficult to show that $\mathcal{V}$(A) is the union of some of the edges of B. Then the following result holds (see [1]).

Theorem 6 *If* $\mathcal{V}(A) \subseteq \mathbf{H}^n$ *then* $\mathcal{C}(A)) \subseteq \mathbf{H}^n$

Thus, in particular, if we want to know if a given convex polyhedron $P = \mathcal{C}(A)$ is contained in $\mathbf{H}^n$ or not, we only need to investigate whether $\mathcal{V}(A)$ is contained in $\mathbf{H}^n$ or not. The number of edges of P which are contained in $\mathcal{V}(A)$ assuming a more or less uniform distribution of edges over the boundary of P is roughly half of the numbrer of edges contained in A. Thus in many cases, it should be simpler to apply theorem 6 than theorem 3. Finally we can combine theorem 6, and corollary 1 and we obtain the following.

Corollary 2 *If $\mathcal{V}(A) \subseteq H^n$ then $\mathcal{T}(\mathcal{C}(A)) \subseteq H^n$.*

5. Some generalizations in the context of entire functions

An entire function is a function $f(z)$, that is analytic on the whole complex plane **C**. We are interested in the set $\mathcal{E}$ of real entire functions of order $\lambda = 0$. (See [4] for reference on entire functions).

According to Hadamard theoren the order $\lambda(f)$ and the genus $h(f)$ of a given entire function, are related by the following formula: $h(f) \leq \lambda(f) \leq h(f) + 1$. Therefore $h(f) = 0$ implies $\lambda(f) = 0$. The Weiertrass theorem implies in this case that if $\gamma_k, (\gamma_k \neq 0), k = 1, \ldots$ are the nonzero zeroes of a given nonconstant $f \in \mathcal{E}$, then $\sum_k (1/|\gamma_k|) < \infty$ and the following factorization holds

$$f(z) = Cz^q \prod_k (1 + z/\gamma_k).$$

In particular, if the number of zeroes γ_k is finite, $f(z)$ is a polynomial. In this context the Hermite-Bieler and Kharitonov theorems have a natural obvious generalization.

Finally let us mention that in [9], [18], [27] which are only a few of the many available references many results related to generalizations of the Hermite-Bieler property have been proved.

6. References

1. Alvarez, M. and Cendra, H. *Stability of Convex Polytopes: It Suffices to check the Edges ... How many of them ?* Preprint. (1994).
2. Barmish, B. et al. *A survey of extreme point results for robustness of control system*, Automatica, Vol. 29(1993), 13-35.
3. Bartlett, A. C., Hollot C. V. and L. *Root Location of an Entire Polytope of Polynomials: It Suffices to Check the Edges*, Math. of Control Signal and Systems, Vol. 1(1987), 61-71.
4. Boas, R. P. *Entire Functions* . Ac. Press (1954).
5. Bose, N. K. and Shi, Y. Q. *A Simple Proof of Kharitonov's Generalized Stability Criterion* IEEE, Trans. on Circuits and Sys., Vol. CAS 34 N_o10. (1987)
6. Cendra, H., Torresi, A., Desages, A. and Orofino, S. *Stability Radius for a Real Hurwitz Polynomial*, Preprint (1992).
7. Cendra, H., Torresi, A., Desages, A. *Convexity Properties of the Set of Hurwitz Polynomials* Preprint (1992).

8. Cendra, H., Itovich, G., Salthú, R. y Torresi, A. *Determinación de puntos de bifurcación de Hopf*. Actas del Primer Congreso Dr. Antonio A. R. Monteiro, Bahía Blanca (1991), 39-50.
9. Chevotarev, N. G. and Meiman N., N. *The Routh-Hurwitz Problem for Polynomials and Entire functions*. Trudy Matematiches Kogo instituta im V. A. Steklova $N°26$. Akademia Nauk SSSR, Moscow and Leningrad (1949), (Russian).
10. Desages, A., Castro, L. and Cendra, H. *Distance of a Complex Coefficient Stable Polynomial from the Boundary of the Stability Set*. Multidimensional Systems and Signal Processing, 2(1991), 189-210.
11. Gantmacher, F.R. *The Theory of Matrices*, Vol. II, Chelesa Pub. Co. (1959).
12. Guckenheimer,J. and Holmes, P. *Nonlinear Oscillations, Dynamical Sistems, and Bifurcations of Vector Fields*, Applied Math. Science **42**, Springer-Verlag, NY (1986)
13. Hinrichsen, D. and Prithard, A.J. *An Aplication of State Space Methods to Obtain Explicit Formulae for Robustness Measure of Polynomials*, Robustness in Identification and Control, Eds. M. Milanese-Plenum. N.Y. (1969).
14. Holmes, P. *Bifurcation to Divergence and Flutter in Flow-Induced Oscilations: A Finite-Dimensional Anlysis.* J. Sound of Vibration, **53**(1977), 471-503.
15. Huang, L., nad Hollot, C. V. Results on Positive Pair of Polynomials and their Applications to the Construction of Stability Domains, Int. J. of Control, Vol. 46(1987), 153-159.
16. Kharitonov, V. L. *Asymptotic Stability of an Equilibrium Position of a Family of Systems of Linear Differential Equations*. Differensial'nye Uravneniya, 14, 11 (1978).
17. Marden, M. *The Geometry of the Zeroes of a Polynomial in a Complex Variable*, AMS, (1949).
18. Meiman, N. N. *On the Problem of Hermite-Hurwitz for Integer Trascendental Functions*, Doklady Akad. Sci. 40(1943), 46-49 (Russian).
19. Moiola, J., Castro, L., Cendra, H. and Desages, A. *Degenerate Bifurcations and Catastrophe Sets via Frecuency Analysis*, Dynamics and Stability of Systems, Vol. 5(1990).
20. Reartes, W., Desages, A. and Mandolesi, P. *D-Stability Margins for Systems with Structured Uncertainties*. Anales de la 5° Reunion de Trabajo en Procesamiento de la Información y Control. Tucumán, Argentina (1993).
21. Robledo, C., Desages, A. and Cendra, H. *On the Distance to the In-*

stability Border of Hurwitz Polynomials with Coefficients that Depend Affinely on m Parameters and a Particular Convexity Property of H^n . Multidimensional Systems and Signal Processing 3, (1992), 45-61.

22. Soh, C. B., Berger, C. S. and Dabke, K. P. *On the stability properties of polynomials with perturbed coefficients* IEEE Trans. Automat. Cont., Vol. AC-30(1985), 1033-1036.
23. Soh, C. B., Berger, C. S. and Dabke, K. P. *Addendum to on the stability properties of polynomials with perturbed coefficients* IEEE Trans. Automat. Cont., Vol. AC-32(1987), 239-240.
24. Tsypkin, Y. Z., Fu, M. Y. *Robust stability of time-delay with an uncertain time-delay constant* . International Journal of Control, Vol. 57(1993), 865-879.
25. Vicino, A. *Some results on robust stability of discrete time systems* IEEE Trans. Automat. Cont.,vol. AC-33(1989), 844-847.
26. Yanuskevsky, R. T. *Robust Control of Delay Feedback Systems with Bounded Uncertainty*, SIAM Journal on Matrix and Applications, 14(1993), 978-990.
27. Zhbko, A. P. and Kharithonov, V. L. *Stability of Families of Quasipolynomials of Neutral Type*, Automation and Remote Control. Vol. 54 (1993), 85-99.

THE SPRING-PENDULUM SYSTEM AND THE RIEMANN EQUATION

RICHARD.C. CHURCHILL
Department of Mathematics, Hunter College and Graduate Center, CUNY
695 Park Avenue, New York, NY 10021, USA
E-mail: richc@cunyvm.cuny.edu

JOAQUIN DELGADO *
Departamento de Matemáticas, Universidad Autónoma Metropolitana-Iztapalapa
Apdo Postal 55-534, México DF 09340, MÉXICO
E-mail: jdf@xanum.uam.mx

and

DAVID.L. ROD†
Department of Mathematics and Statistics, University of Calgary
Calgary, Alberta, T2N 1N4, CANADA
E-mail: rod@acs.ucalgary.ca

1. Introduction

The planar spring-pendulum system is a model Hamiltonian system studied by many authors (see, e.g., [A-D] and [B-B]). With appropriate normalizations (see [A-D] for the full derivation) the motion in polar coordinates is governed by the Hamiltonian

$$H_\lambda(r,\theta,p_r,p_\theta) = \frac{1}{2}(p_r^2 + \frac{p_\theta^2}{r^2}) + r[(\lambda-1) - \lambda\cos\theta] + \frac{1}{2}r^2, \tag{1}$$

where λ is a parameter satisfying $0 \le \lambda \le 1$ and the standard symplectic form $\omega = dr \wedge dp_r + d\theta \wedge dp_\theta$ is assumed. We will prove, using Ziglin theory and results on the monodromy group of the Riemann equation, that for almost all $\lambda \in (0,1)$ the system is not integrable when regarded as complex analytic. In fact the argument remains valid when $\cos\theta$ in (1) is replaced by any holomorphic even function of θ, defined on a neighborhood of 0 in the θ-axis, having value 1 at $\theta = 0$. Our results imply, in particular, that the corresponding real system can have no rational integral for almost all $\lambda \in (0,1)$.

One can apply Ziglin theory to such problems in various ways. The comprehensive approach involves computing the differential Galois group of an associated linearized equation by means of the Kovacic algorithm (see [KOV] and

---

*Supported by CONACYT 1772-E9210.
†Research supported in part by Natural Sciences and Engineering Research Council of Canada Grant OGP 0008507.

[BCRS]), but this can be tedious in practice. For the Hamiltonian under study the relevant linearized equation reduces to a Riemann equation, and we are able to exploit known facts about the corresponding monodromy group. It is to be noted that the differential Galois group of any Fuchsian equation is the Zariski closure of its monodromy group (for details see [BCRS]).

2. The Normal Variational Equation

Because H_λ in (1) is even in θ and p_θ any local integral curve of the Hamiltonian vectorfield X_{H_λ} with initial value in the (r, p_r)-plane, where $(\theta, p_\theta) = (0, 0)$, will be contained in that plane. Choose such a non-equilibrium solution of energy zero and let $\Gamma \subset \{(r, p_r)\} \simeq \{(r, 0, p_r, 0)\}$ denote the image of the associated maximal analytic continuation. Then Γ is a Riemann surface, with time t a local parameter, which by (1) is contained in the rational curve defined by the equation $p_r^2 = r(2-r)$.

The variational (i.e., linearized) equations of X_{H_λ} along Γ decouple and are given in linearized coordinates $(\xi_1, \xi_2, \sigma_1, \sigma_2)$ by

$$\begin{cases} \dot{\xi}_1 = \sigma_1, & \dot{\xi}_2 = \dfrac{1}{r^2}\sigma_2, \\ \dot{\sigma}_1 = -\xi_1, & \dot{\sigma}_2 = -r\lambda\xi_2, \end{cases} \tag{2}$$

where $\cdot = d/dt$. The differential equation for (ξ_2, σ_2) is called the (reduced) normal variational equation (NVE).

The projection $(r, p) \mapsto z = r/2$ carries Γ into the complex plane $\mathbb{C}$, and along Γ we have $dr/dt = p_r$, $dp_r/dt = -r + 1$, whence $dz/dt = p_r/2$. We can therefore use z as the time-parameter for the induced equation (on $\mathbb{C}$), which on setting $\xi_2 = y$ in the NVE is seen to be the (Fuchsian) Riemann equation (see (6) and (7) in the next section)

$$y'' + \left[\frac{5/2}{z} + \frac{1/2}{z-1}\right] y' + \left[\frac{\lambda/2}{z^2} - \frac{\lambda/2}{z(z-1)}\right] y = 0, \quad ' = \frac{d}{dz}. \tag{3}$$

This passage from the NVE to (3) can be viewed as reduction by a lift of the symmetry $(r, p_r) \mapsto (r, -p_r)$ of Γ to the (reduced) normal bundle of this Riemann surface. More precisely, in differential-geometric terms the NVE can be identified with a connection ∇ on the normal bundle, and the symmetry can be lifted in such a way as to preserve this connection. Equation (3) is the result of reducing ∇ by the lifted group action. For details of this geometrical approach see [BCRS, Proposition 4.9], [C-R-S, Theorem 7.15], or [R-S, Theorem 1.6].

Choose a basepoint $z_0 \in \mathbb{C}\backslash\{0,1\}$ for the purpose of computing circuit matrices of (3) and let $G_\lambda \subset Gl(2,\mathbb{C})$ denote the resulting monodromy group. For an arbitrary subgroup $G \subset Gl(2,\mathbb{C})$ a rational function $f = p/q$ on $\mathbb{C}^2 = \{z = (z_1, z_2)\}$ is said to be G-invariant if $g^* f = f$ for all $g \in G$. For example, $f(z) = z_1$ is invariant under

$$G = \left\{ \begin{pmatrix} 1 & 0 \\ c & 1 \end{pmatrix} : c \in \mathbb{C} \right\} \subset Sl(2,\mathbb{C}).$$

When G admits such a non-constant rational function we refer to G as a *Ziglin group.*

Our approach to proving nonintegrability can now be explained.

Proposition 1 ((Ziglin)) *Assume the Hamiltonian vectorfield X_{H_λ} admits a meromorphic integral F defined on some neighborhood (in $\mathbb{C}^4$) of Γ and functionally independent of H_λ. (F is allowed to 'blow-up' along Γ.) Then G_λ is a Ziglin group. In particular, when G_λ is not Ziglin the vectorfield X_{H_λ} is not integrable.*

Proof: See [Z] or [BCRS, Chapter 1]. Q.E.D.

Proposition 1 is only a special case of Ziglin theory. The survey [BCRS] details the general case with applications to two degree of freedom Hamiltonians, and [C-R-S] is concerned with the three degree of freedom case.

3. On the Monodromy Group of the Riemann Equation

The normal form of a second order Fuchsian equation

$$y'' + c_1(z)y' + c_2(z)y = 0 \tag{4}$$

on $\mathbb{C}\cup\{\infty\}$ is

$$y'' + c(z)y = 0, \quad c(z) = c_2 - (c_1/2)^2 - (c_1)'/2. \tag{5}$$

When (4) is achieved through symmetry reduction from a linearization of a Hamiltonian system, as described immediately following (3), the monodromy group of (4) is Ziglin if and only if this is the case for (5). More precisely, let G^0 denote the connected component of the identity of the differential Galois group G of (4). Then the monodromy groups of (4) and (5) are simultaneously Ziglin or non Ziglin if G^0 is a subgroup of $Sl(2,\mathbb{C}) = Sp(1,\mathbb{C})$. (For details see [BCRS, Proposition 4.25] or [R-S, Proposition 1.15].) From the fact that fundamental solution matrices of (5) have constant determinant (as can be seen by straightforward differentiation) the monodromy group of this second equation is seen to lie in $Sl(2,\mathbb{C})$, which is a considerable computational advantage.

Proposition 2 *Let G be any subgroup of $Sl(2,\mathbb{C})$. Then either:*

(a) G is reducible, i.e., conjugate to a subgroup of

$$\left\{\begin{pmatrix} a & 0 \\ b & a^{-1} \end{pmatrix} : a \in \mathbb{C}\backslash\{0\},\ b \in \mathbb{C}\right\};$$

(b) G is a DP-group, i.e., is conjugate to a subgroup of $D \cup P$, where

$$D = \left\{\begin{pmatrix} a & 0 \\ 0 & a^{-1} \end{pmatrix}\right\}_{a\in\mathbb{C}\backslash\{0\}} \quad \text{and} \quad P = \left\{\begin{pmatrix} 0 & b \\ -b^{-1} & 0 \end{pmatrix}\right\}_{b\in\mathbb{C}\backslash\{0\}};$$

(c) G is finite; or

(d) the Zariski closure of G is $Sl(2,\mathbb{C})$.

Proof: Take the Zariski closure of G and use the result in [KOV, p. 7] or [KAP, pp. 31-32]. Q.E.D.

We remark that the cases in Proposition 2 are not disjoint, e.g., a diagonalizable group is both reducible and DP, and might well be finite. However, (d) holds if and only if (a)-(c) fail.

An element $g \in Sl(2,\mathbb{C})$ is (non)-resonant when its eigenvalues are (not) roots of unity.

Proposition 3 *Assume a subgroup $G \subset Sl(2,\mathbb{C})$ contains a nonresonant element. Then*

(a) G is infinite, and

(b) G is Ziglin if and only if G is a DP-group.

Proof: (a) is clear; for (b) see [B-C, Theorem 3.5]. Q.E.D.

Now specialize (4) to the Riemann equation

$$y'' + \left[\frac{1-\eta_1-\mu_1}{z} + \frac{1-\eta_2-\mu_2}{z-1}\right] y' + \left[\frac{\eta_1\mu_1}{z^2} + \frac{\eta_2\mu_2}{(z-1)^2} + \frac{\eta_3\mu_3-\eta_1\mu_1-\eta_2\mu_2}{z(z-1)}\right] y = 0, \qquad (6)$$

in which the complex parameters η_j, μ_j are arbitrary modulo the condition $\Sigma(\eta_j + \mu_j) = 1$. Note that the (nonunique) choices

$$\begin{cases} \eta_1 = -\frac{3}{4} + \frac{1}{4}\sqrt{9-8\lambda}, & \mu_1 = -\frac{3}{4} - \frac{1}{4}\sqrt{9-8\lambda} \\ \eta_2 = \frac{1}{2} & \mu_2 = 0 \\ \eta_3 = 2 & \mu_3 = 0 \end{cases} \tag{7}$$

give (3).

Proposition 4 *For the monodromy group $G \subset Sl(2,\mathbb{C})$ of the normal form of (6) the following assertions hold.*

(a) G is reducible if and only if at least one of the four numbers $(\eta_1 + \eta_2 + \eta_3)$, $(\eta_1 + \eta_2 + \mu_3)$, $(\eta_1 + \mu_2 + \eta_3)$, and $(\mu_1 + \eta_2 + \eta_3)$ is an integer.

(b) G is generated by the three circuit matrices g_j, $j = 1, 2, 3$, corresponding (in order) to the singularities $0, 1$, and ∞, subject to the single relation $g_1 g_2 g_3 = id$.

(c) $t_j = trace(g_j) = -2\cos\pi(\eta_j - \mu_j)$, $j = 1, 2, 3$.

(d) The generator g_k is nonresonant if and only if the number $(\eta_k - \mu_k)$ is not rational. In particular, G contains a nonresonant element (and is therefore infinite) if at least one of the numbers $(\eta_j - \mu_j)$ is not rational.

Proof: (a) is classical (see, e.g., [B-C, Corollary 2.28]), (b) and (c) are standard (see Section 2 of [B-C], and in particular Example 2.14(b)), and (d) is immediate from (c). Q.E.D.

Proposition 5 *Let G denote the monodromy group of the normal form of the Riemann equation (6). Suppose that*

(a) none of the four numbers $(\eta_1 + \eta_2 + \eta_3)$, $(\eta_1 + \eta_2 + \mu_3)$, $(\eta_1 + \mu_2 + \eta_3)$, and $(\mu_1 + \eta_2 + \eta_3)$ is an integer, and that

(b) one of the numbers $(\eta_j - \mu_j)$, say $(\eta_k - \mu_k)$, is not rational.

Then a necessary condition for G to be a Ziglin group is that $(\eta_j - \mu_j)$ be an odd multiple of 1/2 for $j \neq k$.

Proof: By (a) and Proposition 4(a) the group G is not reducible, and by (b) and Proposition 4(d) the element g_k is nonresonant. Assuming G to be Ziglin we may then assume $G \subset D \cup P$ by Proposition 3(b).

From the relation $g_1g_2g_3 = id$ of Proposition 4(b) we see that if two g_j are in D then the third must also belong to this set, contradicting irreducibility. Since elements in P have spectra $\{i, -i\}$ they are resonant, and g_k must therefore lie in D. But the above reasoning then implies that the other two generators must lie in P and thus have trace zero. The expression for this trace in Proposition 4(c) then gives the result. Q.E.D.

In the case of equation (3) we compute from (7) that

$$\begin{cases} \eta_1 - \mu_1 = \frac{1}{2}\sqrt{9-8\lambda} \\ \eta_2 - \mu_2 = \frac{1}{2} \\ \eta_3 - \mu_3 = 2, \end{cases} \tag{8}$$

$$\begin{cases} (\eta_1 + \eta_2 + \eta_3) = \frac{1}{4}(7 + \sqrt{9-8\lambda}) \\ (\eta_1 + \eta_2 + \mu_3) = \frac{1}{4}(-1 + \sqrt{9-8\lambda}) \\ (\eta_1 + \mu_2 + \eta_3) = \frac{1}{4}(5 + \sqrt{9-8\lambda}) \\ (\mu_1 + \eta_2 + \eta_3) = \frac{1}{4}(7 - \sqrt{9-8\lambda}), \end{cases} \tag{9}$$

$$\begin{cases} t_1 = -2\cos(\frac{\pi}{2}\sqrt{9-8\lambda}) \\ t_2 = 0 \\ t_3 = -2. \end{cases} \tag{10}$$

We see that hypotheses (a) and (b) of Proposition 5 are satisfied for $(9 - 8\lambda) \neq (rational)^2$. Since $(\eta_3 - \mu_3) = 2$ is not an odd multiple of $1/2$ we conclude that G_λ is not Ziglin for all such values of λ, whence from Proposition 1 that X_{H_λ} cannot be integrable. Here integrability is in the sense of that proposition (see the discussion in [R, Remark 1(b) on p. 265], or that in [R-S, Remark 1.16(b)], for a clarification of these concepts). At $\lambda = 0$ the differential equations for (1) uncouple; hence this particular case is integrable.

4. References

[A-D] M. Alvarez and J. Delgado, The spring-pendulum system, in Hamiltonian Systems and Celestial Mechanics, Advanced Series in Nonlinear Dynamics, vol. 4, 1-13 (E.A. Lacomba and J. Llibre, eds.), World Scientific, Singapore, 1993.

[B-B] R. Broucke and P. Baxa, Periodic solutions of a spring-pendulum system, Cel. Mech. **8** (1973), 261-267.

[B-C] A. Baider and R.C. Churchill, On monodromy groups of second-order Fuchsian equations, SIAM J. Math. Anal., Vol. 21, No. 6 (1990), 1642-1652.

[BCRS] A. Baider, R.C. Churchill, D.L. Rod and M.F. Singer, On the infinitesimal geometry of integrable systems, in Proceedings of the Fields Institute Workshop "Mechanics Days," to be published by the AMS.

[C-R-S] R.C. Churchill, D.L. Rod and M.F. Singer, Group-theoretic obstructions to integrability, to appear in Ergodic Theory and Dynamical Systems.

[KAP] I. Kaplansky, An Introduction to Differential Algebra, second ed., Hermann, Paris, 1976.

[KOV] J. Kovacic, An algorithm for solving second order linear homogeneous differential equations, J. Symbolic Comput., **2** (1986), 3-43.

[R] D.L. Rod, On a theorem of Ziglin in Hamiltonian dynamics, in Hamiltonian Dynamical Systems, (K. Meyer and D.E. Saari, eds.), Contemp. Math. **81**, 259-270, American Mathematical Society, Providence, Rhode Island, 1988.

[R-S] D.L. Rod and B.D. Sleeman, Complexity in spatio-temporal dynamics, to appear in Proceedings A, Royal Soc. of Edinburgh.

[Z] S.L. Ziglin, Branching of solutions and non-existence of first integrals in Hamiltonian mechanics (I), Functional Anal. Appl. **14** (1982), 181-189.

STABILITY OF ALMOST CIRCULAR PERIODIC ORBITS IN HILL'S PROBLEM

JOAQUIN DELGADO*
E-mail: jdf@xanum.uam.mx

and

ERNESTO A. LACOMBA*
E-mail: lace@xanum.uam.mx

*Departamento de Matemáticas, Universidad Autónoma Metropolitana–Iztapalapa,
Av. Michoacán y La Purísima, Col. Vicentina,
Apdo. 55534, C.P. 09340, México, D.F., México.*

ABSTRACT

It is shown that the Hill's problem in Celestial Mechanics has invariant tori for large values of the Jacobian constant, and also that Hill's periodic orbits are orbitally stable. Levi Civita regularization and a scaling of coordinates transforms the problem into a perturbation of two harmonic oscillators in the plane with equal frequencies. Normalization using Lissajous variables permits to compute the Poincaré map on a surface of section and apply Moser's twist theorem to show stability. An explicit interpretation of Marsden-Weinstein reduction shows that the reduced flow is a Hamiltonian flow on a sphere with the standard symplectic structure. The results are similar to those obtained by Kummer for the restricted 3-body problem.

1. Introduction

In this paper we consider some aspects of Hill's problem in Celestial Mechanics. This is a simplified model of the restricted 3-body problem, which was initially proposed by G.W. Hill around 1878 to describe the lunar motion about the earth under the sun's perturbative effect. Even though this model is a simplification of the real situation, the precise description of some of the periodic orbits in this case can be given if corrections are taken into account due to the inclinations of the orbits, the effect of the earth's shape, etc. This idea appeared as Poincare's *main problem* of lunar theory. Among other things, Hill himself computed a family of almost circular periodic orbits, which is now known as Hill's family, and includes the lunar motion as one of its orbits. Other recent applications concern the study of the motion of Saturn's co-orbital satellites [15]. The mathematical problem is a two degrees of freedom Hamiltonian system where the Hamiltonian up to a constant factor is known as Jacobi's constant.

---

*Research partially suported by CONACYT grant 1772-E9210

We study here Hill's problem for large values of the Jacobi's constant. Levi Civita regularization eliminates the singularity keeping the canonical form of the equations, and permits to state the problem as a small perturbation of two harmonic oscillators with equal frequencies. We prove the existence of invariant tori and also the stability of Hill's periodic orbits.

These results follow by application of Moser's twist theorem for the Poincaré map of perturbed two-degrees of freedom Hamiltonian systems. Since the unperturbed hamiltonian is degenerate, we cannot expect to apply Moser's theorem in a straightforward manner. The main tool is the use of normal forms and Lie transforms using Lissajous coordinates introduced by Deprit [4]. For the stability of Hill's solutions we have to transform to Poincaré coordinates because the circular orbits from which they are born in the unperturbed problem are not well represented in Lissajous coordinates. As a geometrical interpretation we also prove that elimination of the elliptic anomaly while keeping constant its conjugate momentum originates intrinsically a reduced Hamiltonian system in a reduced phase space which is symplectomorphic to a two dimensional sphere with the standard area element.

We remark that there are generalizations of Hill's family of periodic solutions. They were found by Brown [3] in the restricted 3-body problem using Hill's method, and by Moulton [11] and Perron [12] in the plane 3-body problem using the Poincaré continuation method.

After this paper was written, the authors were advised that there is a significant overlapping of this paper with the results obtained by Kummer for the restricted 3-body problem in [7]. The variables he uses are closely related to the Lissajous variables of Deprit considered here. At the end of the paper we compare our results with Kummer's.

In fact, for large values of the Jacobian constant, the restricted 3-body problem in the bounded component of Hill's region containing the small primary and Hill's problem in the bounded component have some similarities. Kummer [7] considered for the planar restricted 3-body problem a regularized Hamiltonian formally differing from ours only by higher order terms in the positions. It is for this reason that the normalization and the reduction of the Hamiltonian system gave essentially the same results.

Using McGehee's blow up, the singularity is substituted by an invariant torus. Introducing special canonical variables one can show also the existence of invariant tori by computing the Poincaré map and applying Moser's twist theorem on an appropriate surface of section. Nevertheless in this approach the Poincaré map is not defined on the intersections of the invariant manifolds with the considered surface. The advantage of a prior regularization and normalization, which is the approach of the present paper, is that collisions correspond to regular points for the vector field.

2. Formulation and regularization

Hill's problem is a limiting case of the restricted 3-body problem. Consider two primaries of masses μ and $1-\mu$ moving in circular orbits about their center of mass. We want to describe the motion of a third massless particle (satellite or asteroid) under the gravitational effect of the two primaries. Passing to a rotating frame where the primaries are at rest, we get the (circular) restricted 3-body problem. Assume now that one of the primaries is sensibly smaller than the other one ($\mu \approx 0$), and the satellite moves so close to the primary of mass μ, that the gravitational effect of the other primary can be considered as a secondary perturbation. If we make a scaling of coordinates putting the primary of greater mass at infinity, we get in the limit the *Hill's problem*. This is a Hamiltonian system with two degrees of freedom whose Hamiltonian is

$$H = (p_1^2 + p_2^2)/2 + q_2 p_1 - q_1 p_2 - \frac{1}{\sqrt{q_1^2 + q_2^2}} - q_1^2 + \frac{q_2^2}{2}.$$

Since H is a first integral, we can restrict the study to manifolds of constant H value. Actually it is customary to write $H = -C/2$, where C is known as the Jacobian constant. The projection into configuration space (q_1, q_2) of a Jacobian level is called the Hill's region, and it is symmetric with respect to the origin. See Fig. 1. For $C > 3^{4/3}$ there are two unbounded components and one bounded component containing the singularity at the origin, the latter being diffeomorphic to a punctured disk. Actually, since we are interested in large values of C we will study the bounded component of the case $C > 3^{4/3}$, specially when $C \to \infty$ and the component shrinks to the origin. The details can be seen in [13], [14].

The singularity at the origin can be regularized by using Levi-Civita transformation: in complex notation $q = q_1 + iq_2$, $p = p_1 + ip_2$, we let $q = x^2$, $p = y/(2\bar{x})$ where $\bar{x}$ is complex conjugation, together with a reparametrization $dt/d\tau = 4|x|^2$ of the solutions over a fixed Jacobian level. Since we are interested in the case $C >> 1$, let us introduce the small parameter $\varepsilon = (4C)^{-3/2}$ and make a new scaling in the configuration coordinate $x \to \varepsilon^{1/3} x$, and the time $\tau \to \varepsilon^{1/3}\tau$. The new Hamiltonian becomes

$$\overline{H} = (|y|^2 + |x|^2)/2 + 2\varepsilon|x|^2(x_2 y_1 - x_1 y_2) + 4\varepsilon^2 |x|^2 (4x_1^2 x_2^2 - x_1^4 - x_2^4), \qquad (1)$$

always restricted to the level surface $\overline{H} = 4$. Note that for small ε, the Hamiltonian looks like a perturbation of two harmonic oscillators in the plane with

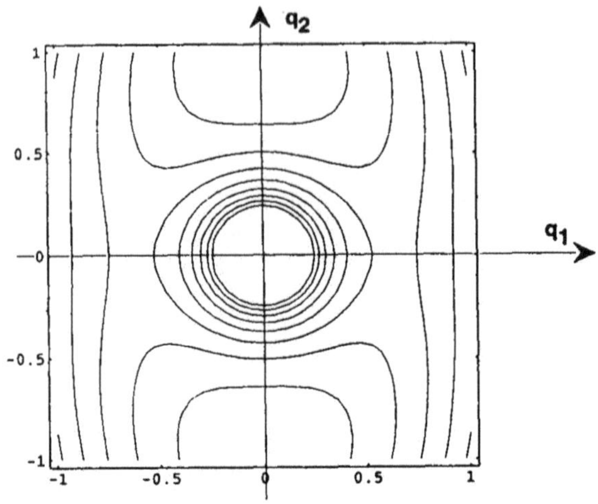

Figure 1: Hill's regions.

equal frequencies so that the unperturbed part is very degenerate.

3. Coordinates of Lissajous and Poincaré

The perturbations of the 1 : 1 resonant Hamiltonian $H_0 = (|y|^2 + |x|^2)/2$, which is the leading term of the regularized Hamiltonian for Hill's problem, can be studied by transforming to angle-action variables l, g, L, G, as introduced by Deprit [5] under the name of Lissajous coordinates. They are related to cartesian canonical coordinates by the formulae

$$\begin{aligned}
x_1 &= b\cos l\cos g - \sigma a\sin l\sin g \\
x_2 &= \sigma a\sin l\cos g + b\cos l\sin g \\
y_1 &= -b\sin l\cos g - \sigma a\cos l\sin g \\
y_2 &= \sigma a\cos l\cos g - b\sin l\sin g
\end{aligned}$$

where $L = (a^2 + b^2)/2$, $G = \sigma ab$, $a \geq b \geq 0$ and $\sigma = \pm 1$. These coordinates are interpreted as follows: given the inital conditions (x, y), the solutions to Hamilton's equations for H_0 project into configuration x-space as ellipses centered at the origin. Coordinates a and b represent, respectively, the major and minor semiaxes of the ellipse, g is the argument of the pericenter (an angle) and l is what is known as the "elliptic anomaly", measured counterclockwise from the

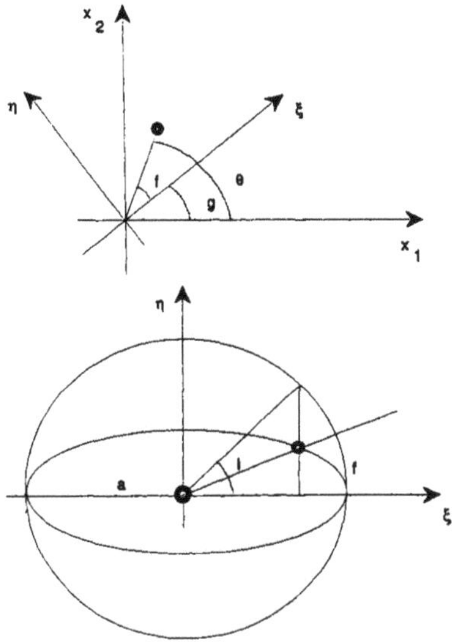

Figure 2: Geometric interpretation of Lissajous coordinates l, g.

periaxis if $G > 0$ and clockwise if $G < 0$. It is clear that (l, g) and $(l+\pi, g+\pi)$ must be identified, because they define the same point in the ellipse. See Fig. 1. We can esily verify the following facts:

1. $H_0 = L$.

2. g is not defined for circular orbits $G = \pm L$.

3. l, g are angular variables and $|G| < L$.

4. $G = x_2 y_1 - x_1 y_2$ is the angular momentum, so $G = 0$ corresponds to collinear orbits.

5. Since H_0 is integrable, the phase space for fixed value of L is foliated by tori, invariant under the hamiltonian flow of $H_0 . H_0$ is degenerate and isoenergetically degenerate.

We remark that near the circular orbits $G = \pm L$ we can switch to Poincaré variables, $(l, g, L, G) \to (\lambda, \varphi, \Lambda, \rho)$, where

$$\begin{aligned} \lambda &= l \pm g, \quad \Lambda = L, \\ \varphi &= \mp g, \quad\ \ \rho = L \pm G. \end{aligned} \tag{2}$$

4. Normalization procedure

When we express the unperturbed part H_0 in Lissajous coordinates, its degeneracy becomes evident, and we cannot apply KAM theory in a straightforward way. Given a Hamiltonian depending on a small parameter, the normalization procedure permits to eliminate "fast" variables to any prescribed order, by means of succesive canonical transformations. The classical normalization theory of Von Zeipel has as a disadvantage the use of generating functions involving both the old and new variables. Using instead the Lie transforms, we can express the new coordinates completely in terms of the old ones in an explicit way. In addition, inversion of the formulas is performed just by changing the sign of the generator, so computations can be carried out in a recursive algorithm. The complete theory of the Lie transforms was introduced by Deprit [4]. It is enough to give here the basic ideas. Let

$$H = H_0 + \varepsilon H_1 + \frac{\varepsilon^2}{2!} H_2 + \cdots,$$

where H_i depends on the Lissajous coordinates (l, g, L, G). The expression of the normalized Hamiltonian to order n will be obtained as

$$H = \overbrace{K_0 + \varepsilon K_1 + \frac{\varepsilon^2}{2!} K_2 + \cdots + \frac{\varepsilon^n}{n!} K_n}^{K} + R$$

which is a function of some new variables (l', g', L', G') such that

1. The variables with primes are related to the original variables via a *Lie transform* (which is a canonical transformation)

$$\begin{array}{ll} l' = E_W(l) & L' = E_W(L) \\ g' = E_W(g) & G' = E_W(G) \end{array}$$

2. The generator

$$W = W_1 + \varepsilon W_2 + \frac{\varepsilon^2}{2!} W_3 + \cdots$$

   is chosen so as to satisfy the condition $\{K_i, K_0\} = 0$, for $i = 1, \ldots, n$.

3. $K_0 = L'$ and is a first integral for K, because of 2.

4. $R = \mathcal{O}(\varepsilon^{n+1})$

We shall omit the primes in the variables in what follows in order to simplify the notation.

The elliptic anomaly will be taken as the fast variable, since $\dot{l} = 1 + \mathcal{O}(\varepsilon)$, while the time variation of the other variables is $\mathcal{O}(\varepsilon)$. The advantage of doing the computations in Lissajous coordinates is that the homological equations

$$\{H_0, W_m\} + I_m = K_m$$

where I_m and K_m are given, become simply

$$\frac{\partial W_m}{\partial l} = I_m - K_m.$$

These are easily solved by integration with respect to l. No integration is needed in fact, if one can give in advance the type of terms in sine and cosine of integer linear combinations of the angles l and g that will appear.

According to the scheme developed in [4], the normalized Hamiltonian up to terms of order 2 is given by

$$H = L - 2\varepsilon LG + \frac{\varepsilon^2}{2!}L\left(-5L^2 - G^2 - 15(L^2 - G^2)\cos 4g\right) + \mathcal{O}(\varepsilon^3), \quad (3)$$

5. The Poincaré map as a twist

In this section we consider a Poincaré map for a convenient surface of section of the flow. We show that it is a perturbation of a twist map so that the existence of invariant tori is asserted by Moser's twist theorem.

We will take l as indendent variable in (3), since $\dot{l} = 1 + \mathcal{O}(\varepsilon)$. This requires to solve the following implicit equation

$$H(l, g, L(l, g, G, \varepsilon), G, \varepsilon) = K(l, g, L(l, g, G, \varepsilon), \varepsilon) + \mathcal{O}(\varepsilon^3) = 4$$

for L to order $\mathcal{O}(\varepsilon^2)$ and solve Hamilton's equations

$$\begin{aligned} \frac{dg}{dl} &= -\frac{\partial L}{\partial G}, \\ \frac{dG}{dl} &= \frac{\partial L}{\partial g}, \end{aligned}$$

for the Hamiltonian $-L$. We have explicitly

$$L = 4 + 8G\varepsilon + \varepsilon^2\left(160 + 18G^2 + \underbrace{30(16 - G^2)\cos 4g}\right) + \cdots \quad (4)$$

The term underlined by the bracket can be killed by a new Lie transform. Then the Poincaré map for the section $l \equiv 0(\mathrm{mod}\ 2\pi)$ can be computed as

$$\begin{aligned} g_1 &= g_0 - 16\pi\varepsilon - 72\pi\varepsilon^2 G_0 + \mathcal{O}(\varepsilon^3) \\ G_1 &= G_0 + \mathcal{O}(\varepsilon^3) \end{aligned} \tag{5}$$

Because of the term $16\pi\varepsilon$, one can expect most of the periodic orbits to have long periods for small ε. Notice that Moser's twist theorem [10] cannot be applied yet. This difficulty can be overcome by using a a time-dependent canonical transformation in the Hamiltonian function $L(g, G, l, \varepsilon)$ in (4). We change from coordinates (g, G) to coordinates (ϕ, F) with l as the independent variable. We take $W(g, F, l) = (g - 8\varepsilon l)F$, so the following transformation of coordinates and Hamiltonian

$$\phi = \frac{\partial W}{\partial F}, \qquad G = \frac{\partial W}{\partial g}, \qquad L' = L + \frac{\partial W}{\partial l}$$

takes the Poincaré map into the desired form

$$\begin{aligned} \phi_1 &= \phi_0 + 2\pi - 36\pi\varepsilon^2 F_0 + \mathcal{O}(\varepsilon^3), \\ F_1 &= F_0 + \mathcal{O}(\varepsilon^3). \end{aligned}$$

Theorem 1 *For sufficient large values of the Jacobian constant C, there exist tori that are invariant by the regularized flow in Hill's problem. The complement of this invariant tori form a set of measure tending to zero with $C \to \infty$.*

For $\varepsilon = 0$, $F = G = 0$ defines the set of collinear orbits through the origin which under the Levi Civita transformation correspond to ejection-collision orbits. Now, if we take the set $|F| < \delta$ in the space of action variables, where δ is some fixed positive constant, then from Moser's theorem, for $\varepsilon \approx 0$, the set of ejection-collision orbits are "trapped" between invariant tori. For this to occur it is sufficient to choose a value of F such that $|F| < \delta$ and satisfying the diophantine condition in the hypothesis of Moser's theorem. In fact this proves some kind of stability in the set of ejection-collision orbits, so they cannot wander in the whole phase space. For an ejection-collision orbit it does not make sense to ask about stability but one can say more for orbits which are solely of ejection or of collision.

Corollary 1 *An ejection (collision) orbit defined for all positive (all negative) times is foward (backward) orbitally stable, provided that the Jacobian constant is taken large enough.*

6. Stability of Hill's periodic orbits

Lissajous variables are not defined for circular orbits; therefore to study the stability of orbits springing out from them by perturbation it is very convenient to switch to Poincaré variables.

We will show that Hill's family of almost circular periodic orbits are analytic continuations of the direct and retrograde circular orbits that exist for $\varepsilon = 0$, and then we will show that they are orbitally stable. We will work out the details only for the case of direct circular orbits.

Passing to Poincaré coordinates (2) in the normalized Hamiltonian (3) we get

$$\begin{aligned} H &= \Lambda - 2\varepsilon(\Lambda - \rho)\Lambda + \\ &\quad \frac{\varepsilon^2}{2!}\Lambda(-6\Lambda^2 + 2\Lambda\rho - \rho^2 - 15(2\Lambda\rho - \rho^2)\cos 4\varphi) + \mathcal{O}(\varepsilon^3) \end{aligned}$$

Solving the equation $H = 4$ to order 2 in ε, we get

$$\begin{aligned} \Lambda &= 4 + 8\varepsilon(4 - \rho) + \\ &\quad \varepsilon^2(704 - 208\rho + 16\rho^2 + 30(8\rho - \rho^2)\cos 4\varphi) + \mathcal{O}(\varepsilon^3). \end{aligned} \tag{6}$$

Transforming to cartesian canonical coordinates $\xi = \sqrt{2\rho}\cos\varphi$, $\eta = \sqrt{2\rho}\sin\varphi$ and taking λ as independent variable, we get

$$\begin{aligned} \frac{d\xi}{d\lambda} &= 8\varepsilon\eta + \mathcal{O}(\varepsilon^2), \\ \frac{d\eta}{d\lambda} &= -8\varepsilon\xi + \mathcal{O}(\varepsilon^2). \end{aligned}$$

The Poincaré map is then given as

$$\begin{pmatrix} \xi_1 \\ \eta_1 \end{pmatrix} = \begin{pmatrix} \cos 8\pi\varepsilon & -\sin 8\pi\varepsilon \\ \sin 8\pi\varepsilon & \cos 8\pi\varepsilon \end{pmatrix} \begin{pmatrix} \xi_0 \\ \eta_0 \end{pmatrix} + \varepsilon \begin{pmatrix} R_1(\xi_0, \eta_0, \varepsilon) \\ R_2(\xi_0, \eta_0, \varepsilon) \end{pmatrix}.$$

Let us consider for a moment the more general fixed point equation

$$x = A(\varepsilon)x + \varepsilon^2 F(x, \varepsilon) \tag{7}$$

where $x \in \mathbb{R}^n$, A is a $n \times n$ real matrix, and ε is a small parameter. We suppose that A is a C^1 close to the identity matrix

$$A = I + \varepsilon a_1 + \cdots,$$

and F is of class C^1 in all the $n+1$ variables. Expanding the equation (7) and canceling out the common factor ε we get

$$0 = a_1 x + \varepsilon F(x, \varepsilon).$$

For $\varepsilon = 0$ we have the solution $x = 0$; therefore if $\det a_1 \neq 0$ then by the implicit function theorem, there exists a unique solution depending smoothly on ε that vanishes for $\varepsilon = 0$.

In our case, the matrix a_1 is given by

$$a_1 = 8\pi \begin{pmatrix} 0 & -1 \\ 1 & 0 \end{pmatrix},$$

whose determinant is not zero. We have shown the following result

Theorem 2 *There exist two families of periodic orbits which are analytical continuations of the limiting circular periodic orbits when the Jacobian constant* $C \to \infty$.

We have mentioned that these orbits are known as Hill's periodic orbits. We will now prove their stability .

Theorem 3 *For C large enough there exist invariant tori arbitrarly close to Hill's periodic orbits. In particular they are orbitally stable.*

Proof: Eliminating the angular variable φ from (6) by means of a Lie transform to order 2 in ε, we get

$$\widetilde{\Lambda} = 4 - 8\varepsilon\rho + \varepsilon^2(-208\rho + 16\rho^2) + \mathcal{O}(\varepsilon^3).$$

Let us make a canonical change of coordinates $(\varphi, \rho) \to (\phi, F)$ with generating function $W(\varphi, F)$, asumming that

$$\phi = \frac{\partial W}{\partial F} = \varphi + 8\varepsilon\lambda, \quad \rho = \frac{\partial W}{\partial \varphi}.$$

Solving the first equation with trivial constant of integration gives $W = (\varphi + 8\varepsilon\lambda)F$. Then $\rho = F$ and the transformed Hamiltonian becomes

$$\begin{aligned} K &= \widetilde{\Lambda} + \frac{\partial W}{\partial \lambda} \\ &= \widetilde{\Lambda} + 8\varepsilon F \\ &= 4 + \varepsilon^2(-208F + 16F^2) + \mathcal{O}(\varepsilon^3). \end{aligned}$$

The computation of the Poincaré map for the section $\lambda = 0 \bmod 2\pi$ gives

$$\begin{aligned}\phi_1 &= \phi_0 + \varepsilon^2(-208 + 32F) + \mathcal{O}(\varepsilon^3)\\ F_1 &= F_0 + \mathcal{O}(\varepsilon^3).\end{aligned}$$

The term containing ε^2 in the first equation has a derivative different from zero at the origin $F = 0$. The result now follows from Moser's twist theorem.

□

7. The reduced flow

We will give an intrinsic description of the reduction process described in Section 4. This reduction of the Hamiltonian system was considered already by [8], since it depends only on the unperturbed Hamiltonian. After the normalization up to a fixed order the Hamiltonian can be written as $H = K + R$ where K is in normal form with respect to $H_0 = (y_1^2 + y_2^2 + x_1^2 + x_2^2)/2$, that is $\{K, H_0\} = 0$ where $\{\cdot,\cdot\}$ denotes the Poisson bracket. The Hamiltonian flow for H_0 is an action of the reals on phase space $P = \mathbb{R}^4 \setminus \{0\}$ which in complex coordinates $z_k = x_k + iy_k$, $k = 1, 2$ can be written

$$e^{i\theta}(z_1, z_2) = (e^{i\theta}z_1, e^{i\theta}z_2).$$

Since the flow is 2π-periodic, the last formula defines an action $S^1 \times P \to P$ which preserves the standard symplectic form $(2i)^{-1}(dz_1 \wedge d\bar{z}_1 + dz_2 \wedge d\bar{z}_2)$. Then the momentum map for this symplectic action is simply H_0. The reduced space $P_L = H_0^{-1}(L)/S^1$ is diffeomorphic to the projective space CP^1 which in turn is diffeomorphic to a sphere in $\mathbb{R}^3$. By the Marsden-Weinstein reduction theorem [1], there exists a unique symplectic form ω_L in P_L such that $\pi^*\omega_L = i^*\omega$, where π is the projection, i the inclusion and * denotes the pullback:

$$\begin{array}{ccc} H_0^{-1}(L) & \overset{i}{\hookrightarrow} & (\mathbb{C}^2, \omega) \\ \pi \downarrow & & \\ (P_L, \omega_L) & & \end{array}$$

Theorem 4 *There exists a symplectomorphism between the reduced space (P_L, ω_L) and $S_L \subset \mathbb{R}^3$, the sphere of radius L in $\mathbb{R}^3$ with the standard volume form times the constant factor $2L$.*

We will denote the classes of P_L by square brackets $[z_1, z_2]$. Let $\varphi_1, \varphi_2 \colon P_L \to \mathbb{C}$ be the proyective charts in CP^1 given by

$$\varphi_1([z_1, z_2]) = z_2/z_1, \qquad \varphi_1([z_1, z_2]) = z_1/z_2,$$

whenever defined. The respective inverses are given by

$$\lambda \mapsto \frac{\sqrt{2L}}{\sqrt{1+|\lambda|^2}}[1,\lambda], \qquad z \mapsto \frac{\sqrt{2L}}{\sqrt{1+|z|^2}}[z,1]$$

and the change of coordinates is given by $z = 1/\lambda$.

Lemma 1 *The local representatives of the symplectic form ω_L in the atlas φ_1, φ_2 are given by*

$$\omega_L^{\varphi_1}(\lambda) = \frac{2L}{(1+|\lambda|^2)^2}\frac{d\lambda\wedge d\bar{\lambda}}{2i}, \qquad \omega_L^{\varphi_1}(z) = \frac{2L}{(1+|z|^2)^2}\frac{dz\wedge d\bar{z}}{2i}$$

Proof: Let $f(\lambda,\bar{\lambda}), g(\lambda,\bar{\lambda})$ be C^∞ functions in $\mathbb{C} = \mathbb{R}^2$ and let

$$F(z_1,z_2,\bar{z}_1,\bar{z}_2) = f(z_2/z_1, \bar{z}_2/\bar{z}_1), \qquad G(z_1,z_2,\bar{z}_1,\bar{z}_2) = g(z_2/z_1,\bar{z}_2/\bar{z}_1),$$

then by definition of the Poisson bracket of f and g with respect to the symplectic form ω_L, we have

$$\{f,g\}_\lambda = \{F,G\}_{z_1,z_2}, \qquad \lambda = \frac{z_2}{z_1}, \bar{\lambda} = \frac{\bar{z}_2}{\bar{z}_1}.$$

Here, $\{F,G\}_{z_1,z_2}$ is the Poisson bracket with respect to the standard symplectic form ω in $\mathbb{C}^2$,

$$\{F,G\}_{z_1,z_2} = 2i\left(\frac{\partial F}{\partial \bar{z}_1}\frac{\partial G}{\partial z_1} - \frac{\partial F}{\partial z_1}\frac{\partial G}{\partial \bar{z}_1} + \frac{\partial F}{\partial \bar{z}_2}\frac{\partial G}{\partial z_2} - \frac{\partial F}{\partial z_2}\frac{\partial G}{\partial \bar{z}_2}\right).$$

By the chain rule, we have

$$\begin{aligned}\{f,g\}_\lambda &= 2i\left(\frac{\partial f}{\partial\bar{\lambda}}\frac{\partial g}{\partial\lambda} - \frac{\partial f}{\partial\lambda}\frac{\partial g}{\partial\bar{\lambda}}\right)\left(\frac{\partial\bar{\lambda}}{\partial\bar{z}_1}\frac{\partial\lambda}{\partial z_1} + \frac{\partial\bar{\lambda}}{\partial\bar{z}_2}\frac{\partial\lambda}{\partial z_2}\right)\\ &= \left(\frac{|z_2|^2}{|z_1|^4} + \frac{1}{|z_1|^2}\right)\{f,g\}_\lambda\\ &= \frac{(1+|\lambda|^2)^2}{2L}(f,g)_\lambda,\end{aligned}$$

where $(f,g)_\lambda$ denotes the Poisson bracket with respect to the canonical form $(2i)^{-1}d\lambda\wedge d\bar{\lambda}$ in $\mathbb{C} = \mathbb{R}^2$. In the same way, given $f(z,\bar{z})$ and $g(z,\bar{z})$, C^∞ functions, and using the local chart $z = z_1/z_2$, we must define

$$\begin{aligned}F(z_1,z_2,\bar{z}_1,\bar{z}_2) &= f(z_1/z_2,\bar{z}_1/\bar{z}_2),\\ G(z_1,z_2,\bar{z}_1,\bar{z}_2) &= g(z_1/z_2,\bar{z}_1/\bar{z}_2),\end{aligned}$$

so that by definition of the Poisson bracket $\{\cdot,\cdot\}_z$,

$$\{f,g\}_z = \frac{(1+|z|^2)^2}{2L}(f,g)_z.$$

Lemma 2 *Let $\{\cdot,\cdot\}$ and $(\cdot,\cdot)$ be two Poisson structures associated to two symplectic forms α and β on M. Supose there exists a never vanishing function k on M such that $\{f,g\} = k(f,g)$, for any smooth functions f and g. Then $\alpha = k^{-1}\beta$.*

Proof: Let V_f and W_f be the vector fields (derivations) defined by

$$V_f(h) = \{f,h\}, \qquad W_f(h) = (g,h).$$

The mapping $C^\infty(M) \to T_B(M)$ such that $f \mapsto W_{p,f}$ is surjective since M is C^∞ and compact [6] and also

$$\alpha(V_f,V_g) = \{f,g\} = k(f,g) = k\beta(W_f,W_g),$$

but $V_f(h) = \{f,h\} = k(f,h) = kW_f(h)$ for all $h \in C^\infty(M)$. Therefore $V_f = kW_f$ and $k^2\alpha(W_f,W_g) = k\beta(W_f,W_g)$, and since $W_{(\cdot)}$ is surjective, it follows that $\alpha = k^{-1}\beta$

□

Proof of theorem 4. Let S_L denote the sphere of radius L in $\mathbb{R}^3$. Let $\psi_i\colon \mathbb{C} \to S_L$, $i = 1,2$ be the local charts defined as follows, write $\psi_1(\lambda) = (x_1,x_2,x_3)$, where

$$x_1 = \frac{L(|\lambda|^2-1)}{1+|\lambda|^2}, \quad x_2 = -\frac{L(\lambda+\bar\lambda)}{1+|\lambda|^2}, \quad x_3 = \frac{L(\lambda-\bar\lambda)}{i(1+|\lambda|^2)}.$$

The expression for ψ_2 is obtained by changing λ by $1/z$. ψ_1 is the stereographic projection from the point $(L,0,0)$, except for a minus sign in x_2. Note that we have not chosen as ψ_1 the stereographic projection, because then the change of coordinates $z \to 1/\bar z$ would not be symplectic. For any smooth curve $\lambda(t)$ in $\mathbb{C}$, let $(x_1(t),x_2(t),x_3(t))$ be the image under any of the charts ψ_i in S_L. Then

$$\dot x_1^2 + \dot x_2^2 + \dot x_3^2 = \frac{4L^2}{(1+|\lambda|^2)^2}|\dot\lambda|^2$$

which shows that the atlas ψ_1,ψ_2 amplifies an infinitesimal area in the complex plane by a factor

$$\frac{4L^2}{(1+|\lambda|^2)^2}.$$

Therefore

$$\begin{aligned}\psi_i^*\Omega_L &= \frac{4L^2}{(1+|\lambda|^2)^2}\frac{d\lambda\wedge d\bar\lambda}{2i}\\ &= 2L\omega_L^{\varphi_i}(\lambda)\end{aligned}$$

for $i = 1,2$, where Ω_L is the canonical area element in S_L. In the last expression we have the local representatives of the symplectic form ω_L in the reduced space P_L, except for the constant factor $2L$.

□

Let us write the conformal symplectomorphism as $h: (P_L, \omega_L) \to (S_L, \Omega_L)$. We will give it explicitly in terms of the atlases φ_i in P_L and ψ_i in S_L, $i = 1,2$. The transition mappings are $h_{i,j} = \psi_i \circ h \circ \varphi_j$, for $i,j = 1,2$, and it is enough to give $h_{1,1}$ since there exists the relationship $h_{i,j} = h_{i,k} \circ (\varphi_k^{-1} \circ \varphi_j)$. In fact,

$$\begin{aligned}x_1 &= (|z_2|^2 - |z_1|^2)/2\\ x_2 &= -(\bar z_2 z_1 + z_2\bar z_1)/2\\ x_3 &= (z_2\bar z_1 - \bar z_2 z_1)/2i.\end{aligned}$$

This is a particular form of the Hopf map employed by Kummer in [8] Formula (8). It also appears in [7] Formula (29). If we parametrize z_1, z_2 by Lissajous coordinates we get a parametrization of the sphere S_L except for the poles $G = \pm L$,

$$\begin{aligned}x_1 &= Le\cos(2g)\\ x_2 &= Le\sin(2g)\\ x_3 &= L\eta,\end{aligned}$$

where $\eta = G/L$ and $e = \sqrt{1-\eta^2}$. Introducing the angle i by $\eta = \cos i$, $e = \sin i$, then $(L, 2g, i)$ become the standard spherical coordinates in $\mathbb{R}^3$. The element of area in this coordinates is simply

$$\Omega_L = 2L\, dG \wedge dg,$$

which is precisely the symplectic form ω_L in Lissajous coordinates.

The coordinate functions $x_i, i = 1,2,3$ are elements in $C^\infty(S_L)$, and if the Poisson bracket between them is computed using the parametrizations $x_i(g, G)$ above one gets

$$\frac{\partial x_i}{\partial g}\frac{\partial x_j}{\partial G} - \frac{\partial x_i}{\partial G}\frac{\partial x_j}{\partial g} = \varepsilon_{ijk} 2L x_k,$$

where ε_{ijk} is the sign of the permutation (ijk). Since the symplectic form in S_L is $\Omega_L = 2L\, dG \wedge dg$ it follows that in terms of Poisson brackets, the above relationship becomes

$$(x_i, x_j) = \varepsilon_{ijk} x_k \tag{8}$$

These are precisely the commutation relations in $so(3)$, the Lie algebra of the group of rotations $SO(3)$; but because of the restriction $x_1^2 + x_2^2 + x_3^2 = L^2$, the pair (S_L, Ω_L) should be interpreted as an orbit of the coadjoint action of the Lie algebra $so(3)$ on $SO(3)$.

Now we will give the intrinsic interpretation of the normalization. Let H_0 denote the Hamiltonian of the two oscillators in the plane with equal frequencies, and suppose that K is in normal form with respect to H_0, that is $\{H_0, K\} = 0$. Then K is constant along the orbits of H_0 and so induces a function K_L on each level P_L, and P_L can be identified with a coadjoint orbit of $so(3)$ on $SO(3)$ whith a natural Poisson structure. This orbit in turn is identical to the sphere S_L with the commutation relations (8). Then the normalization is described by the Hamiltonian system (K_L, S_L).

For any smooth function defined on S_L its time variation is given by

$$\dot{f} = (f, K_L).$$

Using the chain rule and the commutation relations (8) we get for the coordinates

$$\begin{aligned}
\dot{x}_1 &= -x_3\frac{\partial K_L}{\partial x_2} + x_2\frac{\partial K_L}{\partial x_3},\\
\dot{x}_2 &= x_3\frac{\partial K_L}{\partial x_1} - x_1\frac{\partial K_L}{\partial x_3},\\
\dot{x}_3 &= -x_2\frac{\partial K_L}{\partial x_1} + x_1\frac{\partial K_L}{\partial x_2}.
\end{aligned}$$

This system is literally identical with Equations (6) of [8] or (38) of [7].

For Hill's problem the regularization is achieved over the constant level of the Hamiltonian equal to 4, see (1). The Jacobian constant is hidden in the parameter ε, so to study the reduced flow, for given ε and L , we have to pick up the curve which satisfies $K_L = 4$ on the reduced space S_L. It would be better if we could vary the energy level of the reduced Hamiltonian over a fixed reduced space and for some fixed value of ε. This can be done if we change the scale as follows. Let $\bar{L} = \varepsilon L$, $\bar{G} = \varepsilon G$, then the relation

$$K(g, L, G, \varepsilon) = 4 \quad \text{become} \quad K(g, \bar{L}, \bar{G}, 1) = 4\varepsilon$$

and the Hamilton equations for $K(g, L, G, \varepsilon)$ are transformed into the Hamilton equations of $\bar{K}(g, \bar{L}, \bar{G}) := K(g, \bar{L}, \bar{G}, 1)$, if we use the slow time $t' = \varepsilon t$. Moreover, the scaling maps the sphere

$$x_1^2 + x_2^2 + x_3^2 = L^2 \quad \text{into} \quad \bar{x}_1^2 + \bar{x}_2^2 + \bar{x}_3^2 = \bar{L}^2$$

where $\bar{x}_i = \varepsilon x_i, i = 1, 2, 3$, and the commutation relations

$$(x_i, x_j) = \varepsilon_{ijk} x_k \quad \text{become} \quad (\bar{x}_i, \bar{x}_j) = \varepsilon_{ijk} \bar{x}_k \tag{9}$$

Therefore, it is enough to express $\bar{K}_{\bar{L}} = \bar{K}(g, \bar{L}, \bar{G})$ in terms of the coordinates $(\bar{x}_1, \bar{x}_2, \bar{x}_3)$ and use the new commutation relations (9).

In what follows we will omit the bars on the variables. For Hill's problem we have computed the normal form up to the fourth order, which expressed in the scaled variables becomes

$$\begin{aligned} K &= L - 2L^2 x_3 + \frac{L^3}{2!}\left(x_1^2 + x_2^2 - 6 - 15(x_1^2 - x_2^2)\right) \\ &+ \frac{L^4}{3!} x_3 \left(-30(x_1^2 + x_2^2) - 60 - 360(x_1^2 - x_2^2)\right) \end{aligned}$$

The level curves $K = const$ are depicted in the following figures for some fixed values of the parameter L, written on their sides. They where done using a Runge-Kutta algorithm of order 4 to solve the Hamilton equations for K and using *MATHEMATICA* ver. 2.1 of Wolfram and coworkers, especially its graphic library. The vertical axis is x_3 and the horizontal axes form the x_1-x_2 plane.

For any value of L there exist two critical points at the north and south poles, which correspond to the almost circular periodic orbits. A great distortion of the invariant curves is observed for relatively small values of L. Particularly, the northern pole seems to lose its stability before the south pole does. We think it would be interesting to investigate numerically these formal properties.

8. Remarks

In this section we discuss the relationship between our results and those obtained by Kummer [7]. We mentioned already in the introduction that he studied the stability of Hill's solutions of the plane restricted 3-body problem. He uses different variables which are closely related to the Lissajous variables of Deprit that we have considered here. We thank the referee for going through the subtleties of comparing both papers. Nevertheless, the correspondence is in part formal, and an explicit discussion of the stability of these periodic solutions for the Hill's problem did not exist in the literature.

Our Hamiltonian defined in the first paragraph of Section 2 is identical with the Hamiltonian in [7] defined by (7), provided we drop his term $O(|x|^3)$ and we set $\mu = \nu = 1$. It is here the first point where the correspondence is formal, since in his Hamiltonian $\mu + \nu = 1$. His variables in terms of ours are obtained by $x = q_1 + iq_2$, $y = p_1 + ip_2$. The Hamiltonian system defined by (7) in [7] on a surface of large negative energy induces a flow which after a change of time variable and the Levi Civita transformation is described by the Hamiltonian system $(K, \Theta = 2Re(\bar{w}dv))$ with K defined in [7] (31). Using the same ideas but a factor $\frac{1}{2}$ in the Levi Civita transformation, we arrive at an identical Hamiltonian system $(\bar{H} = 2K, 2\Theta)$ where our variables are related to those of [7] by $x = 2v$, $y = 2w$, $\varepsilon = \frac{1}{8}\kappa$, so that $2\Theta = Re(\bar{y}dx)$ (where ε is as defined in this paper). Also, the normal form of the two papers is identical. Substituting $\cos^2\phi = (1 + \cos 2\phi)/2$, again with $\mu = \nu = 1$ (formal correspondence) Hamiltonian (51) of [7] becomes

$$K = 2J - 2\kappa J M_3 - \frac{1}{4}J\kappa^2[5J^2 + M_3^2 + 15(J^2 - M_3^2)\cos 2\phi] + \quad h.o.t.$$

More precisely, the normalized Hamiltonian system is $(K, \Theta = Jdu + M_3 d\phi)$. On the other hand, our normalization leads to the Hamiltonian system $(H = 2K, 2\Theta)$, where Lissajous variables of Deprit L, G, g, l and ε are related to the variables used in [7] as follows

$$L = 4J, \quad G = 4M_3, \quad g = \frac{1}{2}\phi, \quad l = \frac{1}{2}u, \quad \varepsilon = \frac{1}{8}\kappa \Longrightarrow 2\Theta = Ldl + Gdg.$$

Indeed, with these substitutions the Hamiltonian $2K$ of [7] becomes H defined here in (3).

A final remark is that only the first non-zero term in a normal form is uniquely determined. Indeed, a second normalization in the problem at hand leads to the conclusion that there exists a normal form (called canonical) *which is a formal function of G and L only.* For instance, the canonical form up to second order is identical with (3) but with the term $15(L^2 - G^2)\cos 4g$ removed. For general questions of the non-uniqueness of the normal form in the 1:1 resonance see [2].

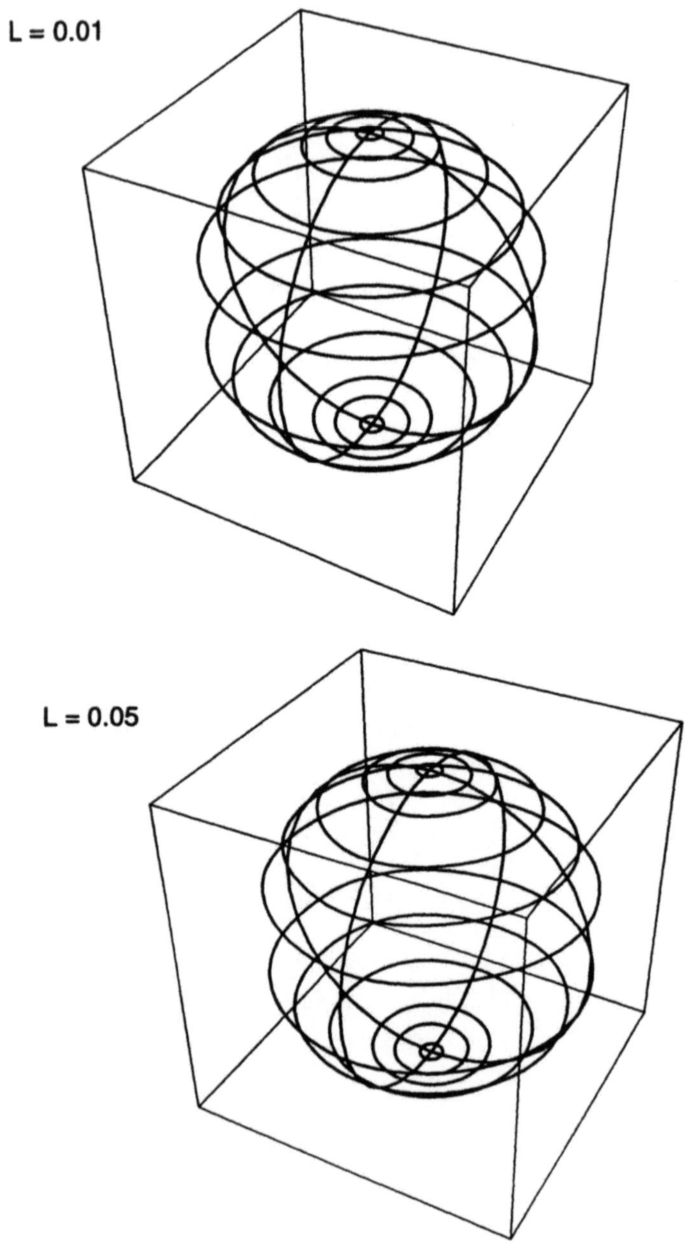

Figure 3

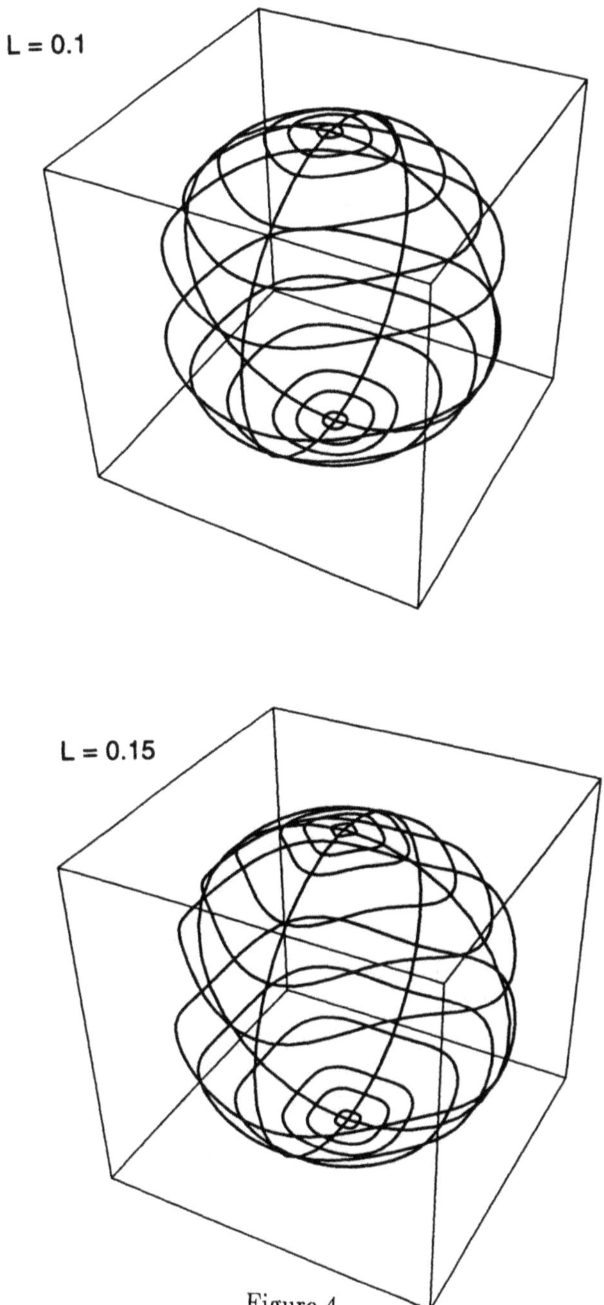

Figure 4

L =0.17

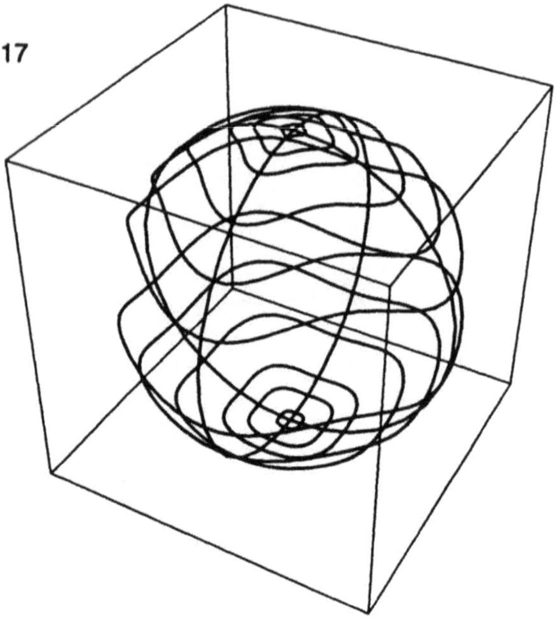

L = 0.18

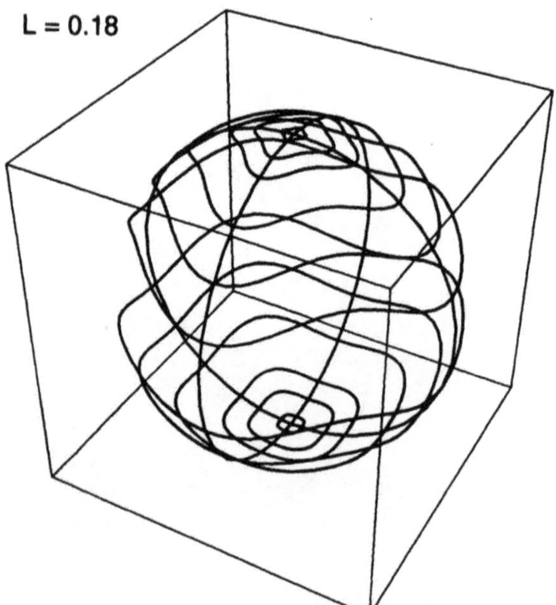

Figure 5

9. References

1. Abraham, A.–Marsden, J. *Foundations of Mechanics,* Benjamin/ Cummings (1979).
2. Baider, A. *Unique normal form for vector fields and Hamiltonians,* J. Differential Eq. 78(1989), 33-52.
3. Brown, E.W. *On the part of the parallactic inequalities in the moon's motion which is a function of the mean motions of the sun and moon,* Am. J. Math. 14(1892), 141-160.
4. Deprit, A. *Canonical transformations depending on a small parameter.*, Cel. Mech. 1 (1969), 12-30.
5. Deprit, A. *The Lissajous transformation I. Basics.* Cel. Mech. and Dyn. Astron. 51(1991), 201-225.
6. Kobayashi, S. and Nomizu, K. *Foundations of differential geometry,* vol. I, Interscience tracts in pure and appl. math. no. 15, 1963.
7. Kummer, M. *On the stability of Hill's solutions of the plane restricted three body problem,* Amer. J. Math. 101(1979), 1333-1354.
8. Kummer, M. *On resonant non-linearly coupled oscillators with two equal frequencies,* Commun. Math. Phys. 48(1976), 53-79.
9. Kummer, M. *A group theoretical approach to a certain class of perturbations of the Kepler problem,* Archive Rat. Mech. and Analysis, 91(1985), 55-82.
10. Moser, J. *Lectures on Hamiltonian systems,* Memoirs of the A.M.S. 81(1986), Providence Rhode Island.
11. Moulton, F.R. *A class of periodic solutions of the problem of three bodies with applications to the lunar theory,* Trans. Amer. Math. Soc. 7(1906), 537-577.
12. Perron, O. *Über eine Schar periodischer Lösungen des ebenen Dreikörperproblems,* Sitzber Bayer, Akad. Wiss., math.-nat. Abt. (1936), 157-176.
13. Whittaker, E.T. *Analytic dynamics of particles and rigid bodies.*, Cambridge Univ. Press, 1927.
14. Szebehely, V.G. *Theory of Orbits.*, Academic Press, N.Y. 1967.
15. Waldvogel, J. and Spirig, F. *The three-body problem with two small masses: a singular perturbation approach to the problem of Sturn's coorbiting satellites,* in Stability of the Solar System and its Minor Natural and Artificial Satellites (V. Szebehely, ed.) Reidel, 1985.

ON THE MÜCKET-TREDER GRAVITATIONAL LAW

FLORIN N. DIACU[1]
Department of Mathematics and Statistics, University of Victoria
Victoria, B.C., Canada, V8W 3P4
E-mail: diacu@math.uvic.ca

ABSTRACT

We consider the 3-body problem with the Mücket-Treder post-Newtonian gravitational law. We show that for certain cases singularities cannot occur, while in the other cases triple collision solutions tend to form Newtonian central configurations near collision.

1. Introduction

The discrepancy between Mercury's observed perihelion advance and the theoretical value obtained in the Newtonian model, led astronomers—during the pre-relativistic years—to attack this problem in two different ways. One was to consider the influence of perturbation forces, the other was to change the attraction law. In the last century there have been several attempts in both directions. For example, Bertrand[2] proposed in 1873 the use of the law of the inverse $2+\alpha$-power of the distance with $\alpha > 0$. In 1895, both Hall[8] and Newcomb[16] found, independently, very close values for α. Hall computed $\alpha = 0.00000016$ by making the theory fit the observed perihelion advance of Mercury, while Newcomb obtained $\alpha = 0.0000001574$, taking into account the secular variations of the inner planets. However, Brown[3] showed in 1903 that the motion of the Moon's perigee is in accord with the Newtonian model and that necessarily $|\alpha| < 4{\cdot}10^{-8}$ in order to make observation and theory match together. As a consequence, Bertrand's proposal proved unsuccessful. (For more details see Hagihara's treatise[7].)

Since relativistic mechanics encounters conceptual difficulties in tackling collision orbits of the n-body problem and leads, in general, to lengthy and cumbersome computations, there still existed post-relativistic attempts to improve Newton's model within the frame of classical mechanics. For example, a succesful attempt has been made by Maneff in the twenties[9,10,11,12]. (This author has recently investigated Maneff's potential[5,6] and found that Maneff's model differs essentially from the Newtonian one not only from the quantitative but also from the qualitative point of view.) Mücket and

[1]Supported in part by NSERC Grant 3-48376

Treder[15] added in 1977 a logarithmic correction term to the Newtonian potential, trying to conciliate the two pre-relativistic directions of work mentioned at the beginning. Their reasoning was as follows. If δ is the distance between two particles, the first two terms in the MacLaurin expansion $\delta^\alpha = 1 + \alpha \ln \delta + (\alpha \ln \delta)^2/2! + (\alpha \ln \delta)^3/3! + ...$, suggest a force proportional to $\delta^{-2}(1 + \alpha \ln \delta)$. If one further considers $|\alpha| < 10^{-8}$, then the perihelion and the perigee motions of the inner planets and the Moon respectively, can both be theoretically explained. However, this could be only a coincidence, and other phenomena might appear hard or impossible to prove within this model. Therefore, in 1991 Mioc and Blaga[14] studied the orbital motion finding some advantages in using Mücket-Treder's potential. Ballinger and Diacu[1] performed in 1993 a qualitative study of the Kepler problem, justifying theoretically the use of this model for the case of two bodies if $\alpha < 0$, and showing that the model doesn't make much sense for astronomy if $\alpha > 0$.

The goal of this paper is to study the qualitative behavior near triple collisions for the 3-body problem with the Mücket-Treder potential. Having a parametrized system of differential equations, small changes of the parameter can lead to drastic changes in the global behavior of solutions. Distortions usually start in the neighborhood of singular solutions, therefore these points should be first checked.

At the beginning we consider $\alpha > 0$ and see that all solutions are globally defined. In other words, singularities (collisions—in particular) cannot appear in this case. Such a situation doesn't correspond to the astronomical reality. Of course, this model might be of interest for other, nonastronomical mathematical models.

Since in this paper we are interested in celestial mechanics, we restrict our attention to the case $\alpha < 0$. We will study orbits passing close to a total collapse. The restriction of focusing on this type of collision and not on a partial one, is not sharp; any particular colliding cluster in the n-body problem can be tackled with the method we use[4]. Nevertheless, the computations are easier to present and comprehend in the total collision case. We apply the technique of McGehee's transformations[13], which allows us to "blow-up" the collision singularity and "paste" instead a—so-called—*collision manifold*. Due to the property of continuity of solutions with respect to the initial data, a study of the flow on this fictitious manifold gives information on solutions passing close to triple collisions. Our computations show that the rest points on the collision manifold depend only on the Newtonian component of the potential and not on the logarithmic correction term or on the value of α. This will imply not only the lack of distortions, but that Mücket-Treder's law stays very close to the Newtonian one in the neighborhood of collisions and tends asymptotically to it when the time variable approaches the collision instant. This is not at all clear from the original equations of motion. From the mathematical point of view, this is the first time when McGehee's transformations are adapted to fit a potential that is neither homogeneous nor quasihomogeneous (i.e. the sum of homogeneous functions).

We can thus draw the conclusion that the model proposed by Mücket and Treder, with $-10^{-8} < \alpha < 0$, doesn't differ qualitatively from the Newtonian one if the particles come close to collisions. The only difference is quantitative and it helps us explain the perihelion advance of Mercury.

2. Equations of Motion

Consider 3 particles of masses $m_i > 0, i = 1,2,3$, in the Euclidean space $\mathbb{R}^3$ having coordinates $\mathbf{q}_i = (q_i^1, q_i^2, q_i^3) \in \mathbb{R}^3, i = 1,2,3$. Denote by

$$M = diag(m_1, m_1, m_1, m_2, m_2, m_2, m_3, m_3, m_3)$$

a square matrix having the above elements on the diagonal and 0 in rest. The 9-dimensional vector $\mathbf{q} = (\mathbf{q}_1, \mathbf{q}_2, \mathbf{q}_3)$ represents the *configuration* of the system and will be regarded as a column vector. The *momentum* of the system is thus defined as $\mathbf{p} = M\dot{\mathbf{q}}$. The equations of motion of the 3-body problem in the Mücket-Treder model will be given by the second order 9-dimensional system of ordinary differential equations

$$M\ddot{\mathbf{q}} = \nabla W(\mathbf{q}), \tag{1}$$

where ∇ is the gradient, $W(\mathbf{q}) = U(\mathbf{q}) + V(\mathbf{q})$,

$$U : \mathbb{R}^9 \setminus \Delta \to \mathbb{R}_+, \;\; U(\mathbf{q}) = \sum_{1 \le i < j \le 3} m_i m_j q_{ij}^{-1},$$

is the Newtonian potential, $\Delta = \bigcup_{1 \le i,j \le 3} \{\mathbf{q} \,|\, \mathbf{q}_i = \mathbf{q}_j\}$ is the *collision set*, and

$$V : \mathbb{R}^9 \setminus \Delta \to \mathbb{R}_+, \;\; V(\mathbf{q}) = \sum_{1 \le i < j \le 3} \alpha m_i m_j (1 + \ln q_{ij}) q_{ij}^{-1},$$

denotes the logarithmic correction term, where α is a real constant and $q_{ij} = |q_i - q_j|$ denotes the Euclidian distance between the particles i and j. The case $\alpha = 0$ corresponds to the classical Newtonian one. Remark that since only theoretical investigations are performed, we have chosen the units such that the gravitational constant equals 1.

The equations (1) can be written as an 18-dimensional first order system

$$\begin{cases} \dot{\mathbf{q}} = M^{-1}\mathbf{p} \\ \dot{\mathbf{p}} = \nabla W(\mathbf{q}) \end{cases} \tag{2}$$

The ten classical *first integrals* of the Newtonian case exist also for the Mücket-Treder model (with obvious changes for the energy integral, only). Thus, from the *momentum* and *center of mass integrals*, one can draw the conclusion that the sets

$$Q = \{\mathbf{q} \mid \sum_{i=1}^{3} m_i \mathbf{q}_i = 0\},\ P = \{\mathbf{p} \mid \sum_{i=1}^{3} \mathbf{p}_i = 0\}$$

are invariant for the equations (2) (i.e. if initial data are in the set then the whole orbit is in the set). By restricting equations (2) to the invariant set $Q \times P$ means, from the physical point of view, that one can translate the origin of the absolute frame to the center of mass of the system of particles, without changing the form of the equations of motion. The *integral of energy* is

$$T(\mathbf{p}(t)) - W(\mathbf{q}(t)) = h,$$

where $T : \mathbb{R}^9 \to [0, \infty)$, $T(\mathbf{p}) = \frac{1}{2}\sum_{i=1}^{3} m_i^{-1}|\mathbf{p}_i|^2$ denotes the kinetic energy and h is the *energy constant*.

Using standard results of the differential equations theory one can show that for any initial conditions $(\mathbf{q}, \mathbf{p})(0) = (\mathbf{q}_0, \mathbf{p}_0) \in (\mathbb{R}^9 \setminus \Delta) \times \mathbb{R}^9$, there exists an analytic solution $(\mathbf{q},\mathbf{p})$ of the corresponding initial value problem, defined on some interval (t^-, t^+) containing 0. Without loss of generality we may study only one of the intervals $(t^-, 0]$ or $[0, t^+)$. Let us choose the second one. The solution can be further extended to a maximal interval $[0, t^*)$, where $0 < t^+ \leq t^* \leq \infty$. Recall that a function is called analytic at some point if there is an interval around that point where the function can be written as a Taylor series. There are several equivalent ways to define solutions with singularities. We adopt here the following one.

Definition. A solution $(\mathbf{q}, \mathbf{p})$ of the equations (2) with initial conditions $(\mathbf{q}_0, \mathbf{p}_0)$, which is defined on the maximal interval $[0, t^*)$, is called regular if it is analytic and if $t^* = \infty$. If the solution is analytic on $[0, t^*)$ but fails to be analytic on $[0, t^*]$, then we say that it has a singularity at t^*.

Notice that, in particular, a singularity can be a collision between at least two particles. In the classical n-body problem there may appear noncollision singularities, which physically correspond to a motion that becomes unbounded in finite time. In the next section we will show that for $\alpha > 0$ this is not the case. In other words all solutions are regular.

3. The Case $\alpha > 0$

In this section we will prove the following global existence result:

Theorem 1. *Any solution* $(\mathbf{q}, \mathbf{p})$ *of the equations (2) with* $\alpha > 0$*, having initial conditions* $(\mathbf{q}_0, \mathbf{p}_0) \in (\mathbb{R}^9 \setminus \Delta) \times \mathbb{R}^9$*, is defined on* $(-\infty, +\infty)$.

In other words the equations (2) with $\alpha > 0$ do not have solutions with singularities. In particular, collisions between particles cannot occur.

Proof of Theorem 1. The idea of the proof is simple. Supposing a solution has a singularity, we will show that the energy relation is not satisfied. For this, take a solution $(\mathbf{q}, \mathbf{p})$ of the equations (2), defined on some interval $[0, t^*)$, with $t^* < \infty$. Let us show first that

$$\lim_{t \to t^*} \min_{i<j} q_{ij}(t) = 0. \tag{3}$$

(This is analog to a result due to Painlevé in the Newtonian case.) For proving it, suppose relation (3) is not true, i.e. $\limsup_{t \to t^*} q_{ij}(t) \geq \gamma > 0$, where γ is a constant. Then, there will exist a sequence $(t_\nu)_{\nu \in \mathbf{N}}, t_\nu \to t^*$, and a constant $\beta > 0$, such that $W(\mathbf{q}(t_\nu)) \leq \beta$, for all $\nu \in \mathbf{N}$. By the integral of energy it follows that $T(\mathbf{p}(t_\nu)) \leq \beta + h$ for all $\nu \in \mathbf{N}$. Therefore there is a constant $\mu > 0$ such that $|\mathbf{p}(t_\nu)| \leq \mu$, for all $\nu \in \mathbf{N}$. In other words $|\mathbf{p}(t_\nu)|$ is uniformly bounded by μ. Therefore $|\mathbf{q}(t_\nu)|$ is also uniformly bounded by some positive constant. This contradicts the fact that the sequence $(\mathbf{q}(t_\nu), \mathbf{p}(t_\nu))$ must become unbounded when t_ν approaches t^*. So t^* is not a singularity, and relation (3) follows.

By (3) we see that there are indices i, j and a sequence $(t_m)_{m \in \mathbf{N}}, t_m \to t^*$, such that $q_{ij}(t_m) \to 0$ when $m \to \infty$. Notice that

$$\limsup_{t \to t^*}[-W(\mathbf{q})] \geq \lim_{\nu \to \infty}[-m_i m_j(1 + \alpha + \alpha \ln q_{ij}) q_{ij}^{-1}] = \infty.$$

This means that along the sequence $(t_m)_{m \in \mathbf{N}}$ the left side of the energy relation becomes unbounded while the right side is constant, a contradiction. Thus the theorem is proved.

This result shows that, from the astronomical point of view, Mücket-Treder's model doesn't make much sense for $\alpha > 0$. Large scale collisions occur in the Univers, so no positive value of α corresponds to reality. We will further see that this is not the case for negative values of α.

4. The Case $\alpha < 0$

We now restrict our attention to negative values of α. The original McGehee transformations[13] will be adapted to Mücket-Treder's gravitational law. This will allow us to "blow-up" the singularity arising from the total collapse of all particles and paste instead a *collision manifold*. Orbits on the collision manifold are fictitious, but their study will give us valuable information on the behavior of solutions coming close to a triple collapse.

Unless otherwise stated, all transformations we consider are analytic functions, bijective, and have the inverse analytic. They will provide equivalent, more convenient, equations of motion. Our first goal will be to eliminate from the equations the singularity arising from the total collapse. For this define

$$\begin{cases} r = (\mathbf{q}^T M \mathbf{q})^{1/2} \\ \mathbf{s} = r^{-1}\mathbf{q} \\ y = \mathbf{p}^T \mathbf{s} \\ \mathbf{x} = \mathbf{p} - yM\mathbf{s} , \end{cases} \tag{4}$$

which are the classical McGehee transformations[13], where ^{T} means the transpose of the column vector $\mathbf{q}$, i.e. the line vector $\mathbf{q}$. Notice that r, y are scalars while $\mathbf{s}$, $\mathbf{x}$ are vectors. (As a rule, we denote scalars by plane letters and vectors by bold ones.) Also observe that

$$\mathbf{s}^T M \mathbf{s} = 1, \ \ \mathbf{s}^T \mathbf{x} = 0$$

$$U(\mathbf{q}) = r^{-1}U(\mathbf{s}), \ \ \nabla U(\mathbf{q}) = r^{-2}\nabla U(\mathbf{s})$$

$$V(\mathbf{q}) = r^{-1}V(\mathbf{s}) + \alpha r^{-1}(\ln r)U(\mathbf{s})$$

$$\nabla V(\mathbf{q}) = r^{-2}\nabla V(\mathbf{s}) + \alpha r^{-2}(\ln r)\nabla U(\mathbf{s}).$$

Since U is a homogeneous function of degree -1, Euler's relation yields

$$\mathbf{q}^T \nabla U(\mathbf{q}) = -U(\mathbf{q}).$$

Using the change of variables (4) and the above relations, equations (2) become

$$\begin{cases} \dot{r} = y \\ \dot{y} = r^{-1}\mathbf{x}^T M^{-1}\mathbf{x} + r^{-2}\mathbf{s}^T \nabla V(\mathbf{s}) - (1+\alpha \ln r) r^{-2} U(\mathbf{s}) \\ \dot{\mathbf{s}} = r^{-1} M^{-1}\mathbf{x} \\ \dot{\mathbf{x}} = -r^{-1} y\mathbf{x} - r^{-1}(\mathbf{x}^T M^{-1}\mathbf{x}) M\mathbf{s} + \nabla U(\mathbf{s})] + (1+\alpha \ln r) r^{-2}[U(\mathbf{s})M\mathbf{s} + \\ \qquad r^{-2}\nabla V(\mathbf{s}) - r^{-2}[\mathbf{s}^T \nabla V(\mathbf{s})] M\mathbf{s} \ . \end{cases} \tag{5}$$

Notice that r is a measure of the distribution of particles in space. For r small, the bodies are close together. Since we are interested in the behavior of solutions coming close to a total collapse, we may assume that r is positive and small. This will allow us to consider further the transformations

$$\begin{cases} v = r^{1/2}(-\ln r)^{-1/2} y \\ \mathbf{u} = r^{1/2}(-\ln r)^{-1/2}\mathbf{x} \ . \end{cases} \tag{6}$$

This new change of variables will transform equations (5) into the following system of differential equations

$$\begin{cases} \dot{r} = r^{-1/2}(-\ln r)^{1/2} v \\ \dot{v} = r^{-3/2}(-\ln r)^{1/2}[\frac{v^2}{2}(1 - \frac{1}{\ln r}) + \mathbf{u}^T M^{-1}\mathbf{u} - \frac{\mathbf{s}^T \nabla V(\mathbf{s})}{\ln r} + \\ \qquad (\alpha + \frac{1}{\ln r}) U(\mathbf{s})] \\ \dot{\mathbf{s}} = r^{-3/2}(-\ln r)^{1/2} M^{-1}\mathbf{u} \\ \dot{\mathbf{u}} = -r^{-3/2}(-\ln r)^{1/2}\{(1 + \frac{1}{\ln r})\mathbf{u}v/2 + (\mathbf{u}^T M^{-1}\mathbf{u} M)\mathbf{s} + \\ \qquad (\alpha + \frac{1}{\ln r})[U(\mathbf{s})M\mathbf{s} + \nabla U(\mathbf{s})] + \frac{1}{\ln r}[\nabla V(\mathbf{s}) + (\mathbf{s}^T \nabla V(\mathbf{s}))M\mathbf{s}]\} \ . \end{cases} \tag{7}$$

In order to eliminate the cumbersome negative powers of r and $-\ln r$ at the beginning of each equation in system (7), we consider the time transformation

$$d\tau = r^{-3/2}(-\ln r)^{1/2} dt. \tag{8}$$

This will introduce the fictitious time variable τ. By abuse of notation we denote the variables with the same letters, and $'$ will mean differentiation with respect to τ. Thus equations (7) become

$$\begin{cases} r' = rv \\ v' = (1 - \frac{1}{\ln r})\frac{v^2}{2} + \mathbf{u}^T M^{-1}\mathbf{u} - \frac{\mathbf{s}^T \nabla V(\mathbf{s})}{\ln r} + (\alpha + \frac{1}{\ln r})U(\mathbf{s}) \\ \mathbf{s}' = M^{-1}\mathbf{u} \\ \mathbf{u}' = -[(1 + \frac{1}{\ln r})\mathbf{u}v/2 + (\mathbf{u}^T M^{-1}\mathbf{u}M)\mathbf{s} + (\alpha + \frac{1}{\ln r})[U(\mathbf{s})M\mathbf{s} + \nabla U(\mathbf{s})] + \\ \qquad \frac{1}{\ln r}[\nabla V(\mathbf{s}) + (\mathbf{s}^T \nabla V(\mathbf{s}))M\mathbf{s}] \;. \end{cases} \tag{9}$$

Using all the above transformations, the energy relation takes the form

$$\frac{1}{2}(\mathbf{u}^T M^{-1}\mathbf{u} + v^2) - \frac{U(\mathbf{s}) + V(\mathbf{s})}{-\ln r} + \alpha U(\mathbf{s}) = \frac{hr}{-\ln r}.$$

Let us finally introduce the transformation

$$\rho = \frac{1}{-\ln r}. \tag{10}$$

System (9) becomes

$$\begin{cases} \rho' = v\rho^2 \\ v' = (1 + \rho)\frac{v^2}{2} + \mathbf{u}^T M^{-1}\mathbf{u} + \rho\mathbf{s}^T \nabla V(\mathbf{s}) + (\alpha - \rho)U(\mathbf{s}) \\ \mathbf{s}' = M^{-1}\mathbf{u} \\ \mathbf{u}' = (\rho - 1)\mathbf{u}v/2 + (\mathbf{u}^T M^{-1}\mathbf{u})M\mathbf{s} - (\rho - \alpha)[U(\mathbf{s})M\mathbf{s} + \nabla U(\mathbf{s})] + \\ \qquad \rho[\nabla V(\mathbf{s}) + (\mathbf{s}^T \nabla V(\mathbf{s}))M\mathbf{s}] \;, \end{cases} \tag{11}$$

and the energy relation changes into

$$\frac{1}{2}(\mathbf{u}^T M^{-1}\mathbf{u} + v^2) - \rho[U(\mathbf{s}) + V(\mathbf{s})] + \alpha U(\mathbf{s}) = h\rho e^{-\frac{1}{\rho}}. \tag{12}$$

Obviously, since a total collapse takes place for $r = 0$, in the new equations this happens for $\rho = 0$. Equations (11) extend naturally to $\rho = 0$, unfortunately this is not immediate for the energy relation (12) because of the term $e^{-\frac{1}{\rho}}$. Let us therefore write relation (12) as

$$\frac{1}{2}(\mathbf{u}^T M^{-1}\mathbf{u} + v^2) - \rho[U(\mathbf{s}) + V(\mathbf{s})] + \alpha U(\mathbf{s}) = hg(\rho), \tag{13}$$

where

$$g(\rho) = \begin{cases} \rho e^{-\frac{1}{\rho}}, & if\ \rho \neq 0 \\ 0, & if\ \rho = 0\ . \end{cases}$$

The function g is analytic for $\rho \neq 0$ and infinitely many times differentiable at 0. Also notice that this is the only way to extend g to $\rho = 0$ and to maintain the infinite differentiability. This makes relation (12) to extend also to the value $\rho = 0$. Since the set $\{(\rho, y, \mathbf{s}, \mathbf{u}) \,|\, \rho = 0\}$ is invariant for the equations (11), we have pasted a manifold to the phase space of system (11). Notice that on this manifold, relation (13) becomes

$$\frac{1}{2}(\mathbf{u}^T M^{-1}\mathbf{u} + v^2) + \alpha U(\mathbf{s}) = 0, \tag{14}$$

and that it doesn't depend on the value of h. The set

$$\mathbf{C} = \{(\rho, y, \mathbf{s}, \mathbf{u}) \,|\, \rho = 0,\ \frac{1}{2}(\mathbf{u}^T M^{-1}\mathbf{u} + v^2) + \alpha U(\mathbf{s}) = 0\}$$

is an infinitely differentiable manifold and will be called the *collision manifold* for the triple collapse. Observe that $\mathbf{C}$ is an invariant manifold for the equations (11). Suppose now that $U(\mathbf{s})$ remains bounded, which implies that "purely" binary collision approaches do not occur. Then this restriction to $\mathbf{C}$ is also compact. Therefore, in the new fictitious time variable τ, the collision manifold $\mathbf{C}$ can be reached only at $\tau = \infty$. Equations (11) restricted to $\mathbf{C}$ take the form

$$\begin{cases} v' = \dfrac{v^2}{2} + \mathbf{u}^T M^{-1}\mathbf{u} + \alpha U(\mathbf{s}) \\ \mathbf{s}' = M^{-1}\mathbf{u} \\ \mathbf{u}' = -\mathbf{u}v/2 + (\mathbf{u}^T M^{-1}\mathbf{u})M\mathbf{s} + \alpha[U(\mathbf{s})M\mathbf{s} + \nabla U(\mathbf{s})]\ . \end{cases} \tag{15}$$

We have thus blown-up the singularity arising from the total collision and pasted to the phase space a fictitious collision manifold. The orbits of system (15), on $\mathbf{C}$, do not have any real meaning or correspondence in the physical space. However, due to the property of continuity of solutions with respect to initial data, a study of these orbits will offer us information on the physical behavior of orbits passing close to a triple collision. Obtaining such information is, in fact, our goal.

Any qualitative study of ordinary differential equations starts with the search for equilibrium solutions (i.e. orbits remaining constant in time). They are given by zeroes of the vector field. Let us solve therefore the equations arising by making the right parts of system (15) equal to zero. Looking at the first one and using the enregy relation (14), we see that this equation is equivalent to $\mathbf{u}^T M^{-1}\mathbf{u} = 0$. This obviously yields $\mathbf{u} = 0$. But in order to fulfill these equations we need $v = \pm(-2\alpha U(\mathbf{s}))^{1/2}$ and $U(\mathbf{s})M\mathbf{s} + \nabla U(\mathbf{s}) = 0$. Consequently we have proved the following fact:

Proposition 2. *Equations* (15) *have equilibrium orbits if and only if the following three conditions are simultaneously fulfilled:*

(i) $\mathbf{u} = 0$

(ii) $v = \pm(-2\alpha U(\mathbf{s}))^{1/2}$

(iii) $U(\mathbf{s})M\mathbf{s} + \nabla U(\mathbf{s}) = 0.$

Before interpreting this result let us give the following:

Definition. A configuration $\mathbf{q} \in \mathbb{R}^9 \setminus \Delta$ of the classical Newtonian 3-body problem is called *central* if there is some nonzero constant σ such that

$$\nabla U(\mathbf{q}) = \sigma M\mathbf{q}. \tag{16}$$

Central configurations play an important role in the study of the n-body problem. It is known that for three bodies there exist five central configurations, three collinear (called Eulerian) and two equilateral (called Lagrangean). There are many important properties connected with the notion of central configuration, and the literature on this topic is growing. We are now interested in just one known asymptotic property, namely:

Theorem 3. *Collision solutions of the Newtonian 3-body problem tend to form a central configuration in the neighborhood of collisions.*

A simple computation will show that using the above McGehee-type transformations, relation (16) takes—in the new variables—the form of relation (*iii*) in Proposition 2. This means that equilibrium solutions on the collision manifold of Mücket-Treder's model are given by Newtonian central configurations.

Continuing the qualitative study of orbits on $\mathbf{C}$, one is looking for periodic orbits. It is easy to see that since (by (14) and (15)) $v' = \mathbf{u}^T M^{-1}\mathbf{u}$, it follows that v is an increasing function (except on equilibrium solutions). Thus we can conclude

Proposition 4. *None of the orbits on the collision manifold is periodic.*

We will now prove the following result:

Theorem 5. *In the Mücket-Treder model, when the time variable tends to the collision instant, any triple collision solution of the 3-body problem approaches a Newtonian central configuration.*

Remark. Proposition 4 is, of course, true for any number n of particles in an n-body problem. Theorem 5, however, could be extended only to a weaker statement, namely by saying that a total-collision solution of the n-body problem, with $n \geq 4$, tends to the set of central configurations. Since very little is known about this set (which might be infinite, and even contain a continuum) we cannot draw the conclusion that the total-collision solution approaches this set with *asymptotic phase*, i.e. by tending to an element of it; it may oscillate between elements without ever reaching any of them.

In order to prove this result we adapt the ideas of McGehee[13] to our case. Denote by $\phi(\tau) = (\rho(\tau), v(\tau), \mathbf{s}(\tau), \mathbf{u}(\tau))$ an orbit of equations (11) and let $\phi_0(\tau)$ be an orbit having the initial conditions $\mathbf{w}_0 = (\rho_0, v_0, \mathbf{s}_0, \mathbf{u}_0)$. Recall that the ω-limit set $L_\omega(\mathbf{w}_0)$ of $\mathbf{w}_0$ is the set of all points $\mathbf{w} = (\rho, v, \mathbf{s}, \mathbf{u})$ such that there is a sequence $(\tau_n)_{n\in\mathbf{N}}, \tau_n \to \infty$, and $\lim_{n\to\infty} \phi_0(\tau_n) = \mathbf{w}$. In other words the ω-limit set of an orbit is the set of points where the orbit "dies". We will make use of the following result of McGehee[13]:

Lemma 6. *If $L_\omega(\mathbf{w}_0)$ is a nonempty compact set, then $L_\omega(\mathbf{w}_0)$ is a single point.*

Proof of Theorem 5. Consider $(\mathbf{q}, \mathbf{p})$ to be a triple-collision solution of equations (2) and let the initial conditions be sufficiently close to the collision instant such that ρ is sufficiently small and the transformations (4), (6), (8) and (10) make sense. Thus, solution $(\mathbf{q}, \mathbf{p})$ is transformed to a solution $\phi = (\rho, v, \mathbf{s}, \mathbf{u})$ of equations (11), and let $\phi_0(\tau)$ be the orbit going through the corresponding initial point $\mathbf{w}_0$. Using McGehee's lemma it is enough to show that $L_\omega(\mathbf{w}_0)$ is a nonempty compact set. Indeed, since there are exactly five rest points, ϕ_0 has to approach one of them, consequently the orbit will tend to one of the five Newtonian central configurations.

In order to see that $L_\omega(\mathbf{w}_0)$ is nonempty and compact, consider first the following notations:

$$S = S(h, \epsilon) = \{\phi \,|\, \rho \leq \epsilon,\ \frac{1}{2}(\mathbf{u}^T M^{-1}\mathbf{u} + v^2) - \rho[U(\mathbf{s}) + V(\mathbf{s})] + \alpha U(\mathbf{s}) = hg(\rho)\},$$

$\epsilon > 0$, constant,

$$\Gamma = \Gamma(h,\epsilon,\mu) = \{\phi \in S \,|\, |v| \le \mu\},\ \mu > 0,$$

$$\Gamma^+ = \Gamma^+(h,\epsilon,\mu) = \{\phi \in S \,|\, v \ge \mu\},$$

$$\Gamma^- = \Gamma^-(h,\epsilon,\mu) = \{\phi \in S \,|\, v \le -\mu\},$$

$$\gamma^\pm = \gamma^\pm(h,\epsilon,\mu) = \{\phi \in \Gamma^\pm \,|\, \rho = \epsilon\},$$

$$\sigma^\pm = \sigma^\pm(h,\epsilon,\mu) = \{\phi \in S \,|\, v = \pm\mu\}.$$

If $[\tau_1,\tau_2]$, with $\tau_1 < \tau_2$, is a time interval in the new fictitious time variable, we call $\phi([\tau_1,\tau_2])$ an *orbit segment*. We say that the orbit segment $\phi([\tau_1,\tau_2])$ is *maximal* in a closed set K, if $\phi([\tau_1,\tau_2]) \subset K$ but $\phi(I)$ is not included in K, for any interval I containing $[\tau_1,\tau_2]$ but larger than $[\tau_1,\tau_2]$. The following two statements are then true:

(i) For ϵ and μ suitably chosen, if $\phi_0([\tau_1,\tau_2])$ is a maximal orbit segment in Γ^+, then $\phi_0(\tau_1) \in \sigma^+$ and $\phi_0(\tau_2) \in \gamma^+$.

(ii) For the same ϵ and μ in (1), if $\phi_0([\tau_1,\tau_2])$ is a maximal orbit segment in Γ^-, then $\phi_0(\tau_1) \in \gamma^-$ and $\phi_0(\tau_2) \in \sigma^-$.

The proof works similar for (i) and (ii), so we will do it only for (i). If $\mathbf{w}(\tau) \in \sigma^+$, then $v = \mu > 0$ and $\rho \le \epsilon$. Using the energy relation (13) and the second equation in (11) we get

$$v' = (\frac{1}{4} + \frac{\rho}{2})v^2 + \frac{1}{2}\mathbf{u}^T M^{-1}\mathbf{u} + \rho(\mathbf{s}^T \nabla V(\mathbf{s}) + V(\mathbf{s}) + h e^{-\frac{1}{\rho}}).$$

Thus for $\mathbf{w}(\tau) \in \sigma^+$ (i.e. for ϵ sufficiently small and a fixed positive μ), we have $v'(\tau) > 0$. From the first equation in (11) we also get $\rho'(\tau) > 0$ when $\mathbf{w}(\tau) \in \sigma^+$. These imply that points on σ^+ are entering Γ^+, so $\phi_0(\tau_2) \in \gamma^+$. Also, points on γ^+ are leaving Γ^+, so $\phi_0(\tau_1) \in \sigma^+$. Statement (i) is thus proved.

Finally, we would like to show that for the above choice of ϵ and μ, since $\rho(\tau) \to 0$ when $\tau \to \infty$, there is a τ_3 such that $\phi_0(\tau) \in \Gamma$ for all $\tau \ge \tau_3$. Suppose that this is not the case, i.e. there exists a $\tau_4 \ge \tau_3$ such that $\phi_0(\tau_4) \in \Gamma^+$. Then, since $\rho(\tau) < \epsilon$, it follows that ρ_0 can never reach γ^+. Thus, by (i) we obtain that $\phi_0(\tau) \in \Gamma^+$ for all $\tau \ge \tau_4$. But $v > 0$ in Γ^+, so $\rho' = v\rho^2 > 0$ for $\tau \ge \tau_4$. This contradicts the fact that $\rho(\tau) \to 0$, when $\tau \to \infty$ and consequently $\phi_0(\tau)$ can not be in Γ^+ for $\tau \ge \tau_3$. In the same way, by (ii), $\phi_0(\tau)$ can not be in Γ^- for $\tau \ge \tau_3$. Consequently $\phi_0(\tau) \in \Gamma$ for $\tau \ge \tau_3$.

Since Γ is compact and any ω-limit set is closed, it follows that $L_\omega(\mathbf{w}_0)$ is nonempty and compact. This completes the proof.

Theorem 5 implies that, asymptotically, Mücket-Treder's law approaches the Newtonian one when the time variable tends to the triple-collision instant. This shows that, from the qualitative point of view, there is no much difference between the behavior of the Newtonian model and that of the Mücket-Treder model, with $10^{-8} < \alpha < 0$, when the particles are close to collisions.

5. References

1. G. Ballinger and F.N. Diacu, *Rom. Astr. Journ.* **3** (1) (1993), 51.
2. G. Bertrand, *Comptes rendus* **174** (1922), 1687.
3. E.W. Brown, *Monthly Notices Roy. Astron. Soc.* **63** (1903), 396.
4. F.N. Diacu, *Diff. Integr. Equations* **5** (1992), 103.
5. F.N. Diacu, *Journ. Math. Phys.* **34** (12) (1993), 5671.
6. F.N. Diacu, Near-Collision Dynamics for Particle Systems with Quasihomogeneous Potentials, *Journ. Differential Equations* (to appear).
7. Y. Hagihara, *Celestial Mechanics*, vol. II, part I, (The MIT Press, Cambridge, Massachusetts, 1972).
8. A. Hall, *Astron. Journ.* **14** (1895), 49.
9. G. Maneff, *Comptes rendus* **178** (1924), 2159.
10. G. Maneff, *Zeitschr. Phys.* **31** (1925), 786.
11. G. Maneff, *Comptes rendus* **190** (1930), 963.
12. G. Maneff, *Comptes rendus* **190** (1930), 1374.
13. R. McGehee, *Inventiones math.* **27** (1974), 191.
14. V. Mioc and P. Blaga, *Rom. Astr. Journ.* **1** (1-2) (1991), 103.
15. J.P. Mücket and H.-J. Treder, *Astron. Nachr.* **298** (1977), 65.
16. S. Newcomb, *The Elements of the Four Inner Planets and the Fundamental Constants of Astronomy*, (Government Printing Office, Washington, D.C., 1895).

RESONANT QUASIMODES OF A THREE-DIMENSIONAL ANHARMONIC OSCILLATOR, LOCALIZED IN THE NEIGHBORHOOD OF CLOSED TRAJECTORIES AND COMPLEX BUNDLES ON THEM

SERGEI Yu. DOBROKHOTOV[*] VICTOR M. OLIVE

and

ANDREI I. SHAFAREVICH[†]

Department of Mathematics, CINVESTAV-IPN
Ap. Postal 14-740 México 07000, D.F.
E-mail: olive@math.cinvestav.mx and dobr@ipm.msk.su

ABSTRACT

A Hamiltonian system in six-dimensional phase space, for an anharmonic oscillator in the resonant case is considered. This system possesses a four dimensional invariant subspace fibered by closed trajectories. It is possible to associate, to each of these trajectories, a quasimode mod $O(h^{3/2})$ (an asymptotic eigenfunction in the quasiclassical approximation; h is the Planck constant) of the associated quantum (Schrödinger) equation. Its construction is based on the construction of an invariant complex Lagrangian manifold, whose intersection with the real phase space coincides with the trajectory considered, and of a complex nondegenerated invariant differential form of maximal rank on the manifold. To obtain quasimodes with this precision it is enough to obtain the linear part of this manifold (and of the invariant form) in the neighborhood of the trajectory. However in this case the asymptotics of the spectrum of the quantum problem is continuous. To achieve more accuracy in the asymptotics on h it is necessary to construct the geometrical object up to infinitesimally small quantities of higher order. In this paper the obstacle to the construction of the cuadratic part is obtained. More precisely, it is shown that the described geometrical object exists if and only if, the corresponding trajectory is also a trajectory of certain additional first integral of the Hamiltonian system.

1. Introduction

In the works of Babich (1968), Dobrokhotov and Maslov (1975, 1983), Ralston (1976), Maslov (1973, 1977), Duistermaat and Guillemin (1975), Belov and

[*]On leave from the Institute of Problems in Mechanics, Russian Academy of Sciences, Moscow 117526 Pr. Vernadskogo 101, Russia
[†]On leave from the Department of Differential Geometry, Mechanico-Mathematical Faculty of the Moscow State University, Moscow 117234, Voroviovye Hills, Russia.

Dobrokhotov (1988, 1992) a way is proposed to construct asymptotic eigenfunctions (quasimodes) of partial differential and pseudodifferential operators with a small parameter. In the most general case these pseudodifferential operators (let us consider that they are selfadjoint) have the form

$$\hat{\mathrm{L}} = \mathrm{L}(-ih\frac{\partial}{\partial x}, x, h).$$

These quasimodes are localized in a neighborhood of closed curves, which are trajectories of the Hamiltonian system given by the Hamilton function $H(p,x) = \mathrm{L}(p,x,0)$. Let us call these trajectories Λ^1. The quasimodes can be constructed when these trajectories $\Lambda^1 : x = X(t),\ p = P(t)$ are orbitally stable in linear approximation, i.e. when all solutions of the linearized system (with periodic coefficients) along Λ^1 are bounded except maybe the conjugated solution (in the sense of the skew–scalar product) to $(\dot{P}, \dot{X})$. In this case we always can construct quasimodes ψ^1 modulo $\mathrm{O}(\mathrm{h}^{3/2})$, i.e.

$$\hat{\mathrm{L}}\psi^1 = \lambda\psi^1 + \mathrm{O}(h^{3/2}).$$

If Floquet exponents β_j of the solutions of the linearized equations and 2π are noncommensurable (i.e. $(\beta, m) + 2\pi k \neq 0$, $m \in \mathbb{Z}^n$, $k \in \mathbb{Z}$) then the construction proposed in these works allows to obtain corrections to the quasimodes up to any order of h. This case is called *nonresonant.* In this case a family of closed trajectories depending precisely on one parameter—the classical energy of the system— exists and this family is quantized.

For the resonant case, in which we are interested, the construction of corrections to quasimodes has not been studied. In this case a family of trajectories depending on more than one parameter can exist and then (one) Bohr–Sommerfeld condition gives a continuous spectrum in the zero approximation, whereas the real spectrum of the operator L can be purely discrete. We consider here an example of this kind.

Let us remember that the quasiclassical approach is based on the solution of a Hamilton–Jacobi equation for a phase S and a transport equation for an amplitude φ of an asymptotic solution (the quasimode is almost everywhere the sum of expressions of the form $\varphi e^{iS/h}$). In the Complex WKB method Hamilton–Jacobi equation is solved not exactly but to order $(ImS)^{3/2}$, i.e.,

$$H(\nabla S, x) = \mathrm{O}((ImS)^{3/2}).$$

From the geometric point of view this procedure is based on the construction of a vector bundle, which base is Λ^1 and each fiber is a complex positive Lagrangian plane in $\mathbb{C}^{2n}$ (see Maslov 1977). *In the nonresonant case this construction is unique* (Dobrokhtov *et. al.* 1984), *but in the resonant case not.* The construction of corrections to a quasimode is connected with the modification of this geometric object. Instead of a vector bundle we obtain a bundle which fibers are certain complex analytic manifolds. The space of the bundle itself is a complex analytic manifold also. Were it possible to solve exactly Hamilton–Jacobi equation then it would be possible to find this complex manifold (invariant with respect to the phase flow of the Hamiltonian system) exactly. As far as we know, there are no nontrivial examples of such manifolds (which are analogous to the Liouville tori of the integrable system). We construct one higher order germ (one "geometric" correction) which makes it possible to obtain quasimodes satisfying

$$\hat{L}\psi = \lambda\psi \ \mathrm{mod}O(h^2).$$

The aim of our report is to illustrate this situation using a very simple (but non–trivial) example. Our example is the *anharmonic oscillator*

$$H(x,p) = p^2/2 + (\omega_1^2 x_1^2 + \omega_2^2 x_2^2 + \Omega^2(x_1,x_2)x_3^2)/2, \qquad x \in \mathbb{R}^3;\ p \in \mathbb{R}^3,$$

the frequencies ω_1 and ω_2 are constant, while Ω is a strictly positive function of x_1, x_2.

2. Resonant quasimodes mod$O(h^{3/2})$.

We consider Schrödinger equation in $L^2(\mathbb{R}^3)$:

$$H(-ih\frac{\partial}{\partial x}, x)\psi = \lambda\psi, \tag{1}$$

where $H(x,p) = p^2/2 + (\omega_1^2 x_1^2 + \omega_2^2 x_2^2 + \Omega^2(x_1,x_2)x_3^2)/2$ and $\Omega^2(x_1,x_2) = a_1^2 x_1^2 + a_2^2 x_2^2 + \omega_3^2$; $\omega_1/\omega_2 = m_1/m_2$, $m_1, m_2 \in \mathbb{Z}$.

We will denote $\omega_0 = \omega_1/m_1 = \omega_2/m_2$ and $T = 2\pi/w_0$. Invariant curves (trajectories) of the Hamiltonian flow in $\mathbb{R}^2$ (with coordinates–(x,p)) lie in the four-dimensional plane $x_3 = p_3 = 0$ and have the form:

$$\begin{aligned} x_i = X_i(t) &\equiv A_i \cos(\omega_i t + \tau_i), \\ P_i = P_i(t) &\equiv -\omega_i A_i \sin(\omega_i t + \tau_i), \\ x_3 = p_3 = 0 = X_3(t) &= P_3(t), \quad i = 1,2. \end{aligned} \tag{2}$$

Here the "amplitudes" A_i and the "phases" τ_i are parameters of the curve and $A_1^2 + A_2^2 \neq 0$. Note that if we replace the variable t by $t^1 = t + \tau_1/\omega_1$, the arguments of sin and cos will be $\omega_1 t^1$ and $\omega_2 t^1 + \omega_0 \xi/\omega_1$, where $\xi = m_1\tau_2 - m_2\tau_1$. Thus the set of closed trajectories is parametrized by three parameters: A_1, A_2, ξ. We will denote the curve (2) by Λ^1; (A, ξ) $(A = (A_1, A_2))$.

In order to obtain quasimodes mod$O(h^{3/2})$ for the equation (1) we have to construct the following geometric object (Maslov 1977) (the result of this section is the explicit construction of this object). Consider the complexification of $\mathbb{R}^6$: $\mathbb{C}^6$. Our object is a vector bundle: $\pi : M \to \Lambda^1(A, \xi)$ with the fiber $\rho \subset \mathbb{C}^6$ satisfying the following properties:

1) Each fiber ρ is a three dimensional complex space Lagrangian with respect to the skew-scalar product in $\mathbb{C}^6$:

$$\{a, b\} = (a_x, b_p) - (a_p, b_x), \quad a, b \in \mathbb{C}^6.$$

Here a_x and a_p are three-dimensional complex vectors; x and p are the components of vector a $((x, p)$ are complex coordinates in $\mathbb{C}^6)$.

2) The fiber $\rho = \pi^{-1}(y)$ $(y \in \Lambda^1)$ contains the complexification ${}^{\mathbb{C}}T\Lambda^1$ of the tangent line to Λ^1 at y;

3) Each fiber ρ is positive:

$$Im\{a, \bar{a}\} > 0 \quad \text{for} \quad a \in \rho, \ a \notin {}^{\mathbb{C}}T\Lambda^1.$$

4) The phase flow of the Hamiltonian system with Hamilton function $H(x, p)$, linearized on the curve Λ^1, brings a fiber onto a fiber.

We now give the construction of all such fiber bundles for our example. The linearized Hamiltonian system has the form:

$$\begin{aligned} \dot{W}_i &= -\omega_i^2 Z_i, \qquad \dot{Z}_i = W_i, \qquad i = 1, 2, \\ \dot{W}_3 &= -\Omega^2(X_1, X_2) Z_3, \qquad \dot{Z}_3 = W_3, \end{aligned} \tag{3}$$

(the unknown functions form a 6 – dimensional complex vector $(Z_1, Z_2, Z_3, W_1, W_2, W_3)$).

From the last two equations in (3) we obtain

$$\ddot{Z}_3 + \Omega^2(X_1, X_2) Z_3 = 0. \tag{4}$$

Consider Floquet solutions of the equation (4), i.e. solutions $Z_3(t)$ satisfying the following condition

$$Z_3(t+T) = \exp\{i\beta T\} Z_3(t). \tag{5}$$

Theorem 1.

1°. The fiber bundle $\pi : H \to \Lambda^1$ with the properties 1)–4) exists if and only if problem (3) – (4) is solvable with real exponent β.
2°. Let this problem be solvable and let β be not a multiple of $2\pi/T$. Then all the fiber bundles E satisfying 1)–4) have the following form:

$$\begin{pmatrix} p \\ x \end{pmatrix} = \alpha_1 a_1 + \alpha_2 a_2 + \alpha_3 a_3,$$

where

$$a_1 = \begin{pmatrix} 0 \\ 0 \\ \dot{Z}_3 \\ 0 \\ 0 \\ Z_3 \end{pmatrix}, \quad a_2 = \tilde{\alpha} \begin{pmatrix} (\sin(\omega_1 t + \tau_1) & - & \gamma\cos(\omega_1 t + \tau_1))(-q_2\omega_1)/A_2\omega_2 \\ (\sin(\omega_2 t + \tau_2) & - & \gamma\cos(\omega_2 t + \tau_2))(q_1\omega_2)/A_1\omega_1 \\ & 0 & \\ (\cos(\omega_1 t + \tau_1) & + & \gamma\sin(\omega_1 t + \tau_1))(q_2/A_2\omega_2) \\ (\cos(\omega_2 t + \tau_2) & + & \gamma\sin(\omega_1 t + \tau_2))(-q_1/A_1\omega_1) \\ & 0 & \end{pmatrix},$$

$$a_3 = (\dot{P}_1,\ \dot{P}_2,\ 0,\ \dot{X}_1,\ \dot{X}_2,\ 0\ )^t, \tag{6}$$

$$q_j = A_j^2 w_j^3/(A_1^2 w_1^3 + A_2^2 w_2^2), \quad \tilde{\alpha} = \sqrt{(A_1^2 w_1^3 + A_2^2 w_2^3)/(w_1 w_2,\ Im\gamma)} \quad Im\gamma > 0.$$

The coefficients $(\alpha_1, \alpha_2, \alpha_3)$ are complex coordinates on the fiber; the set of all fiber bundles is parametrized by three real parameters A_1, A_2, ξ and one complex parameter γ (the base Λ^1 depends on A_1, A_2, ξ and for each fixed base the fiber bundle depends on γ).

3°. Let in addition the following condition be fullfilled (the Bohr – Sommerfeld quantization condition):

$$E = h\beta(\nu_1 + 1/2) + h(\omega_1 + \omega_2)(\nu_2 + 1/2) + h\omega_0\nu_3 + O(h^2). \tag{7}$$

$$E = \frac{1}{T}\oint_{\Lambda^1}(P, dx) = (A_1^2\omega_1^2 + A_2^2\omega_2^2)/2$$

is the classical energy on Λ^1; ν_j $j = 1,2,3$ are some integers; $\nu_j = O(h^{-1})$. Then a quasimode mod$O(h^{3/2})$ of the spectral problem (1) exists, i.e. there is a function ψ_{ν_1,ν_2,ν_3}, such that

$$(H(-ih\frac{\partial}{\partial x}, x) - E)\psi_{\nu_1,\nu_2,\nu_3} = O(h^{3/2})$$

in the L^2– norm.

The proof of this theorem is based on the explicit construction of the quasimodes. It follows from Maslov 1977, Ralston 1976, Belov and Dobrokhotov 1988 and will be given in another publication.

Remark. Equality (7) determines an asymptotic eigenvalue E. It depends on three parameters (A_1, A_2, ξ). So for each value of the vector $\nu = (\nu_1, \nu_2, \nu_3)$ the expression (7) determines an infinite (continuous) set of asymptotic eigenvalues E. From the other hand, the exact spectrum of the operator $H(-ih\partial/\partial x, x)$ is purely discrete; it follows from the Theorem 1 that the distance between two neighbour eigenvalues is $o(h)$. In order to obtain asymptotic eigenvalues with better accuracy, we have to construct higher–order approximations for the quasimode.

3. Complex manifolds and quasimodes mod$O(h^2)$

In order to make our asymptotic solutions more precise, we have to construct "*corrections*" to the fiber bundle M $\to \Lambda^1$ of the previous section. The sense of these corrections is the following. We can imagine the fiber bundle M as part of the tangent bundle to some complex manifold (Λ^1 is the intersection of this manifold with the real subspace $\mathbb{R}^n \subset \mathbb{C}^n$). Note that the existence of the fiber bundle M does not imply the existence of this manifold. Now we have to determine this manifold in the vicinity of Λ^1 with more accuracy, taking into account not only tangent vectors, but also terms of "second order". In coordinates (t, α) this means that we have to take into account cuadratic terms in (α_1, α_2). Locally we can define the "cuadratic part" of our manifold by equations in $\mathbb{C}^6$: $(P,\ X)^t = q(\alpha_1, \alpha_2, t)$. The function q has to satisfy the following conditions:

1°. *Invariance.* There exist functions $\alpha_i'(\alpha_1, \alpha_2, t)$, $i = 1, 2$, and $f(\alpha_1, \alpha_2, t)$ such that $f = O(|\alpha|^2)$ and

$$q(\alpha_1, \alpha_2, t + T + f) = q(\alpha_1', \alpha_2', t), \tag{8}$$

$$|\alpha|^2 = |\alpha_1|^2 + |\alpha_2|^2.$$

The function q satisfies Hamilton equations $\mathrm{mod}O(|\alpha|^3)$.

2°. *Lagrangian property.* For

$$(P, X)^t = q(\alpha_1, \alpha_2, t), \qquad \sum_{i=1}^{3} dx_i \Lambda dp_i = O(|\alpha|^2). \tag{9}$$

3°. $q = (P(t), X(t))^t + \alpha_1 a_1 + \alpha_2 a_2 + O(|\alpha|^2)$.

In order to construct the "amplitude" of the asymptotic solution we also have to construct an "almost–invariant" 3–form on our manifold. In our case the existence of such a 3–form is equivalent to condition 4.

4°. There exist a function $\chi(\alpha_1, \alpha_2)$ and a number μ, such that the function

$$\frac{\chi e^{i\mu t}}{\sqrt{J(\alpha_1, \alpha_2, t)}},$$

where $J = \partial X(\alpha_1, \alpha_2, t)/\partial(\alpha_1, \alpha_2, t)$ satisfies (8).

The result of this section is a neccessary and sufficient condition for the existence of this object.

Theorem 2.

1) The function q satisfying 1°–3° exists for all the values of the parameters A_1, A_2, ξ and γ.

2) The condition 4° is satisfied if and only if the following condition holds:

$$\nabla\beta|_{\Lambda^1} \parallel \nabla H|_{\Lambda^1}. \tag{10}$$

3) If in addition to (10) the Bohr – Sommerfeld condition (7) is fulfilled, then there exists a function $\tilde{\psi}$ such that

$$(H(-ih\frac{\partial}{\partial x}, x) - E)\tilde{\psi} = O(h^2)$$

in the L^2 – norm.

The proof is based on the direct construction of the functions q and $\tilde{\psi}$.

Remark. Equality (10) gives two real conditions on the parameters of the closed curve A_1, A_2 and ξ. Indeed, in the neighborhood of Λ^1 we can introduce "action–angle" variables (I, J, η, ξ), where

$$I = (m_1\omega_1 A_1^2 + m_2\omega_2 A_2^2)/2, \quad J = (n_2\omega_1 A_1^2 - n_1\omega_2 A_2^2)/2, \quad \eta = n_1\tau_1 + n_2\tau_2,$$

$$(n_1, n_2) \in \mathbf{Z}^2 \quad \text{and} \quad n_1 m_1 + n_2 m_2 = 1.$$

In these variables the condition (10) becomes

$$\frac{\partial\beta}{\partial\xi}|_{\Lambda^1} = \frac{\partial\beta}{\partial J}|_{\Lambda^1} = 0.$$

Together with (6) these conditions determine a discrete series of asymptotic eigenvalues E.

Remark.

1. As a result we obtain the following geometric object. We construct a fiber bundle on a closed trajectory Λ^1 (we can treat this trajectory as a complex analytic manifold if we consider complex values of t). Each fiber is a two–dimensional complex analytic manifold (it is defined in $\mathbb{C}^6$ by a system of algebraic equations of the second order).

2. The Complex WKB method (see Maslov 1977 and Belov and Dobrokhotov 1992) makes it possible to construct quasimodes connected not only with closed trajectories, but also with 2–dimensional tori. In our example one can define 2–dimensional tori in $\mathbb{R}^6$ formed by closed trajectories Λ^1. It comes out that one can construct a quasimode associated with a torus only if a Floquet exponent β is constant on the whole torus (i.e. it is equal for all closed trajectories forming the tours) (see Karasev 1987 and Dobrokhotov and Olivé 1995). So the family of 2-D tori is defined by equations in $\mathbb{R}^6$: $x_3 = p_3 = 0$; $H(x,p) =$ const; $\beta =$ const. Consider the two last equations in the complexification of $\mathbb{R}^4 = \mathbb{R}^6 \cap \{x_3 = p_3 = 0\}$. They define some 2-D complex manifold. Consider the intersection of this manifold with the real subspace $\mathbb{R}^4$. If in the points of this intersection $\nabla H \not\parallel \nabla\beta$ then the intersection is a 2-D torus; the tangent planes to the complex manifold in the points of this tours contain the tangent planes to the torus and therefore can not satisfy condition 3) from Section 1 (at each point there are two real lineary independent vectors). Thus the complex manifold under consideration can define the fiber bundle considered in this paper only if in the real points of this manifold $\nabla H || \nabla\beta$ (i.e. eq. (10) is fullfilled). In this case the intersection of this 2-D complex manifold with the real subspace is a 1-D manifold — a closed trajectory Λ^1.

4. Acknowledgment.

The authors thank the Consejo Nacional de Ciencia y Tecnología de México for the support received through the Cátedras Patrimoniales de Excelencia II No 920247 (S. Dobrokhotov) and 940031 (A. Shafarevich).

5. References

1. Babich, V.M. *Zap. Nauchn. Seminarov LOMI*, **9** (1968) Leningrad (in Russian)
2. Belov, V.V., Dobrokhotov, S.Yu. *Dokl. Akad. Nauk SSSR*, **298** (1988) No. 5 1037–1042 (Russian). Translation in *Soviet Math. Dokl.*, **37** (1988) No. 1, 180–185.

3. Belov, V.V., Dobrokhotov, S.Yu. *Theor. Math. Phys.*, **92** (1992) No. 2, 843–868.
4. Dobrokhotov, S.Yu., Maslov, V.P. Sovremennye problemy matematiki. Itogi nauki i tekhniki, No. 5 (1975) VINITI, (Russian). Translation in *J. Sov. Math.*, **5** (1976) 552–605.
5. –ibid– *Sovremennye problemy matematiki. Itogi nauki i tekhniki,* No. 23 (1983) VINITI, (Russian). Translation in *J. Sov. Math.*, **28** (1985) 91–143.
6. Dobrokhotov, S. Yu., Nekhoroshev, N. N., Valiño, B. *Russ. Math. Surveys* **39** (1984) 233-234.
7. Dobrokhotov, S. Yu., Olivé, V.M. *Lissajoux figures, two-dimensional tori and spectral series of the three-dimensional anharmonic oscillator.* (Proc. Conf. *"Asymptotic Spectral Problems"*. Euler Int. Mathematical Inst. Gordon & Breach, 1995).
8. Karasev, M.V. *Quantum reduction to orbits of symmetry algebras and the Ehrenfest problem.* (In Russian) Preprint No. ITF–87–177P. (Akad. Nauk Ukrain. SSR, Inst. Theoret. Phys., Kiev, 1987).
9. Duistermaat, J.J., Guillemin, V.W. *Inventiones math.*, **29** (1975) 39–79.
10. Maslov, V. P. *Operatornye Metody* (Moscow: Nauka) (1973) (in Russian) *Operational Methods* (Moscow: Mir) (1976).
11. Maslov, V.P. *Kompleksnyi metod WKB v nelineinyx uravnenyax* (Moscow: Nauka) (1977) (in Russian) *The Complex WKB Method in Nonlinear Equations.* (Birkhäuser) 1994.
12. Ralston, J.V. *Commun. Math. Phys.*, **51** (1976) 219–242.

NUMERICAL CONTINUATION OF PERIODIC ORBITS FOR THE J_2, J_3, J_4 HARMONICS IN SATELLITE THEORY

ANTONIO ELIPE
Grupo de Mecánica Espacial, Universidad de Zaragoza, 50009 Zaragoza, Spain
E-mail: elipe@msf.unizar.es

and

MARTIN LARA
Real Observatorio de la Armada. San Fernando, 11110 Cádiz, Spain
E-mail: ccastro@czv1.uca.es

ABSTRACT

Taking into consideration the J_2, J_3 and J_4 harmonics in the geopotential, and the polar component of the angular momentum vector as a parameter, we continue numerically natural families of periodic orbits. We find five natural families. As it was expected, the break of the symmetry due to the J_3 harmonic produces bifurcations that do not occur in the symmetric problem.

1. Introduction

In the *main problem* of the artificial satellite theory, the existence of families of periodic orbits have been verified by several authors using different techniques. Thus, Coffey *et al.*[3,5], find the families and their bifurcations by analyzing the flow phase evolution of the reduced Hamiltonian (obtained after applying a Lie–transformation). The same result is obtained by Broucke[2] by means of the Poincaré sections corresponding to the original problem and constructing the periodic solutions through numerical integration of the differential equations.

When the influence of the J_3 harmonic is considered in the geopotential, the Hamiltonian no longer is symmetric, and therefore new families and different bifurcations are expected. In a recent paper, Coffey *et al.*[5] find these families by using the technique of *painting Hamiltonians*[4] in the reduced Hamiltonian containing several zonal harmonics.

We take into consideration the effect of the J_2, J_3 and J_4 and by a different technique, we verify the results given in Coffey *et al.*[5] for the above mentioned harmonics. Essentially, we follow the method given by Deprit and Henrard[8] for continuing numerically natural families of periodic orbits. The method consists of separating the normal displacements along an orbit from the tangential ones, that have a secular character, and by means of a predictor–corrector scheme the evolution of a periodic orbit with respect to a parameter is obtained numerically. However, in our case, some modifications to this algorithm are necessary; namely, the adaptation of the method for variations of a parameter distinct to the energy, and a modification of the algorithm when the velocity is zero or near zero. These modifications are given in Lara *et al.*[9].

Since the zonal problem is invariant under the group of rotations about the polar axis, the polar component of the angular momentum is an integral. This fact allows to reduce the problem to only two degrees of freedom; indeed, when it is expressed in cylindrical coordinates (ρ, z, λ, P, Z, Λ), the coordinate λ is ignorable. Therefore, we find periodic orbits in the two dimensional space (ρ, z), that is, the orbits are periodic in the meridian plane and in general, they are not in the three dimensional space.

The parameter that we choose for the natural family of periodic orbits is precisely the polar component Λ of the angular momentum, although for a better understanding of the graphics, it has been replaced by the inclination angle $I = \arccos(\Lambda/G)$, where G is the norm of the angular momentum.

To begin with, we need some periodic orbits that will be used as initial conditions for the propagation. These have been obtained by means of the Poincaré surface of section of the J_2–problem, and for a value of J_2 big enough to have several periodic solutions easily recognizable from the graphic. Since with our modification to the Deprit and Henrard's algorithm we are able of choosing from a set of several parameters, for a fixed value of Λ, we vary the harmonic J_2 until reach the value adopted (WGS 84). The orbits thus obtained are used as the starting orbits for the variation of Λ in the J_2, J_3 and J_4 problem. We find 5 natural families, of the same type that the ones found by Coffey *et al.*[5]. The numerical integration has been performed by recurrent power series. The stability index is determined by computing the trace of the matrizant.

2. The algorithm

For conservative dynamical systems with two degrees of freedom, the equations of motion may be put as[1]

$$\ddot{x} = 2A\dot{y} + W_x, \qquad \ddot{y} = -2A\dot{x} + W_y, \tag{1}$$

with A and W functions of the coordinates x and y. These equations admits the integral or Painlevé constant

$$C = 2W - (\dot{x}^2 + \dot{y}^2). \tag{2}$$

In the case that the functions A an W depend on a parameter σ_j, any solution of (1) will depend also on this parameter

$$x(t;\sigma_j), \qquad y(t;\sigma_j). \tag{3}$$

Besides, any displacement of (3) due to a variation of the parameter is given by solving the variational equations

$$\begin{aligned} \delta\ddot{x} &= \quad 2A\,\delta\dot{y} + (W_{xx} + 2A_x\dot{y})\,\delta x + (W_{xy} + 2A_y\dot{y})\,\delta y + W_{x\sigma_j} + 2A_{\sigma_j}\dot{y}, \\ \delta\ddot{y} &= -2A\,\delta\dot{x} + (W_{xy} - 2A_x\dot{x})\,\delta x + (W_{yy} - 2A_y\dot{x})\,\delta y + W_{y\sigma_j} - 2A_{\sigma_j}\dot{x}, \end{aligned} \tag{4}$$

that have the variational integral

$$\gamma = W_{\sigma_j} + W_x\,\delta x + W_y\,\delta y - \dot{x}\,\delta\dot{x} - \dot{y}\,\delta\dot{y}. \tag{5}$$

We will deal only with isoenergetic variations ($\gamma = 0$). For a fixed value of the parameter σ_j, and for approximate initial conditions on periodicity, we use Eqs. (4) and (5) for obtaining a periodic solution of (1). With this periodic solution for a given parameter σ_j, we use again (4) and (5) to produce an approximate periodic solution for a variation of the parameter $\sigma_j + \delta\sigma_j$. However, the secular effects of (4) are not easy to handle in this way. This is the reason why Deprit and Henrard[8] elaborated an algorithm that separates purely periodic contributions from secular effects. Essentially, this algorithm consist of formulating the variational equations (4) in its intrinsic form by means of the normal (n) and tangential (p) displacements

$$\begin{aligned} \delta x &= p\cos\phi - n\sin\phi, \\ \delta y &= p\sin\phi + n\cos\phi, \\ \delta\dot{x} &= \dot{p}\cos\phi - \dot{n}\sin\phi - \dot{\phi}\,\delta y, \\ \delta\dot{y} &= \dot{p}\sin\phi + \dot{n}\cos\phi + \dot{\phi}\,\delta x; \end{aligned} \tag{6}$$

the angle ϕ is the inclination of the velocity vector with respect to the axis x.

Deprit and Henrard managed to separate the intrinsic displacements into

$$\ddot{n} + \Theta\,n = -2W_{\sigma_j}\frac{A+\dot{\phi}}{V} - 2A_{\sigma_j}V + W_{y\sigma_j}\cos\phi - W_{x\sigma_j}\sin\phi; \tag{7}$$

where V is the velocity, the coefficient Θ is

$$\Theta = \frac{\ddot{V}}{V} + 2(A+\dot{\phi})^2 + 2A^2 - W_{xx} - W_{yy} - 2V(A_x\sin\phi - A_y\cos\phi) \tag{8}$$

and the quadrature

$$\frac{d}{dt}\left(\frac{p}{V}\right) = 2\frac{A+\dot{\phi}}{V}n + \frac{W_{\sigma_j}}{V^2}. \tag{9}$$

The general solution of (7) is the linear combination

$$n = \alpha n^{\mathrm{I}} + \beta n^{\mathrm{II}} + n^{\sigma_j}$$

of the solutions ($n^{\rm I}$ and $n^{\rm II}$) of the homogeneous equation ($\ddot{n} + \Theta n = 0$) for the initial conditions

$$n^{\rm I}(0) = \dot{n}^{\rm II}(0) = 1, \qquad \dot{n}^{\rm I}(0) = n^{\rm II}(0) = 0$$

whereas n^{σ_j} is the particular solution of the nonhomogeneous equation (7) for the initial value

$$n^{\sigma_j}(0) = \dot{n}^{\sigma_j}(0) = 0.$$

Once calculated the normal displacement, the tangential one is computed from

$$p = \alpha p^{\rm I} + \beta p^{\rm II} + p^{\sigma_j},$$

where $p^{\rm I}$ and $p^{\rm II}$ are the solutions of

$$\frac{d}{dt}\left(\frac{p}{V}\right) = 2\frac{A + \dot{\phi}}{V} n, \tag{10}$$

for $n^{\rm I}$ and $n^{\rm II}$ respectively, and p^{σ_j} is the particular solution of (9) for n^{σ_j} and $p(0) = 0$.

The coefficients α and β of the linear combination are obtained from the initial conditions.

Deprit and Henrard proposed a method for continuing natural families of periodic orbits when the parameter is the energy. When the potential function is of the form

$$W = -\sum_{i>0} \sigma_i F_i(x, y),$$

we extended (Lara *et al.*[9]) this algorithm for whatever parameter σ_j. Let us assume that for a particular value of the parameter $\sigma_{j,0}$, the solution (3) is periodic with period T_0, and let us assume, too, that for small variations of the initial conditions and of the parameter there exists a family of periodic orbits. This family is given by the expansions

$$\begin{aligned} X(t; \sigma_{j,0} + \delta\sigma_j) &= x(t; \sigma_{j,0}) + \delta\sigma_j \left.\frac{\partial x(t; \sigma_j)}{\partial \sigma_j}\right|_{\sigma_j = \sigma_{j,0}} + \ldots, \\ Y(t; \sigma_{j,0} + \delta\sigma_j) &= y(t; \sigma_{j,0}) + \delta\sigma_j \left.\frac{\partial y(t; \sigma_j)}{\partial \sigma_j}\right|_{\sigma_j = \sigma_{j,0}} + \ldots, \end{aligned} \tag{11}$$

and the period is the series

$$T(\sigma_{j,0} + \delta\sigma_j) = T_0 + \delta\sigma_j \frac{\partial T_0}{\partial \sigma_j} + \ldots. \tag{12}$$

By means of our adaptation of the algorithm of Deprit and Henrard, a first order approximation of the initial conditions for the new periodic orbit is given by

$$\begin{aligned} x_0^* &= x_0 + \delta\sigma_i \frac{dx_0}{d\sigma_i}\Big|_{\sigma_i=\sigma_{i,0}}, & \dot{x}_0^* &= \frac{V_0^*}{\tilde{V}_0}\tilde{x}_0, \\ y_0^* &= y_0 + \delta\sigma_i \frac{dy_0}{d\sigma_i}\Big|_{\sigma_i=\sigma_{i,0}}, & \dot{y}_0^* &= \frac{V_0^*}{\tilde{V}_0}\tilde{y}_0, \end{aligned} \tag{13}$$

with

$$V_0^* = \sqrt{2W(x_0^*, y_0^*; \sigma_{i,0} + \delta\sigma_i) - C}, \quad \text{and} \quad \tilde{V}_0 = \sqrt{\tilde{x}_0^2 + \tilde{y}_0^2},$$

where

$$\tilde{x}_0 = \dot{x}_0 + \delta\sigma_i \frac{d\dot{x}_0}{d\sigma_i}\Big|_{\sigma_i=\sigma_{i,0}}, \quad \text{y} \quad \tilde{y}_0 = \dot{y}_0 + \delta\sigma_i \frac{d\dot{y}_0}{d\sigma_i}\Big|_{\sigma_i=\sigma_{i,0}}.$$

The values of the derivatives with respect to the parameter σ_j at an initial value are computed by

$$\begin{aligned} \frac{dx_0}{d\sigma_j}\Big|_{\sigma_j=\sigma_{j,0}} &= \delta x(0) = -\alpha \sin\phi_0, \\ \frac{dy_0}{d\sigma_j}\Big|_{\sigma_j=\sigma_{j,0}} &= \delta y(0) = \alpha \cos\phi_0, \\ \frac{d\dot{x}_0}{d\sigma_j}\Big|_{\sigma_j=\sigma_{j,0}} &= \delta\dot{x}(0) = -F_j(x_0, y_0)\frac{\cos\phi_0}{V_0} + (2A_0 + \dot{\phi}_0)\,\delta y(0) - \beta \sin\phi_0, \\ \frac{d\dot{y}_0}{d\sigma_j}\Big|_{\sigma_j=\sigma_{j,0}} &= \delta\dot{y}(0) = -F_j(x_0, y_0)\frac{\sin\phi_0}{V_0} - (2A_0 + \dot{\phi}_0)\,\delta x(0) + \beta \cos\phi_0, \\ \frac{dT_0}{d\sigma_j}\Big|_{\sigma_j=\sigma_{j,0}} &= \delta T = -\frac{1}{V_0}[\alpha p^{\mathrm{I}}(T) + \beta p^{\mathrm{II}}(T) + p^{\sigma_j}(T)]. \end{aligned}$$

The approximation (13) is improved by means of a corrector procedure (Deprit and Henrard, 1967).

3. The J_2, J_3 and J_4 problem

We shall find natural families of periodic orbits for the zonal problem, but taking into consideration the J_2, J_3 and J_4 harmonics. The potential is

$$\mathcal{V} = -\frac{\mu}{r} + \frac{\mu}{r} \sum_{2\leq n\leq 4} \left(\frac{\alpha}{r}\right)^n J_n P_n(z/r)$$

where μ is the Gaussian constant for the Earth, r is the radial distance, α is the equatorial radius of the Earth, $J_n = -C_{n,0}$ is the zonal coefficient of degree n and $P_n(z/r)$ is the Legendre polynomial of degree n.

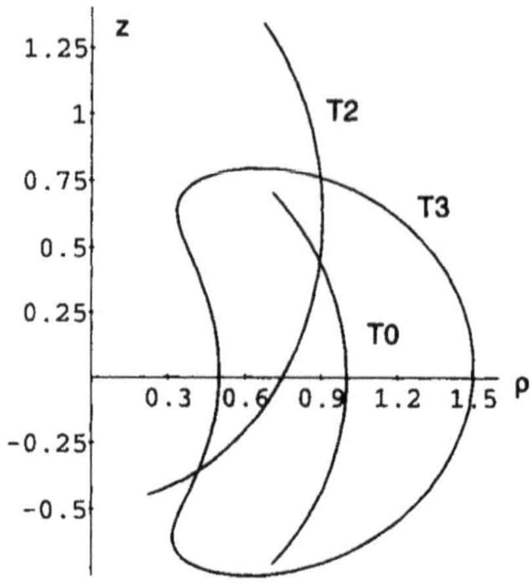

Figure 1: Several periodic orbits considering the J_2, J_3 and J_4 zonals. The orbit $T_0 \in S_0$ is *quasi* – circular; the T_2 ($\in S_2$) and T_3 ($\in S_3$) are eccentric; the argument of the perigee of the orbit T_3 is zero.

When expressed in cylindrical coordinates (ρ, λ, z, P, Λ, Z), the Hamiltonian for this problem is

$$\mathcal{H} = \tfrac{1}{2}(P^2 + \Lambda^2\rho^{-2} + Z^2) + \mathcal{V}(\rho, z).$$

The moment Λ (polar projection of the angular momentum vector) is an integral for its conjugate coordinate λ is cyclic, and therefore, the problem may be considered as a problem with two degrees of freedom (ρ, z) plus a rotation of the meridian plane $(\lambda(t))$. Thus, when we speak about periodic orbits, we mean 'periodic orbits' in the meridian plane.

Following the notation of the previous section, $A = 0$, $x = \rho$, $y = z$, $W = -\frac{1}{2}\Lambda^2\rho^{-2} - \mathcal{V}(\rho, z)$, and the equations of the motion are

$$\ddot{\rho} = W_\rho, \qquad \ddot{z} = W_z, \tag{14}$$

with the Painlevé integral

$$C = 2W - (\dot{\rho}^2 + \dot{z}^2).$$

The corresponding variational equations are $(0 \le i \le 4)$

$$\begin{aligned} \delta\ddot{\rho} &= W_{\rho\rho}\,\delta\rho + W_{\rho z}\,\delta z + W_{\rho\sigma_i}, \\ \delta\ddot{z} &= W_{z\rho}\,\delta\rho + W_{zz}\,\delta z + W_{z\sigma_i}, \end{aligned} \tag{15}$$

that have the variational integral

$$W_\rho \, \delta\rho + W_z \, \delta z - \dot\rho \, \delta\dot\rho - \dot z \, \delta \dot z - F_i = 0, \tag{16}$$

with the possible parameters σ_i and functions F_i given by

$$\begin{array}{llll} \sigma_0 = \frac{1}{2}C, & \sigma_1 = \Lambda^2, & \sigma_i = J_i, & \\ F_0 = 1, & F_1 = \frac{1}{2}\rho^{-2}, & F_i = \frac{\mu}{r}\left(\frac{\alpha}{r}\right)^i P_i(z/r). & (2 \leq i \leq 4) \end{array}$$

To this problem, we apply the Deprit and Henrard's algorithm with the modifications above mentioned.

4. Natural families of periodic orbits

Before starting the task of propagating a natural family, we need a periodic orbit (or equivalently, the initial conditions for such a periodic orbit) to begin with. We approach the problem indirectly: from the *main problem*, we obtain an approximation of a periodic orbit; with the corrector scheme, we find a better approximation of the periodic orbit, which will be used as the starting orbit for the J_2, J_3 an J_4 problem. The numerical values of these coefficients have been borrowed from the World Geodetic System 1984 (WGS 84)

The numerical integration has been done by recurrent power series. This method has been proved to be accurate and stable[7,6]. However, one of the disadvantages of this method is that it must be programmed for each specific problem; for each differential system to integrate, it is necessary to select some auxiliary variables to be used within the algebra of series. We solved this difficulty by elaborating a piece software with an algebraic manipulator that automatically provides the Fortran code for the integration by recurrent power series, simply by introducing the differential system as input. Details of this procedure will be published elsewhere.

The linear stability of each orbit is determined by computing the trace of the resolvent of the homogeneous Hill's equation (7) at the end of one fundamental period, and checking whether $|\mathrm{Tr}(T)| < 2$ (stable orbit), $|\mathrm{Tr}(T)| > 2$ (unstable orbits) or $|\mathrm{Tr}(T)| = \pm 2$ (indifferent stability).

4.1. Main problem: variation of J_2

First, we consider the *main problem*, and by means of Poincaré surfaces of section (Figure 2) we determine the fixed points (which correspond to periodic orbits). However, by using the actual value of J_2 for the Earth, fixed points

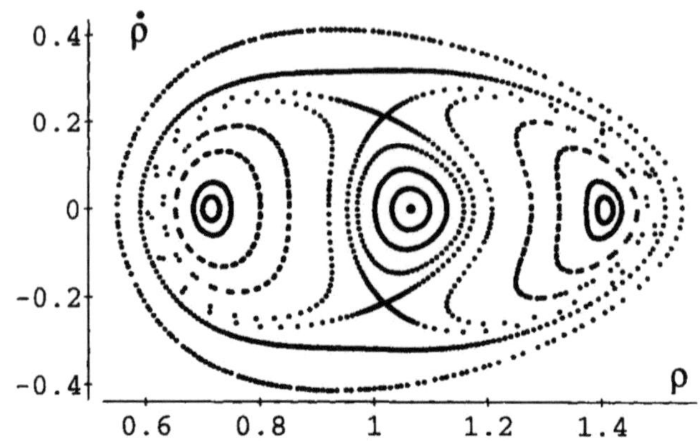

Figure 2: Poincaré surface of section corresponding to the *main problem*. There are 5 periodic trajectories: 3 are stable (corresponding to the elliptic fixed points on the axis $\dot{\rho} = 0$) and 2 are unstable (corresponding to the hyperbolic fixed points). The values of the parameters are $J_2 = 0.2$ and $\Lambda^2 = 0.15$

hardly can be seen from the figure. This is why —in a similar way as Broucke[2] did— we take a bigger value for J_2. In our case, we take $J_2 = 0.2$ and $\Lambda^2 = 0.15$ in order to have in the figure the fixed points.

From the screen on the computer, we estimate the initial conditions for the five periodic solutions that appear in figure 2, that are improved with the corrector procedure. However, these are not the values corresponding to the Earth's J_2, but since we have extended the algorithm for whatever parameter, we may continue the periodic orbits by varying the parameter $\sigma_2 = J_2$ from 0.2 until 0.001. We observe too (Figure 3) that the stability of these periodic orbits does not change along the continuation, that is, there are no bifurcations, as it was expected because of the bi-homogeneity of the potential (Broucke, 1994).

4.2. Zonal problem: variation of Λ

The initial conditions obtained in the previous step for $J_2 = 0.001$ will be used as the starting periodic orbit for the zonal problem (J_2, J_3 and J_4). Now we will continue the natural families by varying the parameter $\sigma_1 = \Lambda^2$; in other words, we consider the inclination I of the orbit as the parameter.

To visualize the evolution of the natural families, we make a plot of the inclination I versus the eccentricity e (Figure 4). Since for the plot no great

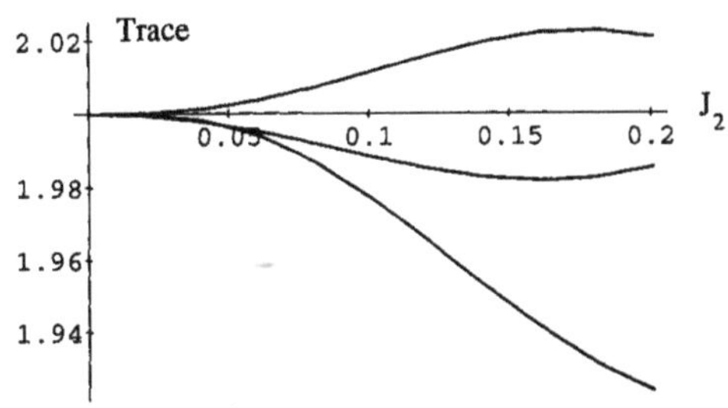

Figure 3: Evolution of the trace of the matrizant with respect to the parameter J_2 for the families of periodic orbits. The stability character of the orbits remains unchanged along the variation of the parameter.

accuracy is required, for each orbit, we represent the averaged values of these elements over the period, already computed during the integration

$$\langle e \rangle = \frac{1}{T}\int_0^T e(t)\,dt, \qquad \langle I \rangle = \frac{1}{T}\int_0^T I(t)\,dt.$$

We find 5 natural families of periodic orbits (Figure 4). The orbits of the family S_0 —emanating from a circular equatorial orbit— remains everywhere almost circular, except in the proximities of the critical inclination, where the eccentricity grows fast. At difference with the *main problem* (Coffey *et al.*[3,5], Broucke[2]), this family does not bifurcate and it is made of stable orbits. The polar family S_1 is also made of stable orbits, that are almost circular; in the neighborhood of the critical inclination, the eccentricity drops to almost 0 and suddenly, it jumps to high values. This family ends at the value $\Lambda^2 = 0.2003$. At this point, another family (S_2) starts. The orbits of this family have high eccentricity, and in the interval ($0.20025 < \Lambda^2 < 0.2003$) these eccentric orbits are unstable. At the value $\Lambda^2 = 0.20025$, a bifurcation occurs; the orbits of the family become stable, and two unstable families (S_3 and S_4) appear.

5. Conclusions

By numerically continuation of natural families, we find five families of periodic orbits in the artificial satellite problem when the influence of the J_2, J_3 and J_4 zonal harmonics is considered. Our numerical work corroborates the results

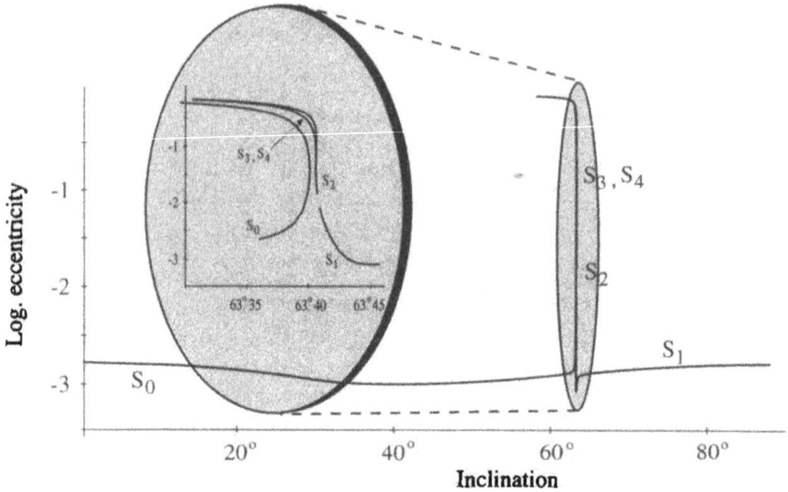

Figure 4: Natural families. The families S_3 and S_4 coincide because of the symmetry along the line $\dot{\rho} = 0$

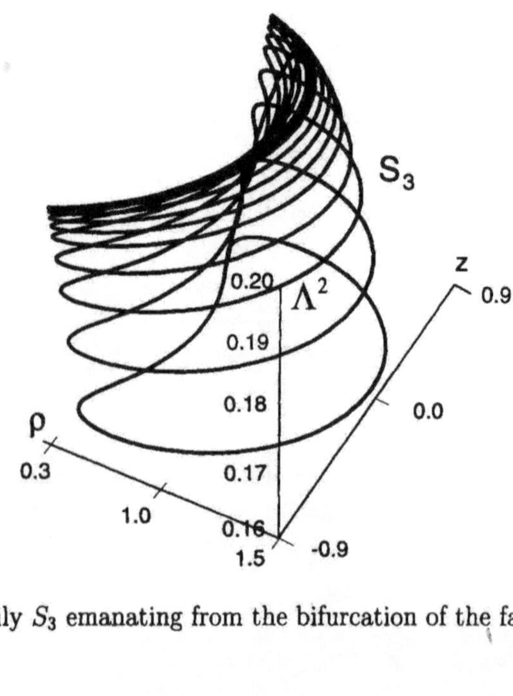

Figure 5: The family S_3 emanating from the bifurcation of the family S_2.

of Coffey *et al.*[5] who found almost periodic orbits for very high inclinations.

6. Acknowledgements

We are grateful to Dr. Deprit, who suggested us the topic. Comments from Professor Broucke have been very useful in our work.

This work has been supported in part by CICYT (Project # ESP91-0919)

7. References

1. G. Birkhoff, G., "Dynamical Systems with two Degrees of Freedom," *Trans. Amer. Math. Soc.* **18** (1917) 199–300.
2. R. Broucke,, "Numerical Integration of Periodic Orbits in the Main Problem of Artificial Satellite Theory," *Celest. Mech.* **58** (1994) 99–123.
3. S. Coffey, A. Deprit and B. Miller, "The critical inclination in artificial satellite theory," *Celest. Mech.* **39** (1986) 365–406.
4. S. Coffey, A. Deprit, E. Deprit and L. Healy, "Painting the phase space portrait of an integrable dynamical system," *Science* **247** (1990) 365–406
5. S. Coffey, A. Deprit and E. Deprit, "Frozen Orbits for Satellites Close to an Earth-like Planet," *Celest. Mech.* **59** (1994) 37–72
6. A. Deprit and J. Price, "The Computation of Characteristic Exponents in the Planar Restricted Problem of three Bodies," *Astron. J.* **70** (1965) 836–846.
7. A. Deprit and R. V. M. Zahar, "Numerical Integration of an Orbit and Its Concomitant Variations," *Z. Angew. Math. Phys.* **17 17** (1966) 425-430.
8. A. Deprit and J. Henrard, "Natural Families of Periodic Orbits," *Astron. J.* **72** (1967) 158-172.
9. M. Lara, A. Deprit and A. Elipe, " Numerical continuation of frozen orbits for the zonal problem," accepted for publication in *Celest. Mech.* (1995)

FAST ROTATIONS OF THE RIGID BODY

FRANCESCO FASSÒ
Università di Padova, Dipartimento di Matematica Pura e Applicata
Via G. Belzoni 7, 35131 Padova, Italy
E-mail: fasso@pdmat1.math.unipd.it

1. Introduction

The aim of this talk is to provide a short description of some recent results on the fast rotations of rigid bodies.[1,2] For simplicity, we consider only the case of a body which has a fixed point and is symmetric (that is, two of the moments of inertia relative to the fixed point coincide). We assume that the forces which act on the body are positional, namely depend on the body configuration alone, and are described by an analytic potential energy V; under these conditions the system is Hamiltonian, with Hamiltonian $K+V$, K being the kinetic energy. A "fast rotation" is any motion in which the modulus of the angular velocity is large, so that $K \gg V$; the study of these motions is of interest both in mechanics and celestial mechanics (motions of planets and satellites, for instance).

A fast rotating rigid body is a small perturbation of the Euler–Poinsot system, that is, the rigid body with a fixed point and no external forces, which is described by the Hamiltonian K and is completely integrable. Therefore, in order to obtain approximate descriptions of these motions, it is possible and natural to resort to the methods of Hamiltonian perturbation theory. As a matter of fact, this approach is common in celestial mechanics,[3] but the existing studies are mostly restricted to first order expansions and are not rigorous. We go instead beyond the formal level, basing our analysis on Nekhoroshev's theory.[4] The reason why we refer to Nekhoroshev's theory rather than to KAM theory is that the rigid body has three degrees of freedom, so that the KAM approach does not provide results for all motions—unless the external forces have special symmetry properties, like in the case of a heavy triaxial body.[5,6,7]

In this way, we obtain rigorous results on the motion of a fast rotating symmetric body which hold on finite but very long time scales, which increase exponentially fast with the (square root of the) angular velocity. Specifically, we first of all show that, on such time scales, the body approximately undergoes a free motion around the instantaneous direction of the angular momentum vector m. Because of the presence of external forces, the angular momentum does not remain (in general) fixed in space. It is however a well known elementary fact (actually, an immediate consequence of the equations of motion, called gyroscopic stiffness) that, in a fast rotation, the angular momentum vector moves *slowly* in space:[8] precisely, denoting by μ the unit vector $m/\|m\|$, one has $\dot{\mu} = \mathcal{O}(\|m\|^{-1})$. We shall provide some informations about the slow spatial motion of the vector m on the considered time scales; in particular, we shall point out that the qualitative properties of the motion of m are deeply different depending on the resonance properties of the motion of the body: in non-resonant motions, m is forced to move in an essentially regular way, while in resonant conditions it can likely perform wide, irregular, chaotic-like movements.

2. The Unperturbed System

Preliminarily, we need to recall a few basic facts about the symmetric Euler–Poinsot system. We shall denote by e_1, e_2, e_3 three orthogonal unit vectors directed like the principal axes of inertia of the body, e_3 being the symmetry axis, and by I_1, $I_2 = I_1$, I_3 the corresponding moments of inertia. Two basic facts about the dynamics of this system are that the angular momentum vector m is constant in space, and that its projection $m_3 = m \cdot e_3$ is constant, too. This implies that, in every motion of the system, the angular velocity w is given by

$$w = \frac{\|m\|}{I_1}\mu + \frac{\eta\, m_3}{I_1} e_3 , \qquad \eta = \frac{I_1}{I_3} - 1 . \tag{1}$$

Equation (1) shows that the body rotates about its symmetry axis e_3 with constant angular velocity $\eta m_3/I_1$ and that, in turn, such an axis uniformly precesses in space about the fixed direction of m, with angular velocity $\|m\|/I_1$ (see figure 1.a). Note that these motions are quasi-periodic with *two* frequencies; this is connected to the fact that Euler–Poinsot is a completely integrable Hamiltonian system which is degenerate: it has three degrees of freedom, but its Hamilton function depends only on two of the three action coordinates. (If $\eta = 0$, namely if the body is spherical, all motions are periodic; in the following, we shall tacitly exclude from our consideration the spherical case).

The degeneracy of the Euler–Poinsot system plays an important role in our study. As a matter of fact, degeneracy strongly influences the long-time dynamics of small perturbations of an integrable system: as a general fact, every action variable which does not enter the Hamilton function is by no means forced to remain almost constant; rather, its evolution is essentially determined by the perturbation. In the present case, as it will be seen, this fact has critical consequences on the long-time properties of the motion of the angular momentum vector. Moreover, degeneracy is connected to the non-existence of a "global" system of action-angle coordinates of the Euler–Poinsot system.[9]

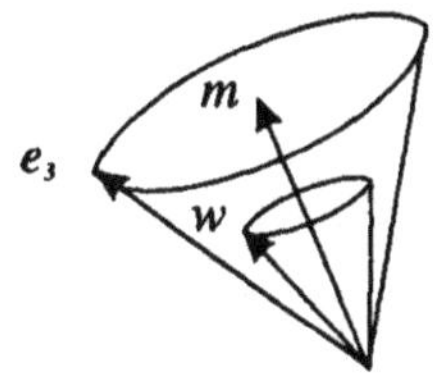

Figure 1.a

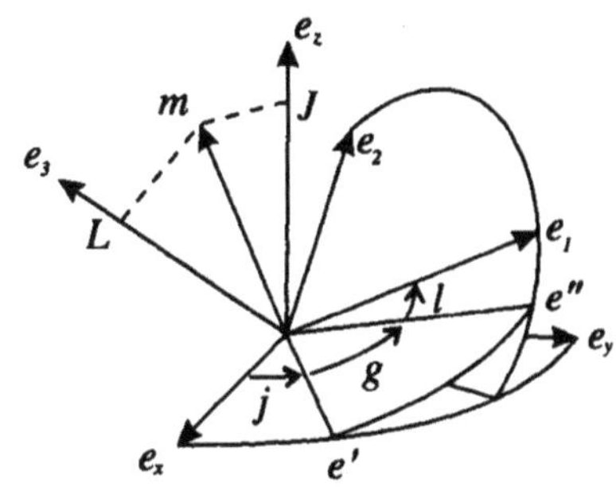

Figure 1.b

The action-angle coordinates of the symmetric Euler–Poinsot system have been known for more than one century.[10,11,12] Their construction depends on the (arbitrary) choice of an orthogonal reference frame $\{e_x, e_y, e_z\}$ fixed in space, with the origin in the fixed point of the body. Let us denote $G = \|m\|$, $L = m\cdot e_3$, $J = m\cdot e_z$, and let the three angles g, l, j be defined as shown in figure 1.b, where e' and e'' are unit vectors, respectively, in the direction of $e_z \times m$ and of $m \times e_3$. Then, as it turns out, (G, L, J, g, l, j) are symplectic coordinates (with 2-form $dG \wedge dg + dL \wedge dl + dJ \wedge dj$). Note that the angles g and l measure, respectively, the precession of the axis e_3 about the angular momentum vector m, and the proper rotation of the body about its symmetry inertia axis e_3. Furthermore, (G, J, j) are (cylindric-like) coordinates for the representative of the angular momentum vector in the spatial base $\{e_x, e_y, e_z\}$. Expressed as a function of the action-angle coordinates, the Hamiltonian $K = (m_1^2 + m_2^2)/(2I_1) + m_3^2/(2I_3)$ of the symmetric Euler–Poinsot system takes the form

$$K(G, L) = \frac{1}{2I_1}\left(G^2 + \eta L^2\right). \tag{2}$$

Hence, Hamilton's equations show that G, J, j (that is, the angular momentum in space) and L (that is, the component of m on the symmetry axis) stay constant, while

$$\dot{g} = \frac{G}{I_1}, \qquad \dot{l} = \frac{\eta L}{I_1}. \tag{3}$$

There are two properties of the action-angle coordinates which should be remarked here. First, the action-angle coordinates are related to the Euler angles and to their conjugate momenta by a canonical transformation which is not point-extended, but is nevertheless a Mathieu transformation. Therefore, the Euler angles (that is, the configuration of the body) are homogeneous functions of degree zero of the actions (G, L, J). This implies that, written in the action-angle coordinates, the potential energy V of a positional force field is a homogeneous function of degree zero of (G, L, J), too; more precisely, if V is analytic, and if one excludes the points where $m = 0$, the local representative of V in such a coordinate system is an analytic function $v(J/G, L/G, g, l, j)$.

Secondly, one should observe that the action-angle coordinates just introduced do not provide a global coordinate system for the whole phase space of the rigid body. The construction of the angles g and l relies indeed on the use of the two nodal lines e' and e'', and at least one of them is not defined on the equilibria ($m = 0$), on the steady rotations about the inertia symmetry axis e_3 (sometimes called "gyroscopic rotations", where m is aligned with e_3), and also on all motions with m parallel to the spatial axis e_z. As it turns out, these singularities have different causes.

In order to understand this situation, one should focus on the foliation of the phase space by the two-dimensional invariant tori.[9] This foliation is an intrinsic object, which is uniquely determined by the dynamics; as equation (2) shows, in fact, a dense subset of these tori support motions with non-resonant frequencies and are thus closures of trajectories. Now, as is easily seen, equilibria and gyroscopic rotations correspond to the singular leaves of such a foliation, which are diffeomorphic, respectively, to points and circles.[9] Therefore, no action-angle coordinates do exist in any neighbourhood of these motions. The situation is quite similar to the equilibrium of the harmonic oscillator or to the circular orbits of the point in a central force field.

On the contrary, motions with m parallel to e_z do not correspond to any singularity of the foliation by the two-dimensional invariant tori; since the direction of the z-axis is completely arbitrary, these motions do not have indeed any special dynamical meaning at all. The geometrical meaning of this singularity is clarified by considering the structure of the set M which is obtained by removing from the phase space the equilibria and the gyroscopic rotations. By construction, the set M is entirely foliated by invariant tori of dimension two, and this foliation is regular; more precisely, it is a fibration.[9] The fact that no system of action-angle coordinates covers all of M (as is clear, since M contains states with m pointing in every spatial direction) just reflects the fact that such a set has not the structure of the product of an open subset of $\mathbb{R}^3$ and of the torus $\mathbb{T}^3$, as the existence of a global set of action-angle coordinates would imply. As a matter of fact, M has not even the structure of a fibration with fiber $\mathbb{T}^3$. Fibrations by invariant three-dimensional tori of subsets of M are created by introducing the local systems of action-angle coordinates, but this procedure is not intrinsic: three-dimensional tori are obtained by grouping together the two-dimensional intrinsic ones, an operation which can be done in different ways but not globally in M. As a matter of fact, the "third" angle j is created by putting cylindrical coordinates (J, j) on the sphere of constant modulus of the angular momentum vector in space (each of which is entirely contained in M), and singularities appear at two poles (the chosen z-axis). By the way, note that one constructs an atlas for the set M by patching together two local systems of action-angle coordinates, relative to two spatial frames with non-parallel z-axes.

Remark: There is a rich and interesting geometry behind this situation, which is typical of degenerate systems. The fact is that the characterization of degenerate systems as completely integrable, based on the Liouville–Arnol'd theorem and on the introduction of action-angle coordinates, does not provide a thorough, global description of them, which is instead obtained by considering the fibration by the invariant tori of the smallest dimension. This fibration has the structure of what in symplectic geometry is called a "bifibration" or a "dual pair". This approach is connected to the so-called "non-commutative" integrability. For a treatment of these facts see refs. 13, 9 and references therein.

Concluding this section, let us observe that the singularity on the equilibria is clearly of no importance for our study (in a fast rotation, m is well away from zero). Fast gyroscopic rotations are instead quite important for applications, but they are outside the range of our approach, which is based on the use of action-angle coordinates. We shall indeed exclude from our consideration a small neighbourhood of these motions, so as to avoid the corresponding coordinate singularities. We shall come back on motions close to gyroscopic rotations in the Conclusions. The role of the singularity on the z–axis will be explained below.

3. Normal Forms

Nekhoroshev's theorem is based on the construction of different normal forms in different regions of the phase space, depending on the resonance properties of the motions of the unperturbed system. The frequencies of the symmetric Euler–Poinsot system depend only on the two functions G and L which, restricted to the subset M of the phase space we are considering here, take values in the set

$$A = \{(G, L) \in \mathbb{R}^2 : \ |L| < G\} . \tag{4}$$

Therefore, the resonance properties of motions are determined by the points of the set A, which will be called the *action space* of the system. More precisely, let us introduce the frequency vector

$$\omega(G, L) = \Big(\frac{G}{I_1}, \frac{\eta L}{I_1}\Big) ,$$

and observe that equations (3) are correspondingly written $(\dot{g}, \dot{l}) = \omega(G, L)$. Then, for any non-zero vector $\nu \in \mathbb{Z}^2$, we define the resonant line R^ν as the set of points $(G, L) \in A$ at which $\omega(G, L) \cdot \nu = 0$.

Figure 2

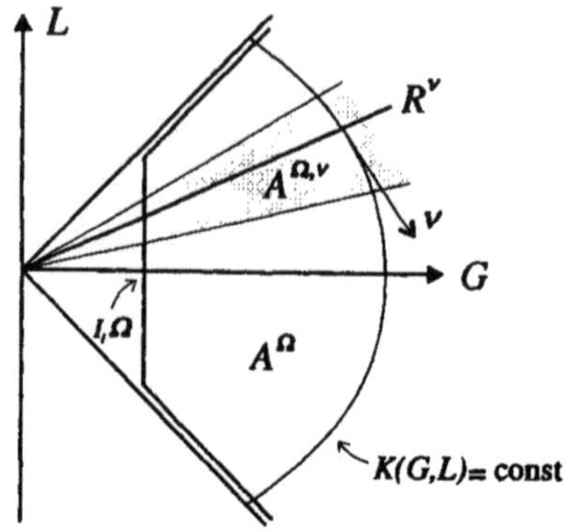

Our perturbation construction depends on the introduction of a (large) parameter Ω, which measures the angular velocity of the body. Precisely, for any given Ω, we restrict ourselves to the subset of the action space where $|\omega(G, L)|_\infty > \Omega$ ($|\cdot|_\infty$ being the l_∞ norm). As already mentioned, in order to avoid the coordinate singularities on equilibria and gyroscopic rotations, we remove from the action space a neighbourhood of its border $G = |L|$; as it turns out, such a neighbourhood can be consistently taken to be a strip of uniform width $\mathcal{O}(1)$, that is, independent of Ω. We denote by A^Ω what remains of the action space after these operations (see figure 2, which is relative to the case $0 < \eta \leq 1$).

As is typical of Nekhoroshev theory, we introduce a cutoff $N = N(\Omega)$, which depends on Ω, and we consider only those resonant lines R^ν for which $|\nu| := |\nu_1| + |\nu_2| \leq N$; as it follows from the proof of Proposition 1 below, the cutoff is conveniently chosen of order $\Omega^{1/2}$. Around each resonant line R^ν, $0 < |\nu| \leq N(\Omega)$, we pick a small "resonant set" $A^{\Omega,\nu} \subset A^\Omega$, defined by the condition

$$|\omega(G, L) \cdot \nu| \leq \frac{\|\omega(G, L)\|}{2N(\Omega)} \qquad \text{for all } (G, L) \in A^{\Omega,\nu}; \tag{5}$$

note that different resonant sets do not overlap (this is proven by observing that the angle between any two integer vectors ν and ν' is not smaller than $|\nu|^{-1}|\nu'|^{-1}$). The complement in A^Ω of the union of all resonant sets $A^{\Omega,\nu}$, $0 < |\nu| \leq N(\Omega)$, constitutes the "non-resonant set", which for notational convenience will be denoted $A^{\Omega,0}$. (The set $A^{\Omega,0}$ should not be thought of as being the set of points which resonate with 0; it contains indeed all resonances of order $|\nu| > N(\Omega)$).

For any given Ω, we shall treat as resonant with a non-zero integer vector ν of norm $|\nu| \leq N(\Omega)$ all motions starting in the "resonant region" $M^{\Omega,\nu}$, which is defined as the subset of the phase space in which the functions G and L take values in $A^{\Omega,\nu}$; we shall instead treat as non-resonant all motions starting in the "non-resonant region" $M^{\Omega,0}$, where G and L take values in $A^{\Omega,0}$. More precisely, we shall construct, up to an exponentially small remainder, a resonant normal form adapted to the ν-resonance in each resonant region $M^{\Omega,\nu}$, and a non-resonant normal form in the non-resonant region $M^{\Omega,0}$.

The construction of these normal forms can be performed by means of essentially standard techniques of Hamiltonian perturbation theory, but one needs to devote some attention to the fact that, because of the coordinate singularities $m \times e_z = 0$, resonant and non-resonant regions $M^{\Omega,\nu}$ are not covered by a single system of action-angle coordinates. The point is that, as will be clear later, because of the degeneracy of the system it may be impossible to control the motion of the angular momentum vector in space; therefore, one cannot know if, and when, m will pass through (or even only near) the singular direction of the z-axis of the chosen spatial reference frame, in which case all predictions obtained from a normal form constructed with reference to the corresponding system of action-angle coordinates loose their validity. This problem, which is typical of degenerate systems, cannot just be disregarded, even though its solution is ultimately simple:[14,15] one can indeed construct a normal form Hamiltonian which is globally defined in the whole resonant, or non-resonant, region $M^{\Omega,\nu}$, and not only within a coordinate domain.

We do not enter here into the details of this construction. However, in order to be able to formulate our results, we need to introduce some notions relative to averaging operations, which have to be performed globally in the phase space (not only locally, within each coordinate system). Let F be a function on M, and let $f^{(1)}$ and $f^{(2)}$ be its local representatives in the two local systems of action-angle coordinates of an atlas of M. Each local representative $f^{(i)}$ can be represented as a Fourier series $\sum_{\nu\in\mathbb{Z}^2} \hat{f}_\nu^{(i)} e^{i\nu\cdot\varphi}$ in the angles $\varphi = (g, l)$. As it turns out,[a] the Fourier series has an intrinsic meaning for the function F, too, in the sense that, for every $\nu \in \mathbb{Z}^2$, the functions $\hat{f}_\nu^{(i)} e^{i\nu\cdot\varphi}$, $i = 1, 2$, are the local representatives of a function defined in M, which will be denoted $\langle F\rangle_\nu$. Therefore, if $\mathcal{L}^\nu$ denotes the subset of $\mathbb{Z}^2$ of all vectors parallel to a given vector

[a]This fact is a simple consequence of the properties of the transition functions between any two local systems of action-angle coordinates, which in turn reflect the "bifibration" structure of the fibration by the invariant two-dimensional tori.[15]

$\nu = (\nu_1, \nu_2) \in \mathbf{Z}^2$, one can define the averaged function $\Pi_\nu F = \sum_{\nu' \in \mathcal{L}^\nu} \langle F \rangle_{\nu'}$; this is precisely the function whose local representatives depend on the angles g and l only through the linear combination $\nu_1 g + \nu_2 l$ (in particular, $\Pi_0 F$ is the average of F over the two-dimensional invariant tori).

We can now state the following

Proposition 1 *Assume that V is real analytic and that Ω is large enough, that is, $\Omega > \Omega_*$ for some constant Ω_* depending only on V and η. Define $N(\Omega) = c\sqrt{\Omega/\Omega_*}$, where c is a constant which depends only on V. Consider any vector $\nu \in \mathbf{Z}^2$ such that $|\nu| \leq N(\Omega)$. Then, there exist a symplectic diffeomorphism $\mathcal{C}$ defined in $M^{\Omega,\nu}$, which is close to the identity, precisely*

$$|\mathcal{C} - \mathrm{id}| = \mathcal{O}\left(\sqrt{\frac{\Omega_*}{\Omega}}\right), \tag{6}$$

and conjugates $K + V$ to

$$K + \Pi_\nu V + \sqrt{\frac{\Omega_*}{\Omega}}\, W + \sqrt{\frac{\Omega_*}{\Omega}} e^{-\sqrt{\Omega/\Omega_*}}\, F \tag{7}$$

where the functions W and F are of order 1 (that is, bounded by constants independent of Ω) and one has, moreover, $W = \Pi_\nu W$.

Sketch of the proof. The proof of this Proposition consists of a computation which is of a standard type in perturbation theory and is conceptually very simple. Nevertheless, as is typical in perturbation theory, there are many details about norms, domains and estimates which make things involved. For shortness, we give here only a sketch of the proof, skipping on details.

The idea of the proof is to construct iteratively a (large) number r of near the identity canonical transformations, each of which moves the perturbation one order up. Let us describe a single step. Consider a real analytic Hamiltonian $K+U+R$, with $U = \Pi_\nu U$. Let us decompose the remainder R in the sum $R^{\leq N} + R^{>N}$, where the "ultraviolet" part $R^{>N}$ contains all the Fourier components $\langle R \rangle_{\nu'}$ of R with $|\nu'| > N$. We aim to construct a near the identity canonical transformation which put $K + U + R^{\leq N}$ into normal form, relative to $\mathcal{L}^\nu$. Using the Lie method, we construct such a canonical transformation as the one-time-map Φ^X of the flow of a Hamiltonian X. Since

$$(K + U + R) \circ \Phi^X = K + U + R^{\leq N} + \{X, K\} + R'$$

with

$$R' = \left[K \circ \Phi^X - K - \{X, K\}\right] + \left[(U + R) \circ \Phi^X - (U + R)\right] + R^{>N}, \tag{8}$$

one sees that, provided X solves the "homological equation"

$$\{K, X\} = R^{\leq N} - \Pi_\nu R^{\leq N}, \tag{9}$$

the new Hamiltonian $(K + U + R) \circ \Phi^X$ has the form $K + U' + R'$, with $U' = U + \Pi_\nu R$. Solving equation (9) is a well known matter; in the considered domain, where the non-resonance condition (5) is satisfied, such an equation has indeed a real analytic solution X, which is given by

$$X = \sum_{\nu' \notin \mathcal{L}} \frac{1}{i\omega \cdot \nu'} \langle R^{\leq N} \rangle_{\nu'} . \tag{10}$$

This function is well defined, since the denominators never vanish and the sum is finite; the fact that it solves (9) is readily verified, for instance by passing to its local representatives in action-angle coordinates. As a consequence, the map Φ^X, and then also the new Hamiltonian $K + U' + R'$, is defined in some subset of the domain where $K+U+R$ and X are defined (Φ^X could leave not invariant such a domain, and one has to remove a neighbourhood of its boundary).

However, going beyond this level, we need to control the domain where the new Hamiltonian is defined and the size of the new remainder R'. We skip on all details concerning these estimates, focusing on a few points. First, in order to make estimates, one needs to pass to the local representatives of functions in two local systems of action-angle coordinates forming an atlas of M. More precisely, one needs to consider analytic extensions of all functions to complex neighbourhoods of their real domains (this allows one to use Cauchy inequality for estimating derivatives). These complex domains can be taken to be unions of polidisks, of some radius σ, centered in the points of the real domains. The existence of such complex extensions is assured by real analyticity, but one needs a precise control on their size (which reduces from step to step). We denote by $|F|_\sigma$ the maximum of the norm of such complex extensions of two local representatives of a function F (one can think of the supremum norm, even though it is not the most appropriate one in perturbation theory; actually, the choice of a norm adapted to Fourier series is crucial for the quality of estimates). The second important point is that one can produce an estimate of R' only in a domain smaller than the domain on which one controls U and R.

Recalling the definition (5) of the resonant set $A^{\Omega,\nu}$, it is not difficult to see that, if Ω is large enough, the solution X of the homological equation satisfies $|X|_\sigma \leq \frac{c_1 N}{\Omega}|R|_\sigma$, where c_1 (as well as $c_2, c_3, \ldots$ appearing below) is a constant independent of Ω, whose precise value depends on the norm used for functions. Estimating the new remainder R' as in (8) is a standard matter: as one can show, for any $0 < \delta < \sigma$ one has

$$|R'|_{\sigma-\delta} \le \left(\frac{c_3 N}{\Omega\delta} + c_4 e^{-c_5 N\delta}\right) |R+U|_\sigma . \tag{11}$$

Here, the first term within the brackets comes from the estimates of the first two groups of terms in (8) (which are remainders of Lie series, of orders respectively two and one). The term $c_4 e^{-c_5 N\delta}$, instead, is produced by the estimate of $R^{>N}$, using the exponential decay of the Fourier components of analytic functions.

Nekhoroshev's method is based on choosing δ and N in such a way to have a constant scaling of the remainder at each step, say of a factor $1/e$, and then iterating the procedure a number r of times which increases as a power of Ω, so as to eventually get a remainder of order e^{-r}. First of all, in order to be able to make r steps with a finite total reduction of the domain, one takes $\delta = c_6\rho/r$, where ρ is the size of the domain of the complex extensions of the local representatives of the original Hamiltonian $K+V$. Then, one chooses $r = c_7 N$ so as to make the second term in the bracket in (11) equal to $1/2e$. After this, the first term in the bracket reduces to $c_8 r^2 \Omega$, and equating it to $1/2e$ leads to $r = \sqrt{\Omega/\Omega_*}$, with $\Omega_* = 2ec_8$.

So, starting with the Hamiltonian $K+V$, one arrives at a Hamiltonian $K + Y + e^{-\sqrt{\Omega/\Omega_*}}Z$, with Y and Z of order 1 and $Y = \Pi_\nu Y$. With some additional attention, one can verify that, as stated in (7), Y has the form $\Pi_\nu V + \sqrt{\Omega_*/\Omega}W$ and that Z is actually of order $\sqrt{\Omega_*/\Omega}$. For this point, as well as for other technical points, like estimate (6), we refer to ref. 2. ∎

4. Non-Resonant Motions

We now apply Proposition 1 to describe the fast rotations of a rigid body with a fixed point. First, we consider non-resonant motions. We denote by $\overline{V}$ the average $\Pi_0 V$ of the potential energy V, which is assumed to be analytic.

Proposition 2. *Consider any $\Omega > \Omega_*$. Then, in any motion of the system described by the Hamiltonian $K+V$ with initial datum at $t=0$ belonging to $M^{\Omega,0}$, the functions G, L and $\overline{V}$ and the map $(\dot g, \dot l) - \omega(G_0, L_0)$ remain constant up to quantities of the order of $\sqrt{\Omega_*/\Omega}$ for all times t such that*

$$|t| \le \mathcal{T}\frac{\Omega_*}{\Omega}\exp\sqrt{\frac{\Omega}{\Omega_*}}, \tag{12}$$

$\mathcal{T}$ being a constant independent of Ω.

Sketch of the proof. We apply Proposition 1 with $\nu = 0$. Denote $(G, L, J, g, l, j) = \mathcal{C}(G', L', J', g', l', j')$ and observe that, since the new angles

(g', l') enter the normal form Hamiltonian (7) only through the exponentially small remainder $\sqrt{\Omega_*/\Omega}e^{-\sqrt{\Omega/\Omega_*}}F$, the new actions G' and L' move of small quantities, of order $(\Omega_*/\Omega)^{3/2}$, on the time scale (12). Note now that, by equation (6), the original actions G and L remain close to the new ones, up to quantities of order $\sqrt{\Omega_*/\Omega}$. Therefore, G and L vary at most of quantities of order $\sqrt{\Omega_*/\Omega}$ on the time scale (12). (This fact is clearly valid on a longer time scale, which grows as $\exp\sqrt{\Omega/\Omega_*}$; the restriction to (12) is made for estimating the variation of $\overline{V}$). The estimate on the variation of the function $\overline{V}$ is obtained by appealing to the energy conservation for the Hamiltonian (7) and observing that, as a consequence of the bound $\mathcal{O}((\Omega_*/\Omega)^{3/2})$ on the variation of G' and L', the kinetic term $K(G', L')$ varies at most of quantities of order $\sqrt{\Omega_*/\Omega}$ on the time scale (12). Finally, let us observe that, in every system of action-angle coordinates, one has $(\dot{g}, \dot{l}) = \omega(G, L) + (\frac{\partial v}{\partial G}, \frac{\partial v}{\partial L})$; the estimate on $\dot{g}$ and $\dot{l}$ follows from here, as one sees by observing that the derivatives of the local representative v of V with respect to the three actions G, L, J are homogeneous functions of degree minus one of (G, L, J), and using again the estimate on the variation of G and L. (As a matter of fact, the derivatives of v diverge on the coordinate singularity $G = |L|$; however, if $G = \mathcal{O}(\Omega)$, they turn out to be $\mathcal{O}(\Omega^{-1/2})$ at distances $\mathcal{O}(1)$ from such a singularity, and this is still enough to obtain the above result). For the details, see ref. 2. ∎

Proposition 2 allows us to describe the motion of the body with some accuracy, on the whole time scale (12). First of all, the body undergoes an approximate Euler–Poinsot motion about the direction of the angular momentum vector (which, as previously noticed, moves slowly in space, with speed $\mathcal{O}(\Omega^{-1})$). Specifically, the frequencies $\dot{g}$ and $\dot{l}$ remain significantly equal to the frequencies of the Euler–Poinsot motion with the same initial conditions, and moreover the angle between m and e_3 remains almost constant, precisely up to quantities of order $(\Omega_*/\Omega)^{3/2}$ (as one deduces from the estimates on the variation of G and L, since this angle is measured by the ratio L/G).

Furthermore, some informations about the slow spatial motion of m can be obtained from the normal form Hamiltonian (7). Roughly speaking, in the approximation in which the exponentially small remainder can be neglected, the new actions G' and L' stay exactly constant, and one is left with an integrable "effective Hamiltonian" $\overline{V}+\sqrt{\Omega_*/\Omega}\,W$ for the coordinates (J', j') alone—that is, for the angular momentum vector in space. More precisely, from the estimates on the variation of G, L and $\overline{V}$ given in Proposition 2, one deduces

$$\left|\overline{v}\Big(\frac{L_0}{G_0}, \frac{J_t}{G_t}, j_t\Big) - \overline{v}\Big(\frac{L_0}{G_0}, \frac{J_0}{G_0}, j_0\Big)\right| = \mathcal{O}\left(\sqrt{\frac{\Omega_*}{\Omega}}\right).$$

This estimate shows that, during the motion, the unit vector $\mu = m/\|m\|$ (for which J/G and j are cylindric coordinates) remains within a small strip, of width $\mathcal{O}(\Omega^{-1/2})$, around a level curve of a function defined on the unit sphere, which is parameterized by the initial value of L/G and has local representatives $\overline{v}(L_0/G_0, J/G, j)$. The motion of the body is illustrated in figure 3.a. Note that, for the whole exponentially long time scale (12), all chaotic effects (if present) are confined to very small oscillations about quite regular motions.

Figure 3.a

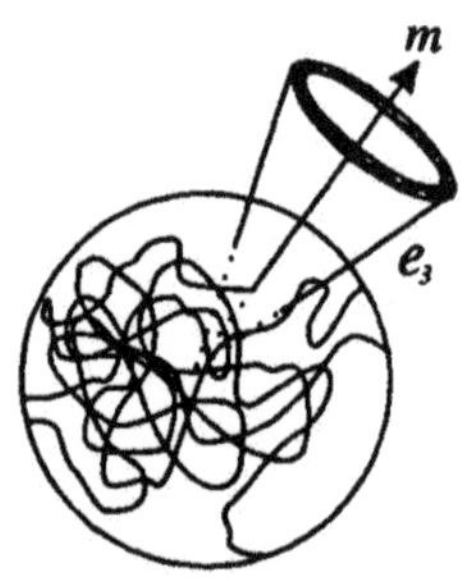

Figure 3.b

5. Resonant Motions

We consider now resonant motions:

Proposition 3. *Let $\Omega > \Omega_*$ and consider a non-zero vector $\nu \in \mathbf{Z}^2$ such that $|\nu| \le N(\Omega)$. Then, in any motion with initial datum belonging to $M^{\Omega+\Omega_1,\nu}$, where Ω_1 is a positive constant independent of Ω and ν, the functions G and L and the map $(\dot{g}, \dot{l}) - \omega(G_0, L_0)$ vary at most of quantities $\mathcal{O}(1)$ (that is, independent of Ω) for all times t as in (12).*

Sketch of the proof. The proof of this Proposition relies on two basic observations. First, the new angles $\varphi' = (g', l')$ enter the term $\Pi_\nu V + \sqrt{\Omega_*/\Omega}\, W$ of the normal form Hamiltonian (7) only through the linear combination $\nu_1 g' + \nu_2 l'$, so that $\frac{\partial}{\partial \varphi'}(\Pi_\nu V + \sqrt{\Omega_*/\Omega}\, W)$ is a vector parallel to ν. Therefore, Hamilton's equations

$$\dot{I}' = \frac{\partial}{\partial \varphi'}\left(\Pi_\nu V + \sqrt{\frac{\Omega_*}{\Omega}}\, W\right) + \mathcal{O}\left(\sqrt{\frac{\Omega_*}{\Omega}}\, e^{-\sqrt{\Omega/\Omega_*}}\right)$$

imply that, for the whole time scale (12), the motion of the new actions $I' =$

(G', L') is flattened on the direction parallel to ν (up to small fluctuations due to the exponentially small remainder). Now, as is easily seen, ν is parallel to the tangent to the level curves of K at the intersection point with the resonant line R^ν, so that ν is in particular transversal to R^ν (see figure 2, and recall that we are assuming $\eta \neq 0$). Second, the energy conservation for the Hamiltonian (7), together with the boundedness of V, implies that, during the motion, the point (G'_t, L'_t) is trapped into the strip between two level curves $K(G', L') = E_1$ and $K(G', L') = E_2$ of the function K, with $E_1 = \mathcal{O}(\Omega^2)$ and $E_2 - E_1 = \mathcal{O}(1)$; it is easy to see that the distance between these two curves is very small, precisely $\mathcal{O}(\Omega^{-1})$ and that the length of any segment parallel to ν contained in such a narrow strip is small, too, precisely $\mathcal{O}(1)$. From these two facts together one deduces that G' and L', and then also G and L by virtue of (6), vary at most of quantities $\mathcal{O}(1)$. For the details, see ref. 2. ∎

Proposition 3 shows that, also in resonant conditions, the body approximately undergoes an Euler–Poinsot motion about the angular momentum vector m, which moves slowly in space. There is however a major difference with respect to the non-resonant case, since now, even neglecting the exponentially small remainder in the normal form Hamiltonian (7), one is left with an effective Hamiltonian which contains also the "slow" angle $\sigma = \nu_1 g' + \nu_2 l'$; more precisely, such an effective Hamiltonian depends on the variables (J', j', σ, S), where S is the linear combination of G' and L' conjugate to σ. Since two-degrees of freedom Hamiltonian systems can likely be non-integrable, it is expected that (at least for not too special V's), on time scales of order Ω (namely long, but very short in comparison with (12)), irregular, large variations of the direction of the angular momentum vector can likely take place. In particular, μ is not forced by the conservation of energy to remain close to a level curve of a smooth function on the unit sphere; it is instead expected to be able to visit a large, genuinely two-dimensional region of the sphere, as illustrated in figure 3.b.

In our opinion, the possibility of these slow chaotic-like motions is a quite interesting fact. Even though a simple analysis indicates that such a behaviour should be expected to be typical,[2] we have no definite results on it. In fact, the question whether such chaotic motions actually take place in systems of (celestial) mechanical interest is a difficult one, which cannot be answered in general, depending on the specific properties of the system and on the initial conditions of the motion.

6. Conclusions

We mention now some extensions of the above results. First of all, we should consider fast motions close to gyroscopic rotations, which have been so far excluded from our analysis but are important for applications. As it turns out, such motions can be studied by the techniques of perturbation theory, but one has to employ different coordinates, which are adapted to the local structure of the foliation by the invariant tori near the singular leaves, and are in fact analogous to Poincaré elements for the circular orbits of Kepler system. Proceeding in such a way, one finds that all of the described results apply to fast motions close to gyroscopic rotations as well.[16,17]

Secondly, the case of a triaxial body can be treated in a quite similar way, and is easy to see that one would obtain quite similar results.[2] The main difference lies in the fact that the component L of the angular momentum vector is not a constant of motion, and one should refer to the true action-angle coordinates of the system.

Finally, the case of a rigid body with no fixed point can be studied by the methods of Hamiltonian perturbation theory, too. However, the coupling with the degrees of freedom of the center of mass may give rise to new, qualitatively different behaviours, like for instance slow chaotic–like motions of m also out of resonance. These behaviours are expected to depend on the specific properties of the system under consideration, and even though a few precise statements can be done,[2] this case is not yet fully understood.

In conclusion, we observe that the approach we have described here provides some general results, and some general a-priori estimates, on the fast rotations of rigid bodies. This approach also suggests the possibility of interesting phenomena, like the presence of slow chaotic-like motions. Concerning the effective existence of such behaviours in specific, significant mechanical systems, our analysis is not conclusive; nevertheless, we think that it can provide a general framework for more specific studies.

References

1. F. Fassò, *Fast rotations of the rigid body and Hamiltonian perturbation theory.* Ph.D. thesis, SISSA, Trieste (1991).
2. G. Benettin and F. Fassò, *Fast rotations of the symmetric rigid body: a study by Hamiltonian perturbation theory.* Preprint (1994).
3. V. Béletski, *Essais sur le mouvement des corps cosmiques* (Mir, Moscow, 1986).
4. N.N. Nekhoroshev, *Russ. Math. Surv.* **32** (1977), 1.
5. V.I. Arnold, *Russ. Math. Surv.* **18** (1963) 9.
6. A.I. Neishtadt, *J. Appl. Math. Mech.* **45** (1982) 766.
7. H. Hanssmann, *Normal forms for perturbations of the Euler top.* Preprint

University of Groningen (1993).
8. J.L. Synge, in *Handbuch der Physik,* **III/1** *Principles of classical mechanics and field theory,* ed. S. Flügge (Springer, Berlin, 1960).
9. F. Fassò, *On the Geometry of the Euler–Poinsot system.* Preprint (1994).
10. J.A. Serret, *Mémoires de l'Académie des Sciences de Paris,* **35** (1866) 585.
11. H. Andoyer, *Cours de Mécanique Céleste* (Gauthier–Villars, Paris 1923).
12. A. Deprit, *Am. J. Phys.* **55** (1967) 424.
13. M.V. Karasaev and V.P. Maslov, *Nonlinear Poisson Brackets. Geometry and Quantization.* Translations of the AMS, **119** (AMS, Providence, Rhode Island, 1993).
14. F. Fassò, in *Hamiltonian Mechanics: Integrability and Chaotic Behaviour,* ed. J. Seimenis (Plenum, New York, 1994), p. 113.
15. F. Fassò, *Hamiltonian perturbation theory on a manifold.* To be published in *Celestial Mechanics & Dynamical Astronomy.*
16. M. Guzzo, *Singolarità virtuali in teoria Hamiltoniana delle perturbazioni e fenomeni giroscopici.* Thesis, Università di Padova (1994)
17. G. Benettin, F. Fassò and M. Guzzo, Paper in preparation.

ON PERTURBED 3D ELLIPTIC OSCILLATORS: A CASE OF CRITICAL INCLINATION IN GALACTIC DYNAMICS

SEBASTIÁN FERRER
Grupo de Mecánica Espacial, Universidad de Zaragoza
Ciudad Universitaria, 50009 Zaragoza, Spain
E-mail: ferrer@msf.unizar.es

and

JORGE GÁRATE
Real Observatorio de la Armada
San Fernando, 11110 Cádiz, Spain

ABSTRACT

We propose a canonical transformation, composition of the rotation from the space frame to the nodal frame, with a Lissajous transformation in the orbital plane, in order to deal with 3D perturbed harmonic oscillators in a 1-1-1 resonance. In this way we enlarge the studies corresponding to planar solutions in the equatorial and meridian planes, considering now orbits in any inclination. The use of this transformation is illustrated by studying an axial symmetric quartic potential where we find a critical inclination in the averaged system, corresponding to a saddle-center bifurcation. Recognizing the fact that, from the Lissajous normalization and the axial symmetry, the double reduced orbit space is S^2, we introduce coordinates which allow to search for the relative equilibria and their local and global bifurcations over the domain of parameters. A complete analysis of the averaged axisymmetrical cubic and quartic potentials is in progress.

1. Introduction

In a phase space made of the Cartesian coordinates $\boldsymbol{x} = (x, y, z)$ and their conjugate momenta $\boldsymbol{X} = (X, Y, Z)$, we are interested in the dynamics defined by the Hamiltonian function

$$\mathcal{H} = \mathcal{H}_2 + \sum_{n\geq 1} \epsilon^n \mathcal{H}_{n+2}(\boldsymbol{x}, \boldsymbol{X}), \tag{1}$$

where $\mathcal{H}_2$ is given by

$$\mathcal{H}_2 = \frac{1}{2}(X^2 + Y^2 + Z^2) + \frac{1}{2}(\omega_1^2 x^2 + \omega_2^2 y^2 + \omega_3^2 z^2), \tag{2}$$

i.e. non linearly coupled oscillators, and ϵ is a small parameter. This is, for instance, the situation corresponding to a Hamiltonian system in the neighbourhood of an equilibrium point[37] with $\mathcal{H}_n$ homogeneous polynomials in $\boldsymbol{x}$

and $\boldsymbol{X}$, where resonant and non resonant situations must be distinguished. For potentials modelling galactic dynamics[26], to which we refer below, the frequencies ω_i more often used are in the resonance ratios 1-1-1 and 1-1-3 or near to them[36]; in this paper we will consider only the first case. About the dynamics of elliptical galaxies, a first aspect to mention is that, although ellipsoidal coordinates are the natural ones to use versus power series expansions in Cartesian coordinates, nevertheless according to de Zeeuw[13], a full treatment of perturbed motion in ellipsoidal variables is not available; that is the reason why we consider Cartesian coordinates. Secondly, as Contopoulos[5] points out, although integrable models have been used on various occasions in galactic dynamics (for instance Stäckel potentials[14]), however most of these models do not treat some resonant phenomena, which are important in realistic models; this explains why resonant integrable cases are receiving now special attention. Finally, studies in galactic dynamics distinguish the rotating or non-rotating character of the galaxy; in this paper we will consider an example of the second class. Still, even in these rather simple models, many questions remain open like: axial symmetry versus triaxial shape of the galaxies as the basic model to consider; or what is the dynamical relation between the core of the galaxy with the halo recently observed in those galaxies. For a very comprehensive report about elliptical galaxies we refer to de Zeeuw and Franx[15].

As we said above, our unperturbed Hamiltonian function will be

$$\mathcal{H}_2 = \frac{1}{2}(X^2 + Y^2 + Z^2) + \frac{1}{2}\omega^2(x^2 + y^2 + z^2), \tag{3}$$

which is called *3D elliptic oscillator* or isotropic harmonic oscillator. Besides the energy integral $\mathcal{H}_2$ and the integral of the angular momentum vector $\boldsymbol{G}$, the dynamical system defined by Eq. (3) has another set of integrals called the Runge-type tensor which components have been defined in the form:

$$A_{i,j} = \frac{1}{2}(X_i X_j + \omega^2 x_i x_j), \quad 1 \leq i,j \leq 3, \tag{4}$$

where $\boldsymbol{x} = (x_1, x_2, x_3)$ and $\boldsymbol{X} = (X_1, X_2, X_3)$. For relations among the integrals as well as the Runge-Lenz vector of the isotropic harmonic oscillator see [18].

Joined with the Kepler problem, the isotropic harmonic oscillator is probably the most studied[18,20,27] dynamical systems. In those studies, different types of variables have been used. Thus, the search for integrable cases have been made in Cartesian variables[23], meanwhile with a differential geometric approach it is convenient to introduce complex variables[24]. But probably, since the pioneering work of Hénon and Heiles[19], most of the work with perturbed systems has been done in cylindrical variables $(\rho, z, \lambda, P, Z, \Lambda)$. In these variables the Hamiltonian Eq. (1), in perturbed 1-1-1 resonance, is given by

$$\mathcal{H} = \mathcal{H}_2 - \mathcal{V} = \frac{1}{2}\left(P^2 + Z^2\right) + \frac{\omega^2}{2}\left(\rho^2 + z^2\right) + \frac{\Lambda^2}{2\rho^2} - \mathcal{V}, \tag{5}$$

where $\mathcal{V}$ stands for the perturbation function. Usually those authors consider not only an axial symmetry $\mathcal{V} \equiv \mathcal{V}(\rho, z)$, but assume also $\Lambda = 0$, i.e. motions in a meridian plane which correspond to the 1-1 resonance case. Here, the effect of $\Lambda \neq 0$ in the dynamics is precisely one of the aspects we want to study.

We propose another set of canonical variables Eq. (9) which we call Nodal-Lissajous variables. The Lissajous transformation, introduced by Deprit[11] is a special set of action-angle variables designed for dealing with perturbations of the degenerate integrable Hamiltonian system made of two linear uncoupled oscillators in the plane with equal frequencies, i.e. in a 1-1 resonance. Here we extend the canonical Lissajous transformation to 3D of freedom, to treat with the case of a 1-1-1 resonance simply by composing of the 2D Lissajous map with the rotation from the fixed Cartesian to the nodal frame and the corresonding transformation of the momenta. A main feature of this transformation is the expresion of the unperturbed Hamiltonian Eq. (3) as one of the momentum coordinates; thus, the solution of the equation that has to be solved in each step of the normalization is simply an average with respect to the fast variable, the elliptic anomaly ℓ. For details on how the Lissajous normalization works to accomplish that see Deprit and Elipe[12].

Although we can use them dealing with perturbed elliptic oscillators of the general form Eq. (1), with $\mathcal{H}_2$ given by Eq. (3), we will restrict to potentials of polynomial form in (x, y, z), which have been widely studied not only in galactic dynamics but also in molecular spectroscopy and several other branches of non-linear dynamics. Several studies with perturbing polynomial potentials $\mathcal{V}(x^2, y^2, z^2)$ in the 1-1-1 resonance have been made, for instance, formulating the problem in action-angles variables and applying the averaging method[13], or by the direct search for periodic orbits[26,35]. In that respect, as Cleary[4] explains: "The modeling of galactic features, such as gas disks, and the construction of self-consistent galatic potentials rely on the completeness of the collection of known orbits families, each of which is generated by a single periodic orbit at each energy." For Hamiltonian systems, as it is well known[3,32], the existence of periodic orbits at low positive energies can be deduced from the existence of nondegenerate critical points of an averaged Hamiltonian on an associated reduced space. As we said before we will follow this approach; for details on the process of reduction by normalization we refer for instance to the recent publications of Marsden[25] and Meyer[29,30].

We know that for 3D isotropic harmonic oscillators the non-zero level sets of the energy (or $\mathcal{H}_2$) are S^5 spheres, and that the first reduction[32] will give us the reduced phase space which is the complex projective space $\mathbb{C}P^2$. Moreover we will assume the axial symmetry mentioned earlier. Then, if there is a second reduction due to an axial symmetry, the reduced space will be a sphere S^2 because the level set of the third component of the angular momentum projecting into $\mathbb{C}P^2$ is a 3-dimensional space S^3, and eliminating the angular variable is equivalent[21] to the Hopf fibration $S^3 \to S^2$. Precisely, the Nodal-Lissajous transformation we propose here (Section 2), with the third component and the modulus of the angular momentum among the conjugate momenta, helps in carrying out the reduction process.

Then, in Section 3, recognizing the fact that the double reduced orbit space is S^2, we still need to introduce coordinates on S^2 which will allow to search for the relative equilibria and its local and global bifurcations over the domain of parameters. In this way we enlarge the studies corresponding to planar solutions in the fundamental plane and meridian planes, considering now orbits with any inclination.

In Section 4, illustrating the use of the proposed transformation, we apply it to the axis symmetric quartic perturbation $(x^2+y^2)z^2$, studied in the meridian plane by Contopoulos and others, but considering here trajectories which are not in the symmetry planes of the potential. We find a critical inclination corresponding to a saddle-center bifurcation of elliptic orbits.

In the Appendix we gather some results concerning cubic and quartic polynomial potential, showing some of the very few known integrable cases. A full study of the parametric bifurcations of the families of axisymmetrical cubic and quartic potentials is in progress.

2. A Nodal-Lissajous transformation

We propose to build up a transformation made of the composition of two canonical transformations: (1), the nodal transformation, which is the rotation from the space frame to the nodal frame; and (2), a Lissajous transformation[11] in the nodal frame. In detail, the process for obtaining this transformation is the following. We assume $\boldsymbol{x}\times\boldsymbol{X}\neq 0$. We first do an homogeneous canonical transformation[10] made of a rotation of coordinates and momenta $\mathcal{N}$: $(x',y',h,X',Y',H)\rightarrow(x,y,z,X,Y,Z)$

$$\begin{aligned} x &= x'\cos h - y'\sin h\cos I, & X &= X'\cos h - Y'\sin h\cos I, \\ y &= x'\sin h + y'\cos h\cos I, & Y &= X'\sin h + Y'\cos h\cos I, \\ z &= y'\sin I, & Z &= Y'\sin I, \end{aligned} \tag{6}$$

where the angle I is determined by $\cos I = H/(x'Y'-y'X')$, $0<I<\pi$. The second transformation $\mathcal{L}_\omega : (\ell,g,L,G)\rightarrow(x',y',X',Y';\omega)$ is a Lissajous transformation

$$\begin{aligned} x' &= s\cos(g+\ell) - d\cos(g-\ell), & X' &= -\omega\,(s\sin(g+\ell)+d\sin(g-\ell)), \\ y' &= s\sin(g+\ell) - d\sin(g-\ell), & Y' &= \omega\,(s\cos(g+\ell)+d\cos(g-\ell)), \end{aligned} \tag{7}$$

where ω is a positive parameter and s,d,η are state functions of the momenta, given by

$$s=\sqrt{\frac{L}{2\omega}(1+\eta)},\qquad d=\sqrt{\frac{L}{2\omega}(1-\eta)},\qquad \eta=\frac{G}{L}. \tag{8}$$

Thus, the composition of Eq. (6) and Eq. (7) gives the transformation Eq. (9):

$$\mathcal{N}\circ\mathcal{L}_\omega : (\ell,g,h,L,G,H)\longrightarrow(x,y,z,X,Y,Z)$$

$$
\begin{aligned}
x &= \frac{1}{2}\left[(1+c)s\cos(\ell+g+h)+(1-c)s\cos(\ell+g-h)\right.\\
&\qquad \left.-(1-c)d\cos(\ell-g+h)-(1+c)d\cos(\ell-g-h)\right],\\
y &= \frac{1}{2}\left[(1+c)s\sin(\ell+g+h)-(1-c)s\sin(\ell+g-h)\right.\\
&\qquad \left.-(1-c)d\sin(\ell-g+h)+(1+c)d\sin(\ell-g-h)\right],\\
z &= \sqrt{1-c^2}\left[s\sin(\ell+g)+d\sin(\ell-g)\right],\\
X &= \frac{\omega}{2}\left[-(1+c)s\sin(\ell+g+h)-(1-c)s\sin(\ell+g-h)\right.\\
&\qquad \left.+(1-c)d\sin(\ell-g+h)+(1+c)d\sin(\ell-g-h)\right],\\
Y &= \frac{\omega}{2}\left[(1+c)s\cos(\ell+g+h)-(1-c)s\cos(\ell+g-h)\right.\\
&\qquad \left.-(1-c)d\cos(\ell-g+h)+(1+c)d\cos(\ell-g-h)\right],\\
Z &= \omega\sqrt{1-c^2}\left[s\cos(\ell+g)+d\cos(\ell-g)\right],
\end{aligned}
\tag{9}
$$

where c, σ are state functions of the momenta, given by

$$c = \frac{H}{G}, \quad \sigma = \eta c = \frac{H}{L}. \tag{10}$$

The Nodal-Lissajous transformation is defined in the domain $\Delta \subset \mathbf{R}^6$

$$\Delta = [0,2\pi)\times[0,2\pi)\times[0,2\pi)\times\{L>0\}\times\{0<G<L\}\times\{0\leq|H|<G\}.$$

Besides, as Deprit explains[11], this transformation is not a Mathieu transformation, because we have the differential identity

$$X dx + Y dy + Z dz - L d\ell - G dg - H dh = d(\tfrac{1}{2}\sqrt{L^2-G^2}\sin 2\ell).$$

In order to see the geometric meaning of the variables (ℓ, g, h, L, G, H), we apply them to the 3D isotropic harmonic oscillator Eq. (3), which in Nodal-Lissajous variables is given by

$$\mathcal{H}_2 = \omega L. \tag{11}$$

Thus, ℓ called elliptic anomaly, is a linear function of time describing the position of the particle in the trajectory which is an ellipse, with the center at the origin of the reference frame, fixed in space. All the other variables (g, h, L, G, H) are constants, defining this ellipse. First h represents the angle of the ascending node of the orbital plane. The inclination of the orbital plane is given by $\cos I = H/G$. We easily compute (a, b), the semi-axes of the ellipse from $a+b=$

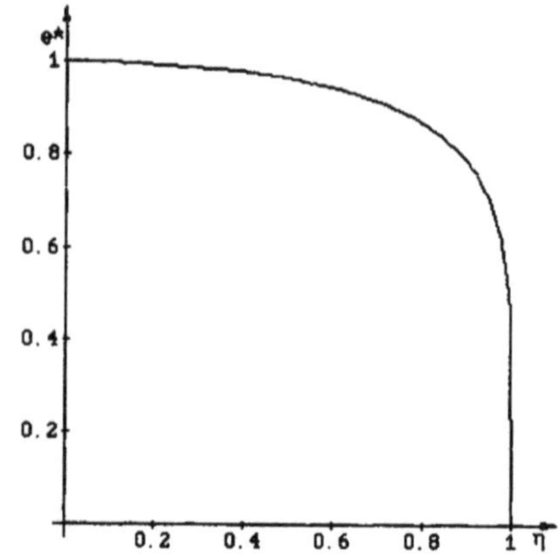

Figure 1: The eccentricity $e^* \equiv e^*(\eta)$ of the ellipse as function of the momenta L, G. Notice the nonsteady variation which keeps very eccentric ellipses until a sudden drop, when η is rather near to one.

$2s$, $a-b=2d$, where s and d are given by Eq. (8); the angle g gives the position of the semi-minor axis reckoning from the ascending node. Then the eccentricity of the ellipse, denoted e^*, is given by (see Fig. 1)

$$e^* = \frac{2\sqrt{sd}}{s+d} = \sqrt{\frac{2\sqrt{1-\eta^2}}{1+\sqrt{1-\eta^2}}}. \tag{12}$$

Finally, the polar coordinates (r, θ) in the orbital plane are given by

$$r^2 = a^2 \sin^2 \ell + b^2 \cos^2 \ell, \qquad \theta = g + f,$$

which tells that, in the unperturbed ellipse trajectory, ℓ differs in $\pi/2$ of the angle called eccentric anomaly (for details see Deprit[11], Fig. 1), and the angle f is given by $r \sin f = a \sin \ell$ and $r \cos f = b \cos \ell$.

3. Lissajous normalization and axial reduction

3.1. On the Lissajous normalization

In dealing with power series perturbations of isotropic harmonic oscillators, the process of Lissajous normalization by averaging has been explained in detail by Deprit[11] and Deprit and Elipe[12]. Moving from two to three dimensions using Nodal-Lissajous variables adds nothing to that, because functions of (h, H) are also in the kernel of the Lie derivative. We only need to remember here what the Lie derivative, defined by the flow $\mathcal{H}_2$ by Eq. (11) acomplishes:

$$\mathcal{L}_{\mathcal{H}_2}(F) = \{F, \mathcal{H}_2\} = \omega \frac{\partial F}{\partial \ell}. \tag{13}$$

Thus, when the perturbation is periodic in the elliptic anomaly, normalizing a perturbed elliptic oscillator is equals to averaging the dynamical system over ℓ. A full description on Lissajous normalization is contained in the above mentioned papers. For instance, the components Eq. (4) of the Runge-Lenz tensor $A_{i,j}$ belong to the kernel. Indeed, we can check immediately that by replacing Eq. (9) in Eq. (4). We get

$$\begin{aligned}
A_{11} &= \omega^2 \left[(d^2+s^2)(1+c^2)\right. \\
&\quad -2ds\,(1-c^2)\cos 2g + (d^2+s^2)(1-c^2)\cos 2h \\
&\quad \left. -ds\,(1-c)^2\cos(2h-2g) - (1+c)^2\, ds\cos(2h+2g)\right], \\
A_{22} &= \omega^2 \left[(d^2+s^2)(1+c^2)\right. \\
&\quad -2ds\,(1-c^2)\cos 2g - (d^2+s^2)(1-c^2)\cos 2h \\
&\quad \left. +ds\,(1-c)^2\cos(2h-2g) + (1+c)^2\, ds\cos(2h+2g)\right], \\
A_{33} &= \frac{\omega^2}{2}(1-c^2)(d^2+s^2+2ds\cos 2g), \\
A_{12} &= A_{21} = \frac{\omega^2}{4}\left[(1-c)^2\, ds\sin(2h-2g) + (d^2+s^2)(1-c^2)\sin 2h\right. \\
&\qquad \left. -(1+c)^2\, ds\sin(2h+2g)\right], \\
A_{13} &= A_{31} = \frac{\omega^2}{2}\sqrt{1-c^2}\,[ds(1-c)\sin(h-2g) - c\,(d^2+s^2)\sin h \\
&\qquad -ds(1+c)\sin(h+2g)], \\
A_{23} &= A_{32} = -\frac{\omega^2}{2}\sqrt{1-c^2}\,[ds(1-c)\cos(h-2g) - c\,(d^2+s^2)\cos h \\
&\qquad -ds(1+c)\cos(h+2g)];
\end{aligned}$$

from those expressions, as they are independent of ℓ, we conclude that they belong to the kernel of the Lie derivative Eq. (13).

Implementing the Lissajous normalization[12],

$$\Psi : (\ell, g, h, L, G, H) \longrightarrow (\ell', g', h', L', G', H')$$

by means of a Lie-Deprit transformation, we have to deal with the differential relation, called by others the homological equation,

$$\mathcal{L}(\mathcal{W}_n) + \mathcal{K}_n = \tilde{\mathcal{H}}_n. \tag{14}$$

Then, the normalization is carried out taking

$$\mathcal{K}_n = \frac{1}{2\pi}\int_0^{2\pi} \tilde{\mathcal{H}}_n\, d\ell, \qquad \mathcal{W}_n = \frac{1}{\omega}\int (\tilde{\mathcal{H}}_n - \mathcal{K}_n)\, d\ell, \tag{15}$$

as the expressions satisfying Eq. (14).

3.2. Axial reduction: Coordinates and differential system in the orbit space S^2

As we said in the indroduction, non-zero level sets of the energy (or $\mathcal{H}_2$) are S^5 spheres. Then, the first reduction will give us the reduced phase space that will be the complex projective space $\mathbb{C}P^2$. If there is a second reduction because of an axial symmetry, the reduced space will be a sphere S^2 because the level set of the third component of the angular momentum projecting into $\mathbb{C}P^2$ is a 3-dimensional space S^3, and eliminating the angular variable is equivalent to the Hopf fibration $S^3 \to S^2$.

Due to the structure of the isotropic harmonic oscillators there is still a discrete symmetry Z_2, the antipodal identification. As a consequence we propose the following set of coordinates for the orbit space

$$\begin{aligned} \bar{I}_1 &= \eta\sqrt{(1-\eta^2)(1-c^2)}\,\cos 2g, \\ \bar{I}_2 &= \eta\sqrt{(1-\eta^2)(1-c^2)}\,\sin 2g, \\ \bar{I}_3 &= \eta^2 - \tfrac{1}{2}(1+\sigma^2) \end{aligned} \tag{16}$$

which verify the equation of the sphere given by

$$\bar{I}_1^2 + \bar{I}_2^2 + \bar{I}_3^2 = \tfrac{1}{4}(1-\sigma^2)^2, \tag{17}$$

with the Poisson algebra

$$\{\bar{I}_i, \bar{I}_j\} = \epsilon_{ijk}\frac{4\eta}{L}\bar{I}_k,$$

where ϵ_{ijk} is the Levi-Civita symbol. The corresponding Liouville equations are

$$\dot{\bar{I}}_i = \{\bar{I}_i, \mathcal{K}\}. \tag{18}$$

Points $(\bar{I}_1, \bar{I}_2, \bar{I}_3,)$ on a sphere $S^2_{L,H}$ represent families of elliptic orbits. From Eq. (16) we have

$$\cos 2g = \frac{\bar{I}_1}{\sqrt{\bar{I}_1^2 + \bar{I}_2^2}}, \quad \sin 2g = \frac{\bar{I}_2}{\sqrt{\bar{I}_1^2 + \bar{I}_2^2}}, \quad \eta = \sqrt{\bar{I}_3 + \tfrac{1}{2}(1+\sigma^2)}, \tag{19}$$

from which we get the orientation of the ellipses in the orbital plane, their eccentricity and the inclination of that plane with respect to the $0xy$ plane. The north pole of $S^2_{L,H}$ corresponds to circular orbits with inclination $\cos I = H/L$. The south pole represents ellipses in the equatorial plane with eccentricity given by $e^* = \sqrt{2\sqrt{1-\sigma^2}/(1+\sqrt{1-\sigma^2})}$.

4. 3D Contopoulos galactic model for low energies and its normalization

In the quartic potential model described by (41), we have as a particular case

$$\mathcal{H} = \mathcal{H}_2 - \epsilon(x^2 + y^2)z^2. \tag{20}$$

The dynamical system defined by Eq. (20) belongs to the family introduced by Andrle[1] in galactic dynamics. Later it has been studied among others by Pullen and Edmonds[34], and more recently by Contopoulos and Kaufmann[8] dealing with escapes.

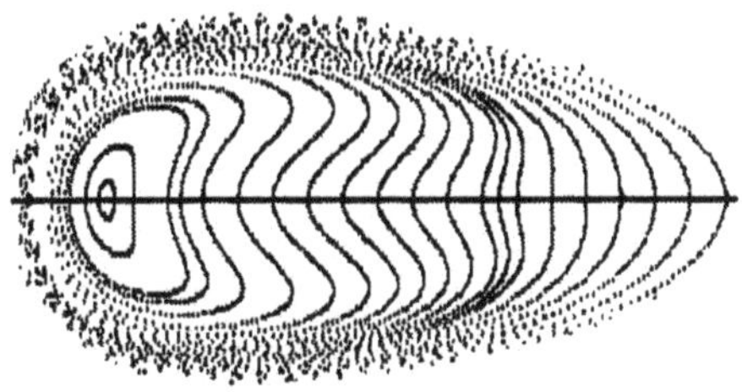

Figure 2: $\epsilon = 0.1$, $\mathcal{H} = 0.5$, $\Lambda = 0.4$. Before the bifurcation

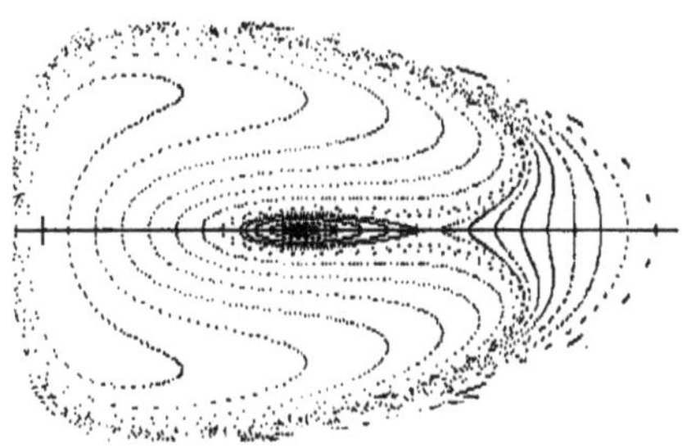

Figure 3: $\epsilon = 0.1$, $\mathcal{H} = 0.5$, $\Lambda = 0.2$. After the bifurcation

We see immediately that the origin is an equilibrium. Besides, as particular solutions, apart from the rectilinear ones corresponding to the three axes and circular motion about the axis of symmetry[9], two planar particular solutions appear: equatorial orbits corresponding to the unperturbed system, and planar polar orbits. Expressing the Hamiltonian Eq. (20) in cylindrical variables we get

$$\mathcal{H} = \frac{1}{2}\left(P^2 + Z^2\right) + \frac{\omega^2}{2}\left(\rho^2 + z^2\right) + \frac{\Lambda^2}{2\rho^2} - \epsilon\,\rho^2 z^2. \tag{21}$$

Thus, the angle λ is ignorable and the three-dimensional problem is reduced to two dimensions in a meridian plane that rotates with angular velocity $\dot{\lambda} = \Lambda/\rho^2$. We get a view of the role of the integral Λ considering the next two Poincaré surfaces of section, Figs. 2 and 3, which illustrate that a bifurcation has occurred. The main features of the orbit of a star is then characterized by its motion toward or away from the galatic center, and above or below the equatorial plane of the galaxy. Here, as we would like to get a better understanding of the evolution of the orbit versus the variation of Λ, in what follows we will use Nodal-Lissajous variables.

4.1. First order normalized Hamiltonian

When we apply our canonical transformation Eq. (9) to the Hamiltonian Eq. (20), we get

$$\begin{aligned}\mathcal{K} &= \omega L + \epsilon \mathcal{K}_1(L, G, H, \ell, g, -)\\ &= \omega L + \epsilon \left(\frac{L}{4\omega}\right)^2 [\mathcal{P}(L,G,H) + \mathcal{Q}(L,G,H,g) + \mathcal{R}(L,G,H,\ell,g)] + \mathcal{O}(\epsilon^2),\end{aligned} \tag{22}$$

where

$$\begin{aligned}&\mathcal{P}(L,G,H) = (1-c^2)\,(1+3c^2)\,(3-\eta^2)\,,\\ &\mathcal{Q}(L,G,H,g) = 12\,ec^2\,(1-c^2)\cos 2g - 3\,e^2(1-c^2)^2\,\cos 4g,\\ &\mathcal{R}(L,G,H,\ell,g) = 4\,[8\,e(1+3c^2)\cos 2\ell - 2\,e^2(1+3c^2)\cos 4\ell\\ &\qquad +8c^2(1+\eta)(2-\eta)\cos(2\ell+2g) - 4\,e\,(1-c^2)\,(1+\eta)\cos(2\ell+4g)\\ &\qquad +8c^2(1-\eta)(2+\eta)\cos(2\ell-2g) - 4\,e\,(1-c^2)\,(1-\eta)\cos(2\ell-4g)\\ &\qquad +4\,ec^2(1+\eta)\cos(4\ell+2g) + (1-c^2)\,(1+\eta)^2\cos(4\ell+4g)\\ &\qquad +4\,ec^2(1-\eta)\cos(4\ell-2g) + (1-c^2)\,(1-\eta)^2\cos(4\ell-4g)]\,,\end{aligned} \tag{23}$$

where, following Miller[31], we have introduced the shorthand $e = \sqrt{1-\eta^2}$. Then, the modulus of the Runge-Lenz vector[18,20], expressed in Lissajous variables, it is $B = L\sqrt{1-\eta^2} = Le$.

Implementing the Lissajous normalization, according to Eq. (15), we have at the first order

$$\mathcal{L}(\mathcal{W}_1) + \mathcal{K}_1 = \mathcal{H}_1.$$

From the form in which Eq. (22) is split, we get immediately as the new Hamiltonian

$$\mathcal{K} = \mathcal{K}_0 + \epsilon \left(\frac{L}{4\omega}\right)^2 (C_0 + C_2\cos 2g + C_4\cos 4g), \tag{24}$$

where

$$C_0 = (1-c^2)(1+3c^2)(2+e^2), \quad C_2 = 12c^2(1-c^2)e, \quad C_4 = -3(1-c^2)^2 e^2,$$

with the generating function given by

$$\begin{aligned}\mathcal{W}_1 = \frac{1}{\omega}\left(\frac{L}{4\omega}\right)^2 \Big[&16\,(1+3c^2)\,\sqrt{1-\eta^2}\sin 2\ell - 2\,(1+3c^2)\,(1-\eta^2)\sin 4\ell \\ &+16c^2(1+\eta)(2-\eta)\sin(2\ell+2g) - 8\,(1-c^2)\,(1+\eta)\,\sqrt{1-\eta^2}\sin(2\ell+4g) \\ &+16c^2(1-\eta)(2+\eta)\sin(2\ell-2g) - 8\,(1-c^2)\,(1-\eta)\,\sqrt{1-\eta^2}\sin(2\ell-4g) \\ &+4c^2(1+\eta)\sqrt{1-\eta^2}\sin(4\ell+2g) + (1-c^2)\,(1+\eta)^2\sin(4\ell+4g) \\ &+4c^2(1-\eta)\sqrt{1-\eta^2}\sin(4\ell-2g) + (1-c^2)\,(1-\eta)^2\sin(4\ell-4g)\Big]\,,\end{aligned} \tag{25}$$

and where primes have been dropped in order to simplify the expressions.

Among the equations of the averaged system we have two quadratures

$$\ell = \int \frac{\partial\mathcal{K}(g(t),L,G(t),H)}{\partial L}\,dt, \qquad h = \int \frac{\partial\mathcal{K}(g(t),L,G(t),H)}{\partial H}\,dt. \tag{26}$$

Therefore, cancelling out the constant factors of the Hamiltonian Eq. (24), and rescaling the time by $d\tau = (L/(4\omega))^2 dt$, the averaged differential system reduces to

$$\frac{dg}{d\tau} = \frac{\partial C_0}{\partial G} + \frac{\partial C_2}{\partial G}\cos 2g + \frac{\partial C_4}{\partial G}\cos 4g, \tag{27}$$

$$\frac{dG}{d\tau} = 2(C_2 + 4C_4\cos 2g)\sin 2g. \tag{28}$$

As soon as we have resolved this system, it still remains to compute two quadratures Eq. (26).

We will not study here the analytic solution of the averaged system. About the existence of what de Zeeuw[13] calls simple periodic orbits: elliptic orbits with both node and perigee stationaries, the answer is negative, or in other words, the averaged orbital plane has a precessional motion given by the second expression in Eq. (26). Thus, focusing on equations Eq. (27) and Eq. (28), what remains is to search for relative equilibria in the averaged orbital plane.

4.2. Relative equilibria, stability and bifurcations

As we have said in the introduction, the orbit space after the second reduction is a sphere S^2. Moreover, due to the structure of the isotropic harmonic oscillators there is still a discrete symmetry Z_2, the antipodal identification. As a consequence using the coordinates Eq. (16) for the reduced orbit space, the

Liouville equations Eq. (18) explicitly are

$$\begin{aligned}
\frac{d\bar{I}_1}{dt} &= \frac{\omega^2 L}{4\eta^3}\left[\alpha_1 + \alpha_2 \bar{I}_1 + \alpha_3 \bar{I}_1^2 + \alpha_4 \bar{I}_2^2\right]\bar{I}_2, \\
\frac{d\bar{I}_2}{dt} &= \frac{\omega^2 L}{4\eta^3}\left[\beta_1 + \beta_2 \bar{I}_1 - \alpha_2 \bar{I}_1^2 + 3\beta_3 \bar{I}_1^3 - \beta_3 \bar{I}_1 \bar{I}_2^2\right], \\
\frac{d\bar{I}_3}{dt} &= \frac{\omega^2 L}{4\eta^3}\left[\gamma_1 + \gamma_2 \bar{I}_1\right]\bar{I}_2,
\end{aligned} \tag{29}$$

with

$$\begin{aligned}
\alpha_1 &= \eta^2\,[3(1+s^2)s^6\eta^6 - 3(3-2s^2)s^4\eta^4 + (3s^8+9s^6+3s^4-6s^2+4)\eta^2 \\
&\quad -3(s^6+18s^4-30s^2+12)], \\
\alpha_2 &= 6s(1-s^2)\eta\,(2(3-2\eta^2)s^2 - 3(1-\eta^2)), \\
\alpha_3 &= -3\eta s(1-s^2)\,(2(2-\eta^2)s^2 - 3(1-\eta^2)), \\
\alpha_4 &= 3(1-\eta^2)s^4\,(2(2-\eta^2)s^2 - 3(1-\eta^2)), \\
\beta_1 &= 6\eta^3(1-\eta^2)(1-s^2)s^3(1-\eta^2+\eta^2 s^2), \\
\beta_2 &= \eta^2\,[3s^6(1+s^2)\eta^6 - 3s^4(3-2s^2)\eta^4 + (3s^8+9s^6-3s^4+6s^2-4)\eta^2 \\
&\quad -3(s^6+18s^4-30s^2+12)], \\
\beta_3 &= (1-\eta^2)s^4\,(2s^2(2-\eta^2) - 3(1-\eta^2)), \\
\gamma_1 &= 12\eta^3(1-\eta^2)^2 s^5(1-s^2), \\
\gamma_2 &= -12\eta^2(1-\eta^2)^3 s^8.
\end{aligned} \tag{30}$$

Here we have made use of the common notation $s^2 = 1 - c^2$, which simplifies some expressions containing square roots. The reader should note that we introduced the shorthand s in the definition of the Nodal-Lissajous transformation, with a different meaning Eq. (8), in accordance with the notation of Deprit[11]. Notice that although $\sigma = H/L$ is not explicit in Eq. (29), it is present through s^2 and η^2.

4.2.1. The south pole as relative equilibrium and its stability

Searching for relative equilibria in the reduced space, we start with the poles. After that, it is simpler look for other roots using Lissajous variables, going back again to the equations Eq. (27) and Eq. (28).

By inspection of the differential system Eq. (29) we see that the south pole, $\bar{I}_1 = \bar{I}_2 = 0$ (and $s = 0$), is an equilibrium. Considering the constraint Eq. (17), at the south pole we have $\delta\bar{I}_3 = 0$. Thus, the variational equations yield the following expressions

$$\frac{d}{dt}\delta\bar{I}_1 = -\frac{L\omega^2\left(3-\sigma^2\right)}{\sigma}\delta\bar{I}_2, \qquad \frac{d}{dt}\delta\bar{I}_2 = \frac{L\omega^2\left(3-\sigma^2\right)}{\sigma}\delta\bar{I}_1, \tag{31}$$

from which we find that the south pole is always stable.

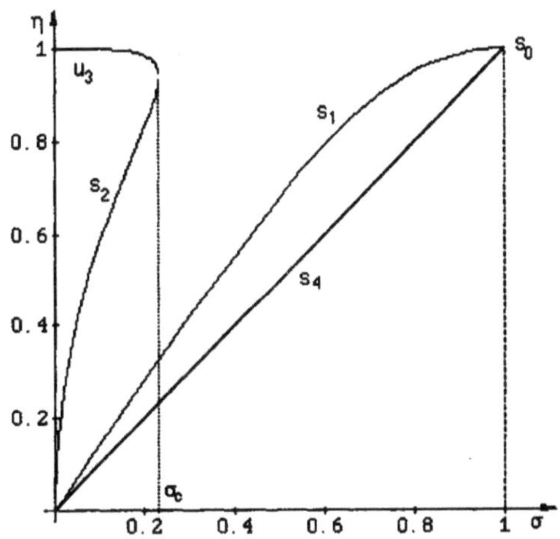

Figure 4: Values corresponding to the equilibria $\eta \equiv \eta(\sigma)$. For any value of $\sigma < 1$ we have two stable equilibria in the orbit space with s_1 corresponding to stable elliptic orbits with periaxis in the line of the nodes and s_4 representing stable elliptic equatorial orbits at the south pole of the orbit space, $S^2_{L,H}$. For $\sigma = \sigma_c$ a saddle-center bifurcation takes place, and for $\sigma < \sigma_c$ we have u_3 corresponding to unstable family of elliptic orbits and s_2 stable family of elliptic orbits, both with periaxis in the line of the nodes.

4.2.2. Orbits with stationary periaxis: Critical inclination

In order to find the other possible equilibria, setting (28) equal to zero, we first get $\sin 2g = 0$, and $\cos 2g = -C_2/(4C_4)$. Then, setting (27) equal to zero and replacing the previous values in it, we get three equations to study

$$g = 0, \pi, \qquad \frac{\partial C_0}{\partial G} + \frac{\partial C_2}{\partial G} + \frac{\partial C_4}{\partial G} = 0, \tag{32}$$

$$g = \frac{\pi}{2}, \frac{3\pi}{2}, \qquad \frac{\partial C_0}{\partial G} - \frac{\partial C_2}{\partial G} + \frac{\partial C_4}{\partial G} = 0 \tag{33}$$

and

$$\frac{\partial C_0}{\partial G} - \frac{\partial C_2}{\partial G}\frac{C_2}{4C_4} + \frac{\partial C_4}{\partial G}\left[2\left(\frac{C_2}{4C_4}\right)^2 - 1\right] = 0. \tag{34}$$

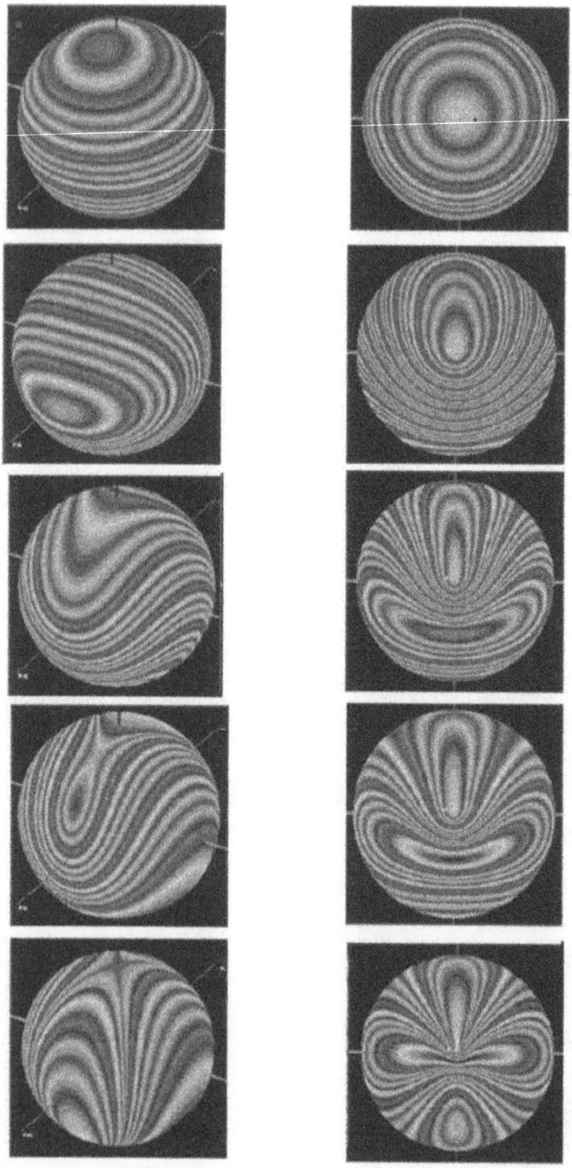

Figure 5: From the top, left column: north pole view for $\sigma = 0.98$, $\sigma = 0.6$ and for $\sigma = 0.27$, $\sigma = 0.22$ just before and after the cusp bifurcation. The last one corresponds to $\sigma = 0.05$. Right column: the corresponding south pole views. (These figures were produced by J.L. Rubio and A. Viartola using the package ESFERAS; for details see [33]).

Let us study each one of these cases. From Eq. (32), looking for roots in η, after some computations we may simplify the problem to the following equation

$$e^4 - 3e^3 + 3e^2 - (1+3\sigma^2)e + 3\sigma^2(1-\sigma^2) = 0. \tag{35}$$

Then, in this quartic we find one or three roots within the interval $0 < \sigma < 1$; we find a double root for $\sigma_0 = 0.2318$. Thus, we have only one root s_1 in the interval $\sigma_0 < \sigma < 1$. For $\sigma_0 = 0.2318$ we have a bifurcation point, with corresponding values for the inclination $I = 75.7$ degrees, and eccentricity $e^* = 0.715124$. Finally for $0 < \sigma < \sigma_0$ we have three roots s_1, s_2 and u_3. We represent the roots in the (σ, η) plane in Figure 4.

From the variational equations

$$\dot{\delta g} = \frac{\partial^2 \mathcal{H}}{\partial g\, \partial G}\delta g + \frac{\partial^2 \mathcal{H}}{\partial G^2}\delta G, \quad \dot{\delta G} = -\frac{\partial^2 \mathcal{H}}{\partial g^2}\delta g - \frac{\partial^2 \mathcal{H}}{\partial g\, \partial G}\delta G,$$

and the corresponding characteristic equation, after some numerical computations we get the stability analysis given in Figure 5.

Finally for the two other cases, when we consider the Eqs. (33) and (34), we find solutions which are out of the domain of our problem.

5. Conclusions and future work

This paper deals with resonant non linearly coupled oscillators with three equal frequencies and their normalization. A composition of nodal and Lissajous canonical variables is proposed, in order to implement efficiently high order Lie-Deprit normalization, needed for breaking the degeneracy which appears at first order in some averaged systems. One feature of the transformation is the use of the third component and the modulus of the angular momentum among the conjugate momenta, which simplifies the study of the effects of the perturbations on the unperturbed ellipses; the variations of the inclination and eccentricity are obtained immediately from Eq. (10) and Eq. (12) respectively.

We consider perturbing potentials with axial symmetry. In that case, from the Lissajous normalization and the axial symmetry, the double reduced orbit space is S^2. We introduce coordinates which allow to look for the relative equilibria and its local and global bifurcations over the domain of parameters. In this way we enlarge the studies corresponding to planar solutions in the equatorial and meridian planes, considering now orbits in any inclination. In particular we apply the transformation to the quartic perturbation $(x^2+y^2)z^2$, studied in the meridian plane by Contopuolos and others. We find a critical inclination corresponding to a saddle-center bifurcation of a family of elliptic orbits.

At present, the study of the generic biparametric cubic potentials Eq. (37) as well as the three parametric quartic model Eq. (41) is in progress, including their relation with the polar case[16]. As far as we know only first order studies have been done; we plan to extend the normalization up to higher orders. One of the aims is to consider the relation between degenerate cases of the averaged system with the integrable cases of the full problem, in particular with the Stäckel integrable potentials. Another related but more ambitious question should be to compare the averaged trajectories with the dust lanes[28,38] observed in several elliptical galaxies, and its relation to the question of axial symmetry versus triaxial shape of the galaxies.

Acknowledgments

We would like to show our gratitude to Prof. Elipe and Drs. Osácar and Palacián for their help and suggestions which have clarified the content of the paper. During his visit to Universidad de Zaragoza, Prof. Broucke showed interest in this work and we owe him the Poincaré surfaces of section.

The research of one of the authors (S. F.) was partially supported by Comisión Interministerial Científica y Técnica of Spain (CICYT ESP91-0919).

Appendix: On cubic and quartic homogeneous potentials

Although the application of Nodal-Lissajous variables are perturbed elliptic oscillators which Hamiltonians have the general form Eq. (1), with $\mathcal{H}_2$ given by Eq. (3), we will restrict to potentials of of polynomial form in (x, y, z). More concretely, we will refer to cubic and quartic homogenous polynomials in the coordinates, focusing in particular in the axisymmetric ones.

From the general form Eq. (1), the cubic family function of (x, y, z) will be represented by

$$\begin{aligned}\mathcal{H}_3 = -\frac{1}{2}\,(&b_1x^3 + b_2y^3 + b_3z^3 + b_4xy^2 + b_5xz^2\\ &+b_6yx^2 + b_7yz^2 + b_8zx^2 + b_9zy^2 + b_{10}xyz)\,.\end{aligned} \tag{36}$$

The case $b_1 = b_2 = b_3 = b_4 = b_6 = b_8 = b_9 = 0$, i.e.,

$$\mathcal{H}_C = -\frac{1}{2}\,(b_5x + b_7y)\,z^2,$$

has been studied by Contopoulos and Barbanis[7]. This model has a second integral

$$\Phi = \frac{1}{2}\left[(b_7X - b_5Y)^2 + (b_7x - b_5y)^2\right].$$

Cubic potentials with an axial symmetry in respect of the third component of the angular momentum are of the form

$$\mathcal{H}_3 = -\left[\alpha z^3 + \beta(x^2 + y^2)z\right]. \tag{37}$$

Indeed, we get that conclusion easily writing down Eq. (36) in cylindrical coordinates, and making zeros the coefficients of terms functions of λ. Hénon-Heiles

Hamiltonian is a member of that family, taking $\alpha = 1/3$, $\beta = -1$, in a meridian plane, $\Lambda = 0$.

With respect to quartic potentials the general expression will be

$$\mathcal{H}_4 = a_1 x^4 + a_2 x^3 y + a_3 x^3 z + a_4 x^2 y^2 + a_5 x^2 z^2 + a_6 x^2 yz + a_7 x^3 y + a_8 xy^2 z$$
$$+ a_9 xyz^2 + a_{10} xz^3 + a_{11} y^4 + a_{12} y^3 z + a_{13} y^2 z^2 + a_{14} yz^3 + a_{15} z^4. \tag{38}$$

One of the few known integrable cases of $\mathcal{H}$ is

$$\mathcal{H}_F = \frac{1}{2}(x^2 + 2y^2 + z^2)^2 - (xz - y^2)^2, \tag{39}$$

which has been found by Fordy *et al.*[17].

The subfamily of $\mathcal{H}$ given by

$$\mathcal{H}_4 = -\frac{1}{2}\left(a_1 x^4 + a_{11} y^4 + a_{15} z^4 + a_4 x^2 y^2 + a_5 x^2 z^2 + a_{13} y^2 z^2\right), \tag{40}$$

defines the type of potentials that models triaxial galaxies in the central region of a few core radii[22]. Again, by expressing Eq. (40) in cylindrical coordinates, quartic potentials with axial symmetry with respect to the third component of the angular momentum are given by the three parameter family

$$\mathcal{H}_4 = -\frac{1}{2}\left[\alpha(x^2 + y^2)^2 + \beta z^4 + \gamma(x^2 + y^2)z^2\right]. \tag{41}$$

Let us note that assuming $\Lambda = 0$, several cases of this family Eq. (41) have been studied. For instance $\mathcal{H}_4 = (\rho^2 + az^2)^2$ was considered by Caranicolas and Barbanis[2]; and the case $\alpha = \beta = 0$, by Contopoulos[6]. In this paper we have focused also on this case but with $\Lambda \neq 0$.

Besides the trivial case $\beta = \alpha$, $\gamma = 2\alpha$, which corresponds to a central potential and has as a third integral the modulus of the angular momentum, there is also an integral[23] when $\beta = \alpha$, $\gamma = 6\alpha$, namely

$$I = \frac{1}{x^2 + y^2}\left\{(xX + yY)Z + 2\left[\omega^2 + 2\alpha(x^2 + y^2 + z^2)\right](x^2 + y^2)z\right\}^2$$
$$+ (xY - yX)^2\left(\frac{1}{x^2 + y^2}Z^2 + 8\alpha z^2\right). \tag{42}$$

References

1. Andrle, P.: 1966, A third integral of motion in a system with a potential of the fourth degree, *Bull. Astron. Inst. of Czechosl.* **17**, 169–175.
2. Caranicolas, N.D. and Barbanis, B.: 1982, Periodic orbits in nearly axisymmetric stellar systems, *Astron & Astrophys* **114**, 360–366.
3. Churchill, R.C., Kummer, M. and Rod, D.L.: 1983, On averaging, reduction, and symmetry in Hamiltonian systems, *J. Diff. Equat.* **49**, 359–414.

4. Cleary, P.W.: 1989, Periodic orbits in the Schwarzschild galactic potential, *Ap. J.* **337**, 108–124.
5. Contopoulos, G.: 1988, Resonant integrable galactic models, in *Integrability in Dynamical Systems* (Eds. Buchler, J.R. *et al.*), Ann. N Y Acad. Sci. **536**, 1–14.
6. Contopoulos, G.: 1990, Asymptotic curves and escapes in Hamiltonian systems, *Astron & Astrophys* **231**, 41–55. See also Caranicolas, N.D.: 1994, 1-1-1 resonant periodic orbits in 3-dimensional galactic-type Hamiltonians, *Astron & Astrophys* **282** 34.
7. Contopoulos, G. and Barbanis, B.: 1985, Resonant systems with three degrees of freedom, *Astron & Astrophys* **153**, 44–54.
8. Contopoulos, G. and Kaufmann, D.: 1992, Types of escapes in a simple Hamiltonian system, *Astron & Astrophys* **253**, 379.
9. Davoust, E.: 1983, Periodic orbits in elliptical galaxies, *Astron & Astrophys* **125**, 101–108.
10. Deprit, A.: 1981, The elimination of the parallax in satellite theory, *Celes. Mech.* **24**, 111-153.
11. Deprit, A.: 1991, The Lissajous transformation. I: Basics, *Celes. Mech. & Dynam. Astron.* **51**, 201-226.
12. Deprit, A. and Elipe A.: 1991, The Lissajous transformation. II: Normalization, *Celes. Mech & Dynam. Astron.* **51**, 227-250.
13. de Zeeuw, T.: 1985, Motion in the core of a triaxial potential, *M.N.R.A.S.* **215**, 731–760.
14. de Zeeuw, T.: 1987, Dynamical models for axisymmetric and triaxial galaxies, in *IAU Symp. 127: Structure and Dynamics of Elliptical Galaxies*, ed. T. de Zeeuw, pp. 271–290, Reidel, Dordrecht.
15. de Zeeuw, T. and Franx, M.: 1991, Structure and dynamics of elliptical galaxies, *Annu. Rev. Astron. Astrophys.* **29**, 239–74.
16. Ferrer, S., and Ibort, A.: 1994, in preparation
17. Fordy, A., Wojciechowski S. and Marshall, I.: 1986, A family of integrable quartic potentials related to symmetric spaces, *Phys. Lett* **113A**, 395–400.
18. Fradkin, D. M.: 1965, Three-dimensional isotropic harmonic oscillator and SU_3, *Am. J. Phys. A* **33**, 207–211.; Fradkin D. M.: 1967, Existence of the dynamic symmetries O_4 and SU_3 for all classical central potential problems, *Prog. Theor. Phys.* **37**, 798–812; Jauch, J.M. and Hill, E.L.: 1940, On the problem of degeneracy in quantum mechanics, *Phys. Rev.* **57**, 641–45; Mukunda, N.: 1967, Dynamical symmetries and classical mechanics, *Phys. Rev.* **155**, 1383–86.
19. Hénon, M. and Heiles, C.: 1964, The applicability of the third integral of motion; some numerical experiments, *Astron. J.* **69**, 73–79.
20. Holas, A. and March N. H,: 1990, A generalization of the Runge-Lenz constant of classical motion in a central potential, *J. Phys. A: Math. Gen.* **23**, 735–749.
21. Ibort A.: 1994, Personal communication.
22. Lake, G. and Norman, C.: 1983, Stellar and gaseous dynamics of triaxial

galaxies, *Ap. J.* **270**, 51–70.
23. Lakshmanan, M. and Sahadevan, R.: 1993, Painlevé analysis, Lie symmetries, and integrability of coupled nonlinear oscillators of polynomial type, *Phys. Rep.* **224**, pp. 93.
24. Marmo, G., Saletan, E.J., Simoni, A. and Vitale, B.: 1985, *Dynamical Systems, A Differential Geometric Approach to Symmetry and Reduction*, (John Wiley & Sons, Chichester)
25. Marsden, J. E.: 1992, *Lectures on Mechanics*, London Math. Soc. Lecture Note Series **174**, (Cambridge Univ. Press, Cambridge).
26. Martinet, L. and de Zeeuw, T.: 1988, Orbital stability in rotating triaxial stellar systems, *Astron & Astrophys* **206**, 269–278.
27. Martínez-y-Romero, R.P., Núñez-Yépez, H.N. and Salas-Brito A.L.: 1992, Closed orbits and constants of motion in classical mechanics, *Eur. J. Phys.* **13**, 26–31.
28. Merritt, D. and de Zeeuw, T.: 1983, Orbital configurations for gas in elliptical galaxies, *Ap. J.* **267** L19–L23.
29. Meyer, K. R.: 1994, Perturbation analysis of nonlinear systems, in *Computer Algebra and Differential Equations*, Ed. E. Tournier, London Math. Soc. Lecture Note Series **193**, pp. 103–139 (Cambridge Univ Press, Cambridge).
30. Meyer, K. R. and Hall, G.R.: 1992, *Introduction to Hamiltonian Dynamical Systems*, Applied Math. Sciences **90**, Springer-Verlag, New York.
31. Miller, B. R.: 1991, The Lissajous transformation. III: Parametric bifurcations, *Celes. Mech & Dynam. Astron.* **51**, 251-270.
32. Moser, J.: 1970, Regularization of Kepler's problem nd the averaging method on a manifold, *Comm. Pure and Appl. Math.* **23**, 609–636.
33. Pascua, P., Rubio J. L., Viartola, A. and Ferrer, S.: 1995, Visualizing relative equilibria and bifurcations by painting Hamiltonians on personal computers, to appear in *Int. J. of Bifurcations and Chaos.*
34. Pullen, R.A. and Edmonds, A.R.: 1981, Comparison of classical and quantal spectra for a totally bound potential, *J. Phys. A: Math. Gen.* **14**, L477–L484.
35. Robe, H.: 1985, Periodic orbits in a triaxial galaxy, *Astron & Astrophys* **142**, 351–354; Periodic orbits in a triaxial galaxy. III. Their stability, idem, 1987, **182**, 202–206.
36. Saaf, A.F.: 1968, A formal third integral of motion in a nearly spherical stellar system, *Ap. J.* **154**, 483–498.
37. Verhulst, F.: 1984, Invariant manifolds and periodic solutions of three degrees of freedom Hamiltonian systems, in *Trends and Applications of Pure Mathematics to Mechanics*, ed. P.G. Ciarlet and M. Roseau, Lecture Notes in Physics **195**, pp. 413-422, Springer-Verlag, Berlin
38. Tohline, J.E., Simonson, G.F. and Caldwell, N.: 1982, Using gaseous disks to probe the geometric structure of elliptical galaxies, *Ap. J.* **252**, 92–101.

BILLIARDS THAT SHARE A TRIANGULAR CAUSTIC

E. GUTKIN
Department of Mathematics, USC, 1042 West 36 Place
Los Angeles, CA, 90089-1113, USA
E-mail: egutkin@math.usc.edu

and

O. KNILL
Division of Physics, Mathematics and Astronomy, Caltech, MS 253-37
Pasadena, CA, 91125, USA
E-mail: knill@cco.caltech.edu

ABSTRACT

We consider a one-parameter family of billiard tables T_L which have a common caustic, the equilateral triangle γ. The billiard tables T_L are constructed geometrically by the string construction, where the length L of the string is the parameter. We study the family of circle homeomorphisms f_L obtained by restricting the billiard map to the canonical invariant circle Γ_L belonging to the caustic and the rotation function $\rho(L) = \rho(f_L)$. We show that the graph of ρ is a devil's staircase. We analyse the passage of a Birkhoff periodic orbits through the caustic as the parameter changes.

1 A Family of Billiard Maps

Let γ be the equilateral triangle with length 1. We construct tables T obtained from γ by the so called *string construction* (see i.e. [9, 10]). One takes an unstretchable string having length $L > 1$, wraps it around γ, pulls it tight at a point, M, and drags it around γ. The point M then traces the table. Variing the parameter $L \in [1, \infty]$, we get a one-parameter family of billiard tables T_L. Each of the tables is composed of arcs of ellipses (Fig. 1.).

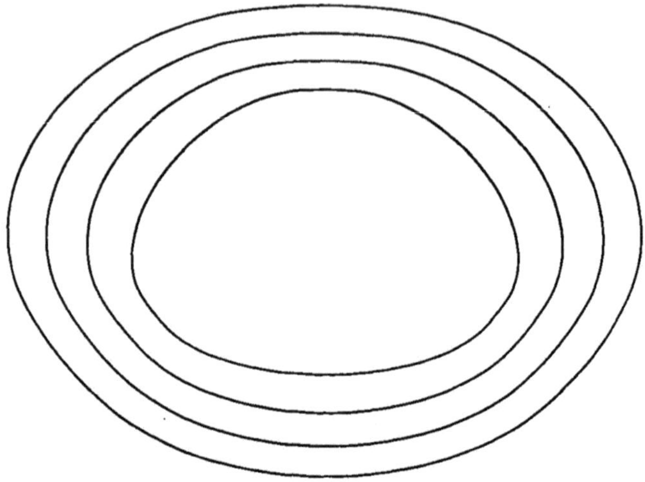

Fig. 1. Four examples of tables given by the string construction at an equilateral triangle.

The corresponding billiard maps, $\phi_L : \Phi \to \Phi$, form a natural one-parameter family of *twist maps*. We investigate this family in the present work.
We denote by Φ the space of unit vectors, with foot points in T_L, directed inward. The space Φ is a closed cylinder, with natural coordinates (s, θ), where $0 \leq s \leq |T_L|$ is the normalized arc-length parameter on T, and $0 \leq \theta \leq \pi$ measures the height in Φ. The cylinder Φ is the phase space for the billiard map, $\phi : \Phi \mapsto \Phi$, of the billiard table T (Fig. 2.).

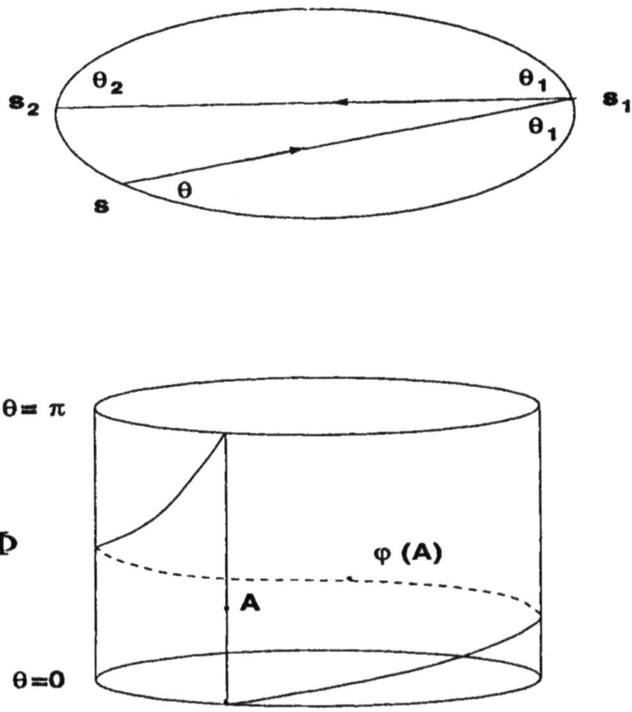

Fig. 2. The billiard map ϕ and the phase space Φ.

By construction, each billiard map ϕ_L has a canonical invariant circle, $\Gamma_L \subset \Phi_L$. It is formed by the rays (i.e., oriented straight lines) supporting γ. Their orientation is induced by the positive orientation of γ. (The opposite choice of orientation yields an other invariant circle $\tilde{\Gamma}_L$). The curve γ is the *caustic* corresponding to the invariant circle Γ_L, for any L [2].

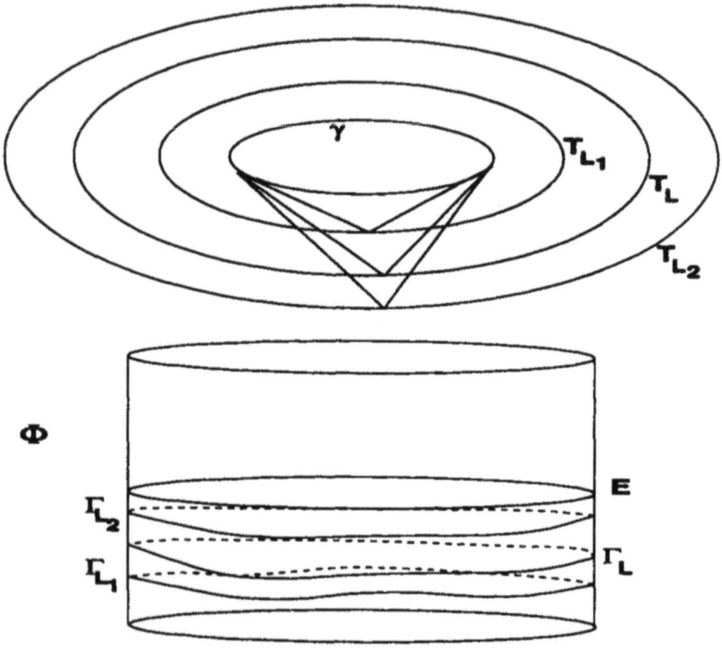

Fig. 3. A family T_L of billiard tables obtained by the string construction. $(L_1 < L < L_2)$. The invariant circles Γ_L are moving up in the phase space Φ.

Thus the family $T(\gamma, L)$, consists of billiard tables with the same caustic γ.

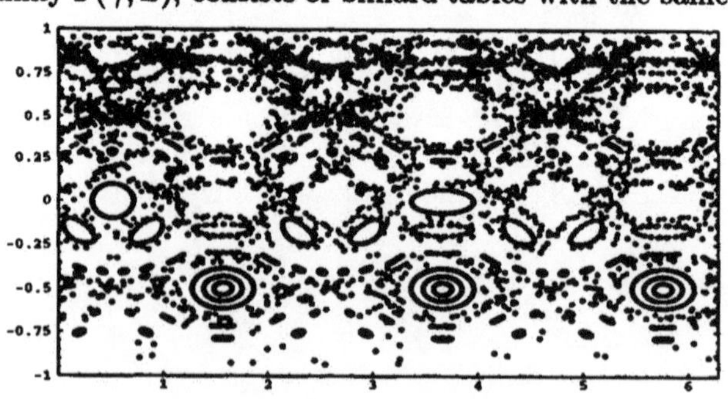

Fig. 4. Some orbits for the parameter $L = 1.25$.

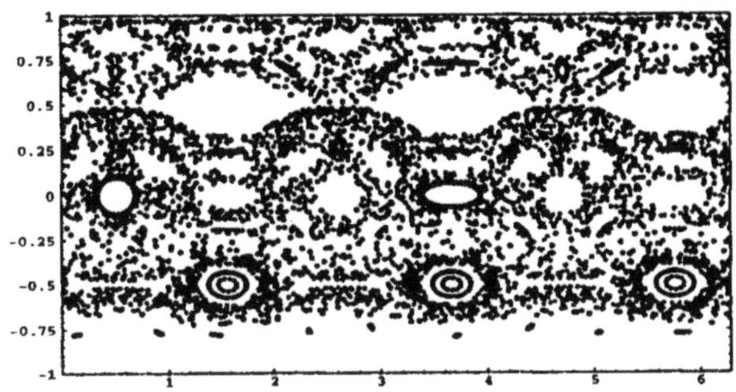

Fig. 5. Some orbits for the parameter $L = 1.33$.

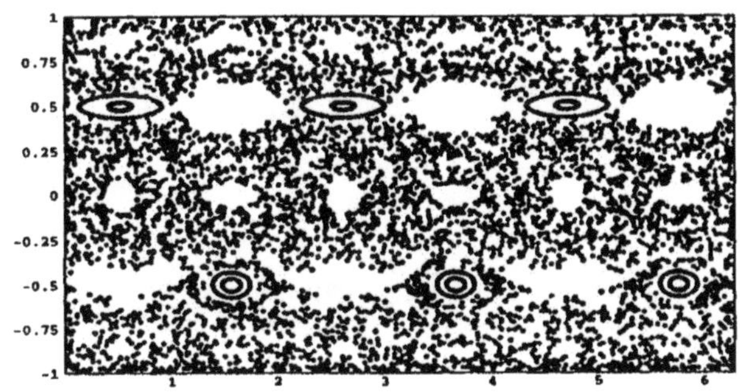

Fig. 6. Some orbits for the parameter $L = 1.38$.

By our choice of Γ_L, the *rotation number* $\rho(L)$ of Γ_L satisfies $\rho(L) \leq 1/2$.

2 A Family of Circle Maps

We denote by $f_L : \Gamma_L \mapsto \Gamma_L$ the circle homeomorphism induced by ϕ_L. We choose a reference direction in $\mathbf{R}^2$, and parametrize Γ_L by the directions, δ, of the supporting rays. With this parametrization, f_L, becomes a family of

Lipschitz homeomorphisms of the *standard circle*, $S^1 = \mathbf{R}/\mathbf{Z}$. The rotation number, $\rho(L) = \rho(f_L)$, is a continuous, nondecreasing function. This function $\rho(L)$ contains significant information about the family $\phi_L, |\gamma| < L$ of twist maps.

Lemma 2.1 *The map $\rho \mapsto \rho(f(L))$ is continuous and we have*

$$\lim_{L\to 1} \rho(L) = 1/3, \ \lim_{L\to\infty} \rho(L) = 1/2 .$$

Lemma 2.2 *Fix $L > |\gamma|$, and let $f_L : S^1 \mapsto S^1, f_L(\delta) = \delta_1$, be the corresponding circle homeomorphism. Denote by $a = a(\delta), b = b(\delta)$ the lengths of the supporting segments (see Fig. 8. for notation). Then $f_L'(\delta) = d\delta_1/d\delta = a/b$. The derivative with respect to L is given by*

$$\frac{\partial f_L}{\partial L} = \frac{\cot\theta}{b} .$$

Proof. See Fig. 7. and Fig. 8.

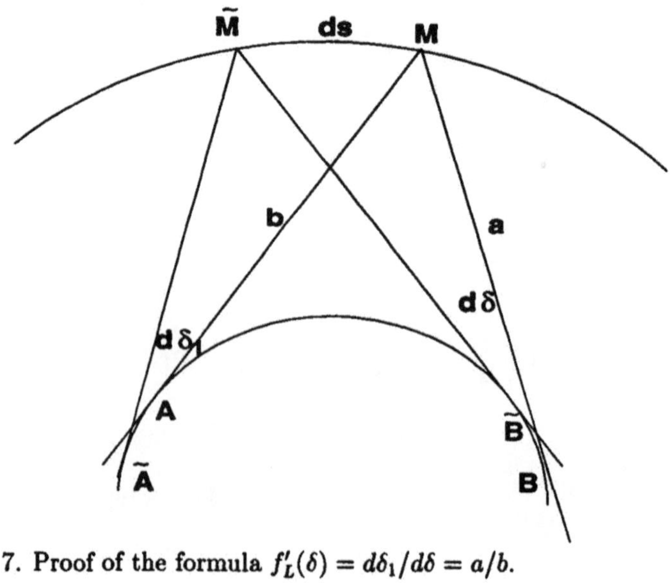

Fig. 7. Proof of the formula $f_L'(\delta) = d\delta_1/d\delta = a/b$.

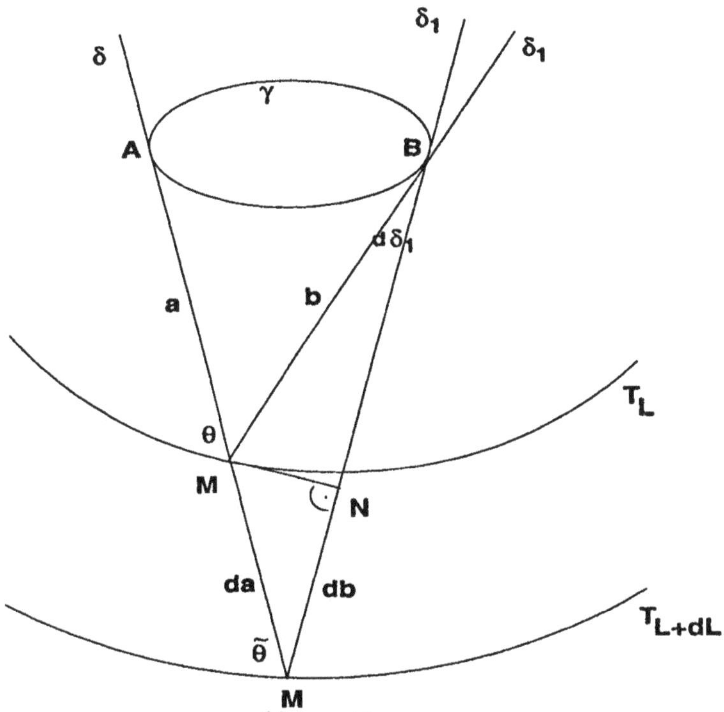

Fig. 8. Proof of the formula $\frac{\partial f_L}{\partial L} = \frac{\cot\theta}{b}$. □

Corollary 2.3 *The one-sided derivatives $f'_L(\delta\pm)$ exist everywhere. The map $f(L)$ is C^1 outside a finite set of δ's.*

Recall that $t \in \mathbf{R}$ is a *point of increase* for a nondecreasing function, g, if $g(t-\epsilon) < g(t) < g(t+\epsilon)$ for all sufficiently small $\epsilon > 0$. Denote with $R(\alpha)$ the standard rotation by $2\pi\alpha$.

Proposition 2.4 *1. ρ is increasing at points L for which $\rho(L)$ is irrational.*
2. ρ is increasing at points L for which $\rho(L) = p/q$, and $f_L : S^1 \mapsto S^1$ is conjugate to the rotation $R(p/q)$ of S^1. In this case, f_L^q is the identity on S^1.

Remark. Case 2. in this proposition, when $\rho(L) = p/q$ is especially important.

3 Devil's Staircase

The graph of a nondecreasing function, $\rho : (1,\infty) \mapsto (\rho(1), 1/2)$, is called a *devil's staircase* if there exists a family of disjoint open intervals, such that their union is dense in $(1,\infty)$, the function ρ is constant on each interval and such that ρ takes different values on different intervals (see, e.g., [6, 8]).

A rational number, $\rho(1) < p/q < 1/2$, is called *exceptional*, if the homeomorphism f_L such that $\rho(L) = p/q$ is conjugate to $R_{p/q}$. For any *non-exceptional* p/q, the set $\rho^{-1}(p/q)$ is an interval, $I_{p/q} = [L^-_{p/q}, L^+_{p/q}]$ with nonempty interior. Using the terminology of circle maps, we call $I_{p/q}$ a *phase locking interval.* If p/q is exceptional, the interval $I_{p/q}$ degenerates to a single point $I_{p/q} = \{L_{p/q}\} = \rho^{-1}(p/q)$. In this case, every point on the invariant curve Γ_L is periodic. From the formula in Lemma 2.2 of f_L we get:

Lemma 3.1 *Let $\delta \in \Gamma_L$ be a periodic point, with period q. Let a_i, b_i, $1 \leq i \leq q$ be the lengths of the associated intervals in the polygon $P = P(\delta)$ (see Fig. 9.). Then the homeomorphism $f_L : S^1 \mapsto S^1$ is differentiable at the fixed point δ, and we have*

$$\frac{df_L^q(\delta)}{d\delta} = \frac{a_1 \cdots a_q}{b_1 \cdots b_q} .$$

This leads to

Corollary 3.2 *1. If the curve Γ_L contains an interval (δ', δ''), of periodic points of period q, then $(f_L^q)'(\delta) = 1$ on (δ', δ'').*
2. Γ_L is p/q-exceptional if and only if

$$a_1 \cdots a_q = b_1 \cdots b_q \tag{1}$$

identically on Γ_L.

By differentiating this relation along a deformation, we can get necessary conditions which have to hold in an exceptional case.
Let $O \subset \Gamma_L$ be a q-periodic orbit, and let $O(t) \subset \Gamma_L, t \in I \subset \mathbf{R}$ be a *deformation* of O. We assume that I is an interval with a nonempty interior, that $0 \in I, O = O(0)$. Let $P(t) = M_1(t) \cdots M_q(t)$ be the polygon corresponding to $O(t)$. We will assume that the deformation $O(t)$, $t \in I$ is differentiable and *nontrivial.* By this we mean that $t \to M_i(t)$ are C^1-functions, that $ds(M_i)/dt = ds_i/dt = \dot{s}_i \geq 0$ for $1 \leq i \leq q$, and that at least one $\dot{s}_j > 0$ for $t \in I$. We say a periodic orbit $O \subset \Gamma_L$ *admits a deformation* if O is contained in an interval of periodic orbits.

Lemma 3.3 *No periodic orbit $O \subset \Gamma_L$ admits a deformation.*

Proof. Assume a deformation $O(t)$ exists. Let $M_i(t)$ be given by arc lengths $s_i(t)$. We have $\dot{a}_i = \cos(\theta_i)\dot{s}_i$ and $\dot{b}_i = -\cos(\theta_i)\dot{s}_i$. Use this and the mirror equation of geometrical optics to get

$$\dot{a}_i/a - \dot{b}_i/b = (1/a + 1/b)\cos(\theta_i)\dot{s}_i = 2\kappa(M_i)\cot(\theta_i)\dot{s}_i .$$

Differentiating $0 = \sum_i \log(a_i) - \log(b_i) = 0$ (obtained from Equation (1)) gives now

$$0 = 2\sum_{i=1}^{q} \kappa(M_i)\cot\theta_i \dot{s}_i$$

which is not possible since $\dot{s}_j > 0$ for at least one j and $\cot\theta_i > 0$ for all $1 \leq i \leq q$. □

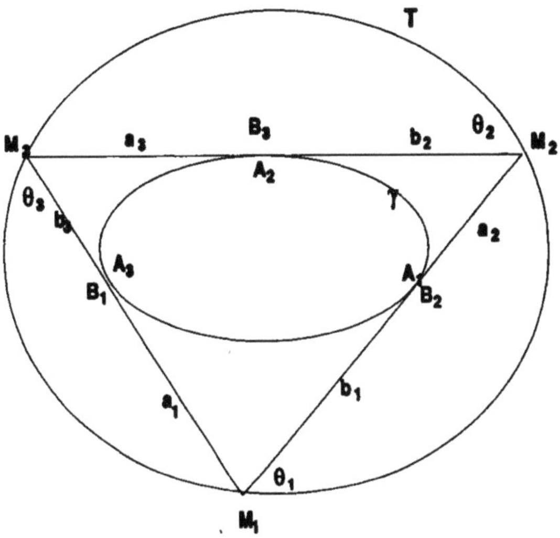

Fig. 9. Proof of Lemma 3.3.

Corollary 3.4 *No $p/q \in (1/3, 1/2)$ is exceptional. The function $L \mapsto \rho(f_L)$ is a devil's staircase.*

This is a special case of a more general theorem, we proven in [3]. There is first of all a geometrical argument which shows that for a generic convex curve γ, the rotation function is a devil's staircase. We also showed that if γ has a flat point of if γ is a polygon, then the rotation function is a devil's staircase.

We show next a numerically computed graph of the rotation function in our case, when the γ is an equilateral triangle. The graph shows the rotation number in dependence on L.

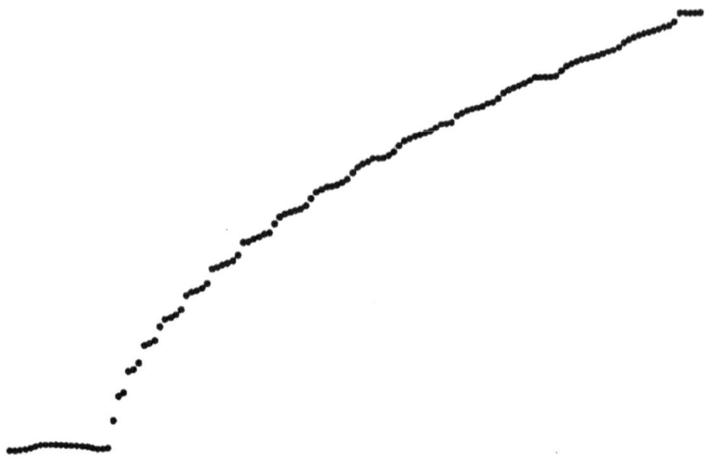

Fig. 10. Numerical computation of the rotation number in dependence of L. The plateau at the beginning corresponds to rotation number 1/3. The end of the initial plateau is given by the value $L = 4/3$.

4 Aubry-Mather Sets and the Invariant Circle

In this section, we analyse the interaction of the moving invariant circle Γ_L with Birkhoff periodic orbits of type p/q if the phase-locking interval $I_{p/q}$ has a nonempty interior.
It is a basic fact for monotone twist maps that for any $p/q \in [0,1]$ at least two Birkhoff periodic orbits of type (q,p) exist (see for example [7, 5]). In the billiard case, one of these orbits is a local maximum of the functional $|O|$, the total length of the trajectory O. Let $M_L(p/q)$ be the set of Birkhoff periodic orbits. Let α be irrational. An accumulation point (in the Hausdorff topology) of sets $M_L(p/q)$ as $p/q \to \alpha$ is called an *Aubry-Mather set* and denoted by $M_L(\alpha)$. Such a set $M_L(\alpha)$ has the property that it is the graph of a Lipschitz continuous function $\phi : K \subset S^1 \to [0,\pi]$ and that a lift $\tilde{\phi}$ of ϕ preserves the order of the covering of M [5].

Let C be a simple, contractible closed curve in Φ avoiding all fixed points of ϕ^q. The *index* $\mathrm{ind}(C,\phi^q)$ of C with respect to ϕ^q is defined as the Brower degree of the map $v : C \to S^1$ $v(x) = (\phi^q(x) - x)/||\phi^q(x) - x||$. The index is a homotopy invariant and does not change if we deform C without intersecting a fixed point of ϕ^q.
If a curve C contains only one fixed point x_0 of ϕ^q, $\mathrm{ind}(x_0,\phi^q) = \mathrm{ind}(C,\phi^q)$ is called the *index* of x_0. If C contains a connected fixed-point set Y_0, we call $\mathrm{ind}(Y_0,\phi^q) = \mathrm{ind}(C,\phi^q)$ the *index of this fixed point set.* If a curve C contains

finitely many fixed points x_i (rsp. connected fixed-point sets Y_i) of ϕ^q, then $\sum_i \mathrm{ind}(x_i) = \mathrm{ind}(C, \phi^q)$.
For L near $|\gamma| = 1$, the invariant circle is near the boundary $\{\theta = 0\}$ of the phase space Φ. For $L \to \infty$, the invariant circle Γ_L moves towards the equator $E = \{\theta = \pi/2\}$ of Φ. Aubry-Mather sets $M_L(\alpha)$ with a fixed rotation number α pass through the moving circle Γ_L.

The passage of $M_L(\alpha)$ with irrational α is easy to describe: since each irrational α is a point of increase of $\rho(L)$, the set of parameters L for which $M_L(\alpha)$ intersects with Γ_L consists of exactly one parameter value $L_\alpha = \rho^{-1}(\alpha)$, for which $\Gamma_L = M_L(\alpha)$. The set $M_L(\alpha)$ is in general a Cantor set, for $L \neq L_\alpha$.

More interesting is when Birkhoff periodic orbits $M_{p/q}$ pass through Γ_L if the phase locking interval $I_{p/q}$ is nontrivial. Since any periodic orbit on Γ_L is a Birkhoff periodic orbit, $I_{p/q}$ is the set of parameters L for which $M_{p/q}$ intersects with Γ_L.

The following theorem is proven in [3] for a general convex caustic γ.

Theorem 4.1 *For $L \to L^-_{p/q}, L < L^-_{p/q}$, a Birkhoff periodic set of index -1 is approaching Γ_L from above, intersects Γ_L at $L = L^-_{p/q}$ with index -1 and leaves it for $L > L^-_{p/q}$ with index 1. For $L \to L^+_{p/q}$, $L < L^+_{p/q}$, a Birkhoff periodic set of index 1 approaches Γ_L from above, hits it at $L = L^+_{p/q}$ and leaves it for $L > L^+_{p/q}$ with index -1. (see Fig. 11.)*

The idea of the proof is as follows: for $L = L^-_{p/q}$ and $L = L^+_{p/q}$, every Birkhoff periodic set on Γ_L is parabolic and has index -1. By the symmetry of the situation, it suffices to study the bifurcation at L^-. For $L < L^-$, there exists no Birkhoff periodic set on Γ_L and all such sets are above Γ_L. Local index conservation implies that such a set must exist nearby. Using properties of the one-parameter family of circle map f_L for $L > L^-$ with L near L^-, we conclude that there exist 2 hyperbolic Birkhoff periodic sets of index -1 on Γ_L. Local index conservation implies that a Birkhoff periodic set of index $+1$ must be nearby. By the global Poincaré index formula, we argue that this set V^- is on the other side of Γ_L.

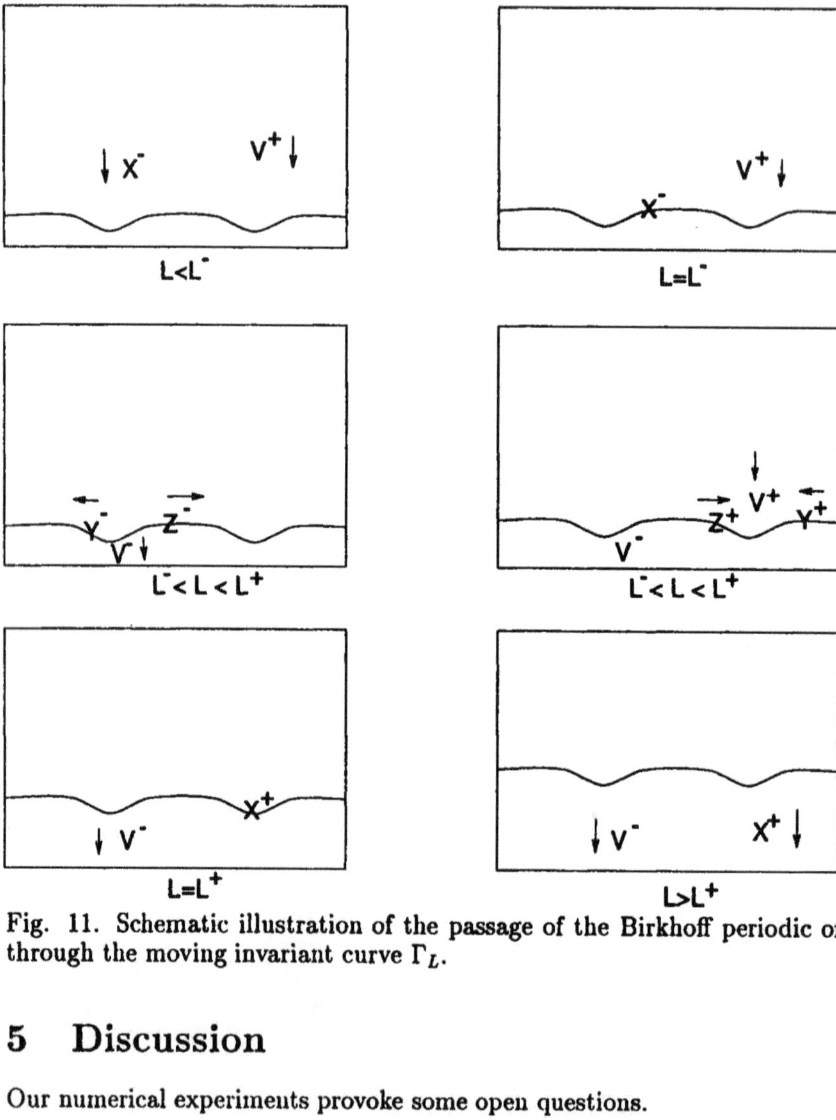

Fig. 11. Schematic illustration of the passage of the Birkhoff periodic orbits through the moving invariant curve Γ_L.

5 Discussion

Our numerical experiments provoke some open questions.

• We were especially interested in the parameter value $L = 4/3$, which is the supremum of all parameter values L for which $\rho(L) = 1/3$. The numerical experiments do not decide clearly, if there exists a second caustics which is different from the canonical caustics Γ_L.

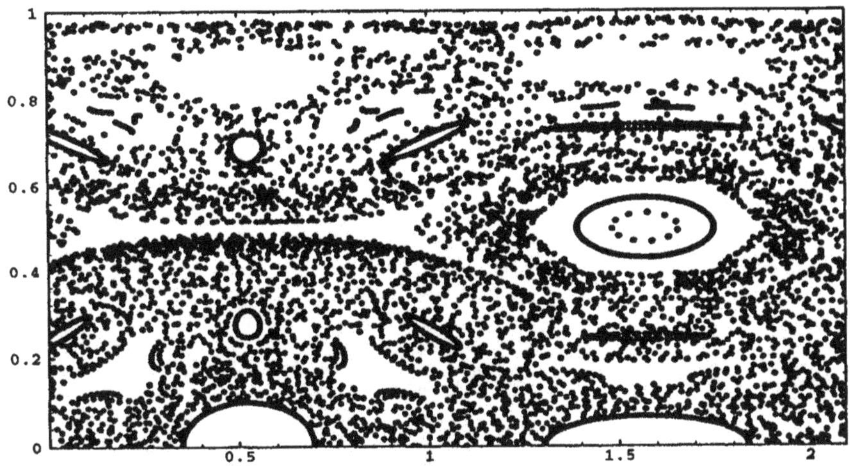

Fig. 12. The same picture as Fig. 5. Some orbits for $L = 1.33$. This time, we identified the dyhedral symmetry of the phase space to see more details. We see the parabolic periodic orbit (see around coordinates $(0.5, 0.5)$) which is on the canonical invariant curve Γ_L. On the same height to the right, there is an elliptic periodic orbit also above Γ_L. The last invariant curve around that iland has a hexagonal shape.

• It looks as if for L in the interior of a phase locking interval, there exists a neighborhood of the canonical invariant circle Γ_L, for which there exists no other invariant circle. One can argue that the stable or unstable manifold of the hyperbolic Birkhoff periodic orbits on the caustics prevent this if they do cross transversely outside the canonically invariant circle.

• The curvature of the tables T_L has discontinuities. Hubacher's result [4] shows that near the boundaries of Φ, there exist no invariant curves. A result of Angenent [1] implies that for any $L > 1$, the topological entropy $h_{top}(\phi_L)$ of the billiard at T_L is positive. It would be interesting to get quantitative results about the topological entropy. Is it monotonically decreasing in L? Is $\limsup_{L\searrow 1} h_{top}(\phi_L) > 0$?

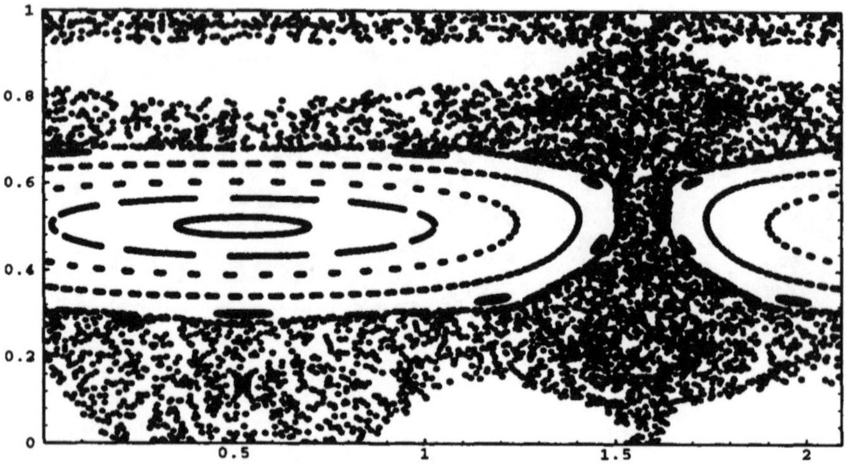

Fig. 13. Some orbits for $L = 1.00016$ for which the billiard table is very close to a triangle. Like in Fig. 12, we have identified the dihedral symmetry. (The arc length parameter displaying the first coordinate is slightly distorted since the program uses a convenient arc length instead of the real arc length).

References

[1] S. Angenent. A remark on the topological entropy and invariant circles of an area preserving twist map. In K.Meyer R.Mc Gehee, editor, *Twist mappings and their Applications*, IMA Volumes in Mathematics, Vol. 44. Springer Verlag, 1992.

[2] E. Gutkin and A. Katok. Caustics for inner and outer billiards. CMP, in press, 1994.

[3] E. Gutkin and O. Knill. Billiards that share a common convex caustic. preprint, 1994.

[4] A. Hubacher. Instability of the boundary in the billiard ball problem. *Commun. Math. Phys.*, 108:483–488, 1987.

[5] A. Katok. Some remarks on Birkhoff and Mather twist map theorems. *Ergod. Th. Dyn. Sys.*, 2:185–194, 1982.

[6] A. Katok and B. Hasselblatt. *Introduction to the modern theory of the smooth dynamical systems.* Cambridge University Press, 1994.

[7] W.D. Neuman M. Brown. Proof of the Poincaré-Birkhoff fixed point theorem. *Mich. Math. J.*, 24:21–31, 1975.

[8] W.de Melo and S.van Strien. *One-Dimensional Dynamics.* A series of Modern Surveys in Mathematics. Springer, Berlin, 1993.

[9] Ya.G. Sinai. *Introduction to ergodic theory.* Princeton university press, Princeton, 1976.

[10] S. Tabatschnikov. Billiards. preprint, Survey paper, 1994.

EQUILIBRIA OF SPHERICAL CHARGE DISTRIBUTIONS

R. CLARISSA HOWISON
Department of Mathematical Sciences, University of Cincinnati
Cincinnati, OH 45221-0025, U.S.A.

and

CHRISTOPHER K. MCCORD
Department of Mathematical Sciences, University of Cincinnati
Cincinnati, OH 45221-0025, U.S.A.
E-mail: chris.mccord@uc.edu

ABSTRACT

We consider the problem of N equally charged particles, confined to the surface of a sphere and moving under their mutual electrostatic forces. This system is of interest physically, and also displays interesting similarities with the N-body problem of celestial mechanics. However, the dynamics of this system are not well-understood. In this paper, we address one of the basic dynamical questions: counting, locating, and determining the stability of the rest points of the system. These rest points correspond to distributions of the charges on the surface of the sphere in which the total repulsive force is directed radially.

1. Introduction

The dynamics of charged particles confined to a surface, and in particular to the surface of a sphere, has long been of interest in physics. The equilibrium configurations of such systems have interpretations in many physical models. (See [Erber & Hockney, 1991] and the references therein.) In recent years, this interest has led to several numerical studies [2,3,5,7], which have sought to locate the energy-minimizing equilibrium configurations of N equally-charged particles, for values of N up to 65. However, there has been almost no mathematical analysis of the problem. In this paper, we use techniques borrowed from the N-body problem of celestial mechanics to investigate the equilibrium distributions of N equally-charged particles (e.g. electrons) confined to the surface of a sphere. We will use Morse theory to give lower bounds on the number of eq, keep iuilibrium distributions, and to give some information about the stability of these equilibria. These results fall far short of a complete description of the equilibrium distributions, but they indicate that topological and dynamical methods can make a useful contribution to the understanding of this problem.

The natural phase space for this problem is

$$\mathcal{S} = \left\{ (q_1, \ldots, q_N) \in \prod_N S^2 \mid q_i \neq q_j \text{ for } i \neq j \right\}.$$

This is an open $2N$-dimensional manifold, in which collisions of the particles are forbidden. Contained in it is an important $N+2$-dimensional submanifold: the manifold of circular distributions

$$\mathcal{C} = \left\{(q_1, \ldots, q_N) \in \mathcal{S} \mid \exists q \in S^2 \text{ such that } q_i \cdot q = 0 \text{ for all } i\right\}.$$

That is, $\mathcal{C}$ consists of all distributions in which all N particles lie on a great circle in the sphere.

The electrostatic potential function $V : \mathcal{S} \to \mathbf{R}$ is given by

$$U(q_1, \ldots, q_N) = -\sum_{i<j} \frac{1}{|q_i - q_j|}.$$

Note that, with this choice of sign, the potential is always negative, so collisions have an infinitely negative potential, and are repelling in the gradient flow $\dot{q} = \nabla V(q)$. It is the critical points of U that we are interested in studying, as these are the equilibrium charge distributions. Note that $q = (q_1, \ldots, q_N)$ is a critical point if and only if $\frac{\partial U}{\partial q_i}(q) = \lambda_i q_i$ for all i. Also note that $\mathcal{C}$ is invariant under the gradient flow.

There is a natural SO_3 action on $\mathcal{S}$: given $A \in SO_3$ and $(q_1, \ldots, q_N) \in \mathcal{S}$, define $A(q_1, \ldots, q_N) = (Aq_1, \ldots, Aq_N)$. Since any nontrivial element of SO_3 fixes exactly two points on S^2, SO_3 acts freely on $\mathcal{S}$ (for $N > 2$). We can then form a smooth, open $(2N-3)$-manifold $S = \mathcal{S}/SO_3$. This contains an $(N-1)$-dimensional submanifold $C = \mathcal{C}/SO_3$. We will refer to points in C as circular distributions, and points in $S \setminus C$ as spherical distributions. To find equilibrium distributions, note that the potential U is clearly constant on SO_3 orbits, so there is a well-defined smooth potential function $V : S \to \mathbf{R}$. Our goal is to describe the critical points of V, which can be thought of as SO_3-families of equilibrium distributions. Specifically, we show

Theorem 1 *There are exactly $\frac{(N-1)!}{2}$ equilibria in C, each with the charges uniformly spaced on the circle. There is one for each (unoriented) ordering of the charges on the circle. These equilibria are attracting in C.*

Theorem 2 *If V is a Morse function, then there are at least $(N-3)(N-2)!$ equilibria in $S \setminus C$.*

In addition to the rotational symmetry, this problem has a rotational symmetry and an exchange symmetry. The SO_3 action described above extends in the obvious fashion to an O_3 action, and the permutation group S_N acts on $\mathcal{S}$ by $\sigma(q_1, \ldots, q_N) = (q_{\sigma(1)}, \ldots, q_{\sigma(N)})$. In fact, the two actions commute with one another, so there is an $O_3 \times S_N$ action on $\mathcal{S}$ which leaves the submanifold $\mathcal{C}$ and the potential function U invariant. Thus there is an action of

$(O_3 \times S_N)/SO_3 \cong \mathbb{Z}_2 \times S_N$ on $\mathcal{S}/SO_3 = S$, and every critical point in S is actually a member of a $\mathbb{Z}_2 \times S_N$-orbit of critical points. Unfortunately, the action of $O_3 \times S_N$ on $\mathcal{S}$ is not free, so $S/(\mathbb{Z}_2 \times S_N)$ is not a manifold, and we cannot apply Morse theory to it directly. However, since all critical points on an S_N orbit have the same Morse index, the Betti number estimates that are used to proof Theorem 2 can also be used to prove

Theorem 3 *If V is a Morse function, then there are at least $N-4$ $O_3 \times S_N$ orbits of spherical equilibria.*

That is, there are at least $N-3$ geometrically distinct equilibrium configurations: one circular; and at least $N-4$ spherical.

We want to draw attention to the very close connection between this problem and the study of central configurations in the N-body problem[6,9]. Central configurations are critical points of the gravitational potential function — a function which differs only by an overall sign. More importantly though, the natural phase spaces for the two problems are different. For planar central configurations, the analogue of $\mathcal{S}$ is an open manifold in S^{2N-2}; here, it is an open manifold in $\prod_N S^2$. However, both the overall form of the problem, and its solution, are very much the same in the two problems: in both problems, we isolate a class of special configurations (circular here; collinear there); use Morse theory on the phase space; and use our knowledge of the special configurations to make estimates on the remaining configurations. Theorems 1 and 2 can be viewed as the analogues of Moulton's and Palmore's results for central configurations[10,12].

The rest of the paper is devoted to the proofs of Theorems 1, 2 and 3. The proof of Theorem 1 is in the next section, the other two are proved in §. The second author would like to thank Bob Brown and Tom Erber for drawing the problem to his attention, and Rick Moeckel for several helpful conversations.

2. Circular Distributions

To find the critical points of V in C, and determine their stability, it is convenient to work with a set of representatives of C in $\mathcal{C}$. The obvious set to consider are the "equatorial configurations"

$$\tilde{C} = \{(q_1, \ldots, q_N) \in S | (0,0,1) \cdot q_i = 0 \, \forall i\}.$$

Every configuration in $\mathcal{C}$ can be rotated to a configuration in $\tilde{C}$ by rotating its normal vector q to the z-axis, and the SO_3 action on $\mathcal{C}$ restricts to an O_2 action on $\tilde{C}$. That is, $C = \tilde{C}/O_2$.

To examine the behavior of U on $\tilde{C}$, it is useful to express the positions q_i in spherical coordinates:

$$x_i = \sin\phi_i \cos\theta_i \quad y_i = \sin\phi_i \sin\theta_i \quad z_i = \cos\phi_i.$$

These are valid coordinates away from $\phi = 0$, and in particular, in a neighborhood of $\tilde{C}$. In these coordinates, $\tilde{C} = \left\{\phi_i = \frac{\pi}{2} \forall i\right\}$, so $\tilde{C} \cong \{(\theta_1, \ldots, \theta_N) | \theta_i \neq \theta_j\}$. We can eliminate the O_2 action on $\tilde{C}$ by setting $\theta_N = 0$ and requiring $\theta_1 < \theta_2$. That is,

$$C \cong \left\{(\theta_1, \ldots, \theta_{N-1}) \in (0,1)^{N-1} | \theta_i \neq \theta_j, \theta_1 < \theta_2\right\}.$$

This space is a disjoint union of $\frac{(N-1)!}{2}$ disks of dimension $(N-1)$. Since V is bounded above, it achieves a maximum on every component of C. In fact, we will show that these maxima are the only critical points of V on C.

To do so, we first express U in spherical coordinates:

$$U = -\sum_{i<j} \{2 - 2\cos(\theta_i - \theta_j)\sin\phi_i \sin\phi_j - 2\cos\phi_i \cos\phi_j\}^{-1/2}.$$

Critical points of U in C are then solutions of

$$\begin{aligned}
\frac{\partial U}{\partial \theta_i} &= \sum_{j \neq i} \frac{\sin(\theta_i - \theta_j)\sin\phi_j \sin\phi_i}{2\sqrt{2}\,[1 - \cos(\theta_i - \theta_j)\sin\phi_i \sin\phi_j - \cos\phi_i \cos\phi_j]^{3/2}} = 0 \\
\frac{\partial U}{\partial \phi_i} &= -\sum_{j \neq i} \frac{\cos(\theta_i - \theta_j)\cos\phi_i \sin\phi_j + \sin\phi_i \cos\phi_j}{2\sqrt{2}\,[1 - \cos(\theta_i - \theta_j)\sin\phi_i \sin\phi_j - \cos\phi_i \cos\phi_j]^{3/2}} = 0.
\end{aligned}$$

If we restrict to $\tilde{C}$ by setting $\phi_i = \pi/2$, these equations reduce to

$$\begin{aligned}
\frac{\partial U}{\partial \theta_i} &= \sum_{j \neq i} \frac{\sin(\theta_i - \theta_j)}{2\sqrt{2}\,[1 - \cos(\theta_i - \theta_j)]^{3/2}} = 0 \\
\frac{\partial U}{\partial \phi_i} &= -\sum_{j \neq i} \frac{0}{2\sqrt{2}\,[1 - \cos(\theta_i - \theta_j)]^{3/2}} = 0.
\end{aligned}$$

That is, any configuration in $\tilde{C}$ satisfies the second inequality, and critical points of U correspond to solutions of the first equality. Before we explicitly locate these, we note that

Lemma 1 *Every critical point of V in C is a local maximum of $V|_C$.*

Proof. To compute the Hessian of V, we set $\theta_N = 0$ and compute $\frac{\partial^2 V}{\partial \theta_j \partial \theta_i} = \frac{\partial^2 U}{\partial \theta_j \partial \theta_i}$ for $i, j \neq 1$. With $\theta_N = 0$, we have

$$\frac{\partial U}{\partial \theta_i} = \frac{\sin(\theta_i)}{2\sqrt{2}\,[1 - \cos(\theta_i)]^{3/2}} + \sum_{1<ij} \frac{\sin(\theta_i - \theta_j)}{2\sqrt{2}\,[1 - \cos(\theta_i - \theta_j)]^{3/2}}$$

and

$$\frac{\partial^2 V}{\partial \theta_i^2} = -\frac{1}{4\sqrt{2}} \cdot \frac{3+\cos\theta_i}{[1-\cos\theta_i]^{1/2}} - \frac{1}{4\sqrt{2}} \sum_{j\neq 1} \frac{3+\cos(\theta_i-\theta_j)}{[1-\cos(\theta_i-\theta_j)]^{3/2}}$$

$$\frac{\partial^2 V}{\partial \theta_j \partial \theta_i} = \frac{1}{4\sqrt{2}} \cdot \frac{3+\cos(\theta_i-\theta_j)}{[1-\cos(\theta_i-\theta_j)]^{3/2}}.$$

Thus the Hessian is given by

$$HV = \begin{bmatrix} -b_2 - S_2 & a_{23} & a_{24} & \dots & a_{2N} \\ a_{32} & -b_3 - S_3 & a_{34} & \dots & a_{3N} \\ \vdots & & & & \\ a_{N2} & \dots & \dots & -b_N - S_N & \end{bmatrix}$$

where

$$b_i = -\frac{1}{4\sqrt{2}} \cdot \frac{3+\cos\theta_i}{[1-\cos\theta_i]^{1/2}}$$

$$a_{ij} = \frac{1}{4\sqrt{2}} \cdot \frac{3+\cos(\theta_i-\theta_j)}{[1-\cos(\theta_i-\theta_j)]^{3/2}}$$

and $S_i = \sum_{\substack{j \\ j\neq i}} a_{ij}$. Note that $b_i, S_i, a_{ij} > 0$ for all i, j. This matrix is symmetric, so its eigenvalues are all real.

By the theorem of Gerschgorin, each eivenvalue of HV must be contained in one of the discs in the complex plane centered at $-b_i - S_i$ of radius $\sum_{\substack{j \\ j\neq i}} |a_{ij}|$.

But

$$\sum_{j\neq i} |a_{ij}| = \sum_{j\neq i} a_{ij} = S_i < S_i + b_i,$$

so each eigenvalue lies in the left hand plane, and being real must be negative. Thus HV is negative definite and any critical point of $V|_C$ is a local maximum. □

This establishes that there are exactly $\frac{(N-1)!}{2}$ critical points in C. To complete the proof of Theorem 1, we must show that these critical points correspond to the equally spaced configuration $\theta_i = \frac{2\pi(i-1)}{N}$ (and its rotations and permutations). But this is simply a matter of observing that, when these values of θ_i are inserted into $\frac{\partial U}{\partial \theta_i}$, we do indeed obtain $\frac{\partial U}{\partial \theta_i} = 0$ for all i.

This completes the proof of Theorem 1. However, there is still one question about the circular configurations that remains unanswered. We have not yet determined their stability in S. We will not fully determine their stability, but we do note that (unlike collinear central configurations in the planar N-body problem) they are not repelling in all directions normal to C. To see this, recall that q is a critical point of U if and only if there exist $\lambda_1, \ldots, \lambda_N$ such that $\frac{\partial U}{\partial q_i}(q) = \lambda_i q_i$ for all i. If $\overline{q} = (\overline{q}_1, \ldots, \overline{q}_N)$ is a circular equilibrium, then (by symmetry) $\lambda_i = \lambda_j \; \forall i, j$. Further, because the potential is repulsive, this single value λ must be positive.

Then the quadratic form associated with the Hessian of U at the point $\overline{q}$ is[11]

$$D^2U(\overline{q})(v,v) = \sum_{i<j} \frac{1}{|\overline{q}_i - \overline{q}_j|^3} \left[|v_i - v_j|^2 - \frac{3(\overline{q}_1 - \overline{q}_j) \cdot (v_i - v_j)}{|\overline{q}_i - \overline{q}_j|^2} \right] - \lambda \sum_j |v_i|^2$$

where $v = (v_i, \ldots, v_N)$ lies in the tangent space $T_{\overline{q}}S$.

Letting v also be normal to $T_{\overline{q}}C$, we have

$$\begin{aligned} D^2U(\overline{q})(v,v) &= \sum_{i<j} \frac{1}{|\overline{q}_i - \overline{q}_j|^3} |v_i - v_j|^2 - \lambda \sum_i |v_i|^2 \\ &= vAv^T - \lambda v I v^T, \end{aligned}$$

where A is the $N \times N$ matrix with entries

$$a_{ii} = \sum_{j \neq i} \frac{1}{|\overline{q}_i - \overline{q}_j|^3}, \quad a_{ij} = -\frac{1}{|\overline{q}_i - \overline{q}_j|^3}.$$

Thus A has eigenvalue 0 associated with eigenvector $v_0 = (1, 1, \ldots, 1)$, and since we know that $\lambda > 0$ we have $D^2U(\overline{q})(v_0, v_0) = -\lambda v_o I v_0^T < 0$.

In other words, electrons placed on the sphere at the same latitude, and at equally spaced longitudes, will be attracted to the "equatorial" circular configuration, moving uniformly along lines of constant longitude. Although numerical calculations indicate that all other directions normal to $\overline{q}$ should be repelling, this has not yet been shown. The index in S of the circular configurations is therefore not determined. However, we know that one of the $N-2$ normal directions is stable, so there are at most $N-3$ unstable directions, and the index is at most $N-3$. This will suffice for our purposes.

3. Spherical Distributions

Our strategy for proving Theorem 2 is to apply Morse theory to estimate the number of critical points of V in S, then use our knowledge of the critical points in C to estimate the number of critical points in $S \setminus C$. We first note that, while S is not compact, V is bounded above on S, and all critical points of S are bounded away from collision. We can apply Morse theory in this setting. An obvious first step is to compute the homology of S. To compute $H_*(S;\mathbb{Q})$, we first note that S can be viewed as a configuration space[4]. Given a topological space X, with Q a set of q points in X, we define the configuration space

$$F_{p,q}(X) = \{(x_1, \ldots, x_p) \in (X \setminus Q)^p \mid x_i \neq x_j \text{ for } i \neq j\}.$$

Lemma 2 *S has the same homotopy type as $F_{N-3,2}(\mathbb{R}^2)$.*

Proof. For any topological space X, there is a natural fibration

$$F_{p-1,q+1}(X) \to F_{p,q}(X) \to X \setminus Q,$$

defined by projection onto the first factor. Thus, we have $F_{N-1,1}(S^2) \to S \to S^2$. The projection $S \to S^2$ is equivariant with respect to the SO_3 action, and the SO_3 action on S^2 is transitive with isotropy SO_2. Thus we can form the fibration of quotient spaces

$$F_{N-1,1}(S^2)/SO_2 \to S \to S^2/SO_3 = *.$$

In other words, $S = F_{N-1,1}(S^2)/SO_2$. But $F_{N-1,1}(S^2) = F_{N-1,0}(S^2 \setminus *)$ can be identified with $F_{N-1,0}(\mathbb{R}^2)$. Indeed, this identification is equivariant with respect to the natural SO_2 actions on $S^2 \setminus *$ and $\mathbb{R}^2$, so we can identify S with $F_{N-1,0}(\mathbb{R}^2)/SO_2$. This space was shown[8] to have the same homotopy type as $F_{N-3,2}(\mathbb{R}^2)$. □

Now, the homology of $F_{N-3,2}(\mathbb{R}^2)$ is well known. Its Poincaré polynomial is

$$(1+2t)(1+3t)\ldots(1+(N-2)t),$$

with $t \in H_1(F_{N-3,2}(\mathbb{R}^2))$. If we let β_i be the ith Betti number of S (i.e. $\beta_i = \dim(H_i(S;\mathbb{Q}))$, then β_i is the coefficient of t^i in this polynomial. There is not a convenient closed form for the β_i's, but the following facts are known (and will be sufficient for our purposes);

1. $\beta_i = 0$ for $i > N-3$, and $\beta_i \neq 0$ for $i \leq N-3$.;
2. The sum $SB = \sum_{i=0}^{N-3} \beta_i = \frac{(N-1)!}{2}$;
3. $\max_i \{\beta_i\} = \beta_{N-3} = (N-2)!$

To formulate the Morse inequalities, let s_i denote the number of critical points of V, in $S \setminus C$, with index i, and let K denote the index in S of the circular critical points. Then the Morse inequalities take the form

$$\frac{(N-1)!}{2} t^K + \sum_{i=0}^{2N-3} s_i t^i = (1+2t)(1+3t)\ldots(1+(N-2)t) + (1+t)Q(t), \quad (1)$$

where $Q(t)$ is a polynomial with positive coefficients.

In particular, the coefficients of t^K are

$$\frac{(N-1)!}{2} + s_K = \beta_K + q_{K-1} + q_K.$$

Since $\beta_K \leq (N-2)!$,

$$\begin{aligned} q_{K-1} + q_K &\geq \frac{(N-1)!}{2} + s_K - (N-2)! \\ &\geq \frac{(N-3)(N-2)!}{2}. \end{aligned}$$

Now, if we evaulate equation 1 at $t = 1$, we have

$$\frac{(N-1)!}{2} + \sum_{i=0}^{2N-3} s_i = \frac{(N-1)!}{2} + 2\sum_{i=0}^{2N-3} q_i,$$

or

$$\begin{aligned} \sum_{i=0}^{2N-3} s_i &= 2\sum_{i=0}^{2N-3} q_i \\ &\geq 2(q_{K-1} + q_K) \\ &\geq (N-3)(N-2)! \end{aligned}$$

Since $\sum_{i=0}^{2N-3} s_i$ is the total number of spherical critical points of V, this completes the proof of Theorem 2.

Theorem 3 follows immediately. Since all of the points in a $\mathbb{Z}_2 \times S_N$ orbit in S have the same index, critical points with different indices must belong to different orbits. The argument above shows that there are at least $N-3$ distinct nonzero Betti numbers , of which only one corresponds to circular equilibria, so there at least $N-4$ distinct $\mathbb{Z}_2 \times S_N$ orbits of equilibria in $S \setminus C$.

4. References

1. R. Brown, J. White. Homology and Morse theory of third configuration spaces. *Ind. J. Math.* **30** (1981) 501 - 512.
2. J. R. Edmundson. The distribution of point charges on the surface of a sphere. *Acta Cryst. A* **48** (1992) 60 - 69.
3. T. Erber, G. Hockney. Equilibrium configurations of N equal charges on a sphere. *J. Phys. A* **24** (1991) 1369 - 1377.

4. E. Fadell, L. Neuwirth. Configuration Spaces. *Math. Scand.* **10** (1962) 111 - 118.
5. L. Glasser, A. G. Every. Energies and spacings of point charges on a sphere. *J. Phys. A* **25** (1992) 2473 - 2482.
6. G. R. Hall, K. R. Meyer. *Introduction to Hamiltonian dynamical systems and the N-body problem.* Springer-Verlag: New York 1992.
7. I. Kogan, A. M. Perelomov, G. W. Semenoff. Charge-distribution in 2-dimensional electrostatics. *Phys. Rev. B* **45** (1992) 12048 - 12087.
8. C. McCord. Planar central configuration estimates in the N-body problem. preprint.
9. R. Moeckel. On central configurations. *Math. Z.* **205** (1990) 499 - 517.
10. F. R. Moulton. The straight line solutions of the problem of N bodies. *Ann. Math. II* Ser. **12** (1910) 1 - 17.
11. F. Pacella. Central configurations of the N-body problem via equivariant Morse theory. *Arch. Ration. Mech. Anal.* **97** (1987) 59 - 74.
12. J. Palmore. Classifying relative equilibria. I *Bull. Amer. Math. Soc.* **79** (1973) 904 - 908; II *Bull. Amer. Math. Soc.* **81** (1975) 489 - 491; III *Lett. Math. Phys.* **1** (1975) 71 - 73.

A GEOMETRIC FRAMEWORK FOR BILLIARDS WITH TIME-DEPENDENT BOUNDARIES

JAIR KOILLER*
Laboratório Nacional de Computação Científica, Rua Lauro Müller 455
Rio de Janeiro, 22290-160, Brazil
and Instituto de Matemática, UFRJ, Cidade Universitária, Rio de Janeiro, Brasil
E-mail: jair@lncc.br

ROBERTO MARKARIAN †
Inst. de Matemática y Estadística "Prof.Ing. Rafael Laguardia"
Fac. de Ingeniería, Montevideo, Uruguay

and

SONIA PINTO DE CARVALHO and SYLVIE OLIFFSON KAMPHORST ‡
Departamento de Matemática, ICEx, UFMG, Belo Horizonte, Brazil

ABSTRACT

We describe the symplectic structure of the time-dependent billiard mapping. The jacobian is explicitly given, with an illustrative example.

1. Introduction

Billiards whose boundaries change in time are important in Physics. We present here some basic mathematical results, hoping to attract further investigations. A few words for motivation: in Nuclear Physics, Bohr's "liquid drop model" [10], assumes that the deforming nucleous is populated by particles moving like billiard balls. In Plasma Physics, time-dependent billiards are toy models for acceleration in magnetic bottles. Time-varying billiard models occur also in Solid State Physics and in some geometric approximations for wave reflections, specially in accoustics [5].

The 1 dimensional case is known as the *Fermi accelerator* [4,8]. When the motion of one wall is prescribed and the other wall is fixed, it gives rise to an area preserving map in the energy × time phase space. Numerical studies, with different periodic wall functions, indicated the existence of invariant curves crossing the time direction for a whole period, thus limiting the maximum energy gain. Chirikov's standard mapping first appeared in this context. Rigorous mathematical proofs of the existence of such invariant curves were given, in

*Partially supported by a fellowship from the Guggenheim Foundation.
†Partially supported by Fac. de Ciencias, Uruguay
‡Partially supported by CNPq, Brasil

some cases, via KAM theory (see, [3], 3.IV, and [6] §4). In the two-dimensional case, using similar methods, M. Levi has given examples of "pulsating soft circular billiards" whose energies stay bounded for all time ([7] §1). Interestingly, in the relativistic version studied by Pustyl'nikov, unbounded growth of the energy is possible [11], [12].

In this work we restrict ourselves to 2-dimensional time-dependent billiards in euclidian space. Everything can be extended to higher dimensions and general metrics. A detailed version will appear elsewhere.

2. Geometric Framework

Consider in $\mathbf{R}^3$ a natural autonomous Lagrangian system $L = T - V(x,y,z)$ with kinetic energy defined by

$$T_\gamma(x',y',z') = \frac{1}{2}(x'^2 + y'^2 + \frac{1}{\gamma^2}z'^2) \tag{1}$$

where γ is a small parameter (and $'$ stands for d/dt). Assume that $\partial V/\partial z$ is uniformly bounded. Then the limiting flow (as $\gamma \to 0$) leaves invariant the hypersurfaces $z' = const.$ In particular, when $z' = 1$, $z = t + const.$ In other words, *time can be viewed as a spatial coordinate whose inertia tends to infinity; in the limit its "motion" becomes uniform.*

In particular, take a "tube" $f(x,y,z) \leq 1$, where f is a smooth function, with $\partial f/\partial z$ bounded. A time-dependent billiard is defined by the potential energy

$$V(x,y,z) = \lim_{N\to\infty} [f(x,y,z)]^N = \begin{cases} 0 & \text{inside the tube} \\ \infty & \text{outside the tube} \end{cases} \tag{2}$$

(for large but finite N, $V(x,y,z) = [f(x,y,z)]^N$ yields a "soft billiard"; formally, the time-dependent billiard problem corresponds to $\gamma \to 0$, $N \to \infty$).

Using the modified metric (1) we can perform time-varying billiard computations in the same way as done for rigid billiards (see eg., [9]). Let $\vec{v}_0$ and $\vec{v}_1$ be the velocity vectors (in the extended phase space $T\mathbf{R}^3$) immediately before and after the collision of the particle with a point $\vec{q}_1$ of the boundary $f(x,y,z) = 1$ (which is now thought as rigid). Then

$$\begin{aligned} \vec{v}_1 &= \vec{v}_0 - 2 < \vec{v}_0, \vec{n} >_\gamma \vec{n} \\ \vec{n} &= \frac{\nabla_\gamma f(\vec{q}_1)}{\| \nabla_\gamma f(\vec{q}_1) \|_\gamma} \end{aligned} \tag{3}$$

where $<,>_\gamma$ is the riemannian metric given by T_γ and ∇_γ its associated gradient operator.

Since

$$\nabla_\gamma f = \left(\frac{\partial f}{\partial x}, \frac{\partial f}{\partial y}, \gamma^2 \frac{\partial f}{\partial z}\right) \tag{4}$$

it follows that, for $\vec{v}_0 = (a_0, b_0, c_0)$,

$$\vec{v}_1 = (a_1, b_1, c_1) \quad = \quad (a_0, b_0, c_0) - \tag{5}$$

$$-2\left[\frac{a_0\frac{\partial f}{\partial x} + b_0\frac{\partial f}{\partial y} + c_0\frac{\partial f}{\partial z}}{\left(\frac{\partial f}{\partial x}\right)^2 + \left(\frac{\partial f}{\partial y}\right)^2 + \gamma^2\left(\frac{\partial f}{\partial z}\right)^2}\left(\frac{\partial f}{\partial x}, \frac{\partial f}{\partial y}, \gamma^2\frac{\partial f}{\partial z}\right)\right]_{|\vec{q}_1} \tag{6}$$

Proposition 1 *Take $c_0 = 1$. Then $\lim_{\gamma \to 0} \vec{v}_1$ exists and is equal to*

$$\vec{v}_1 = (a_1, b_1, 1) \quad = \quad (a_0, b_0, 1) - \tag{7}$$

$$-2\left[\frac{a_0\frac{\partial f}{\partial x} + b_0\frac{\partial f}{\partial y} + \frac{\partial f}{\partial z}}{\left(\frac{\partial f}{\partial x}\right)^2 + \left(\frac{\partial f}{\partial y}\right)^2}\left(\frac{\partial f}{\partial x}, \frac{\partial f}{\partial y}, 0\right)\right]_{|\vec{q}_1} . \tag{8}$$

Conservation of energy holds for any rigid billiard; in particular, for the γ-metric (in the augmented space x, y, z),

$$\| \vec{v}_1 \|_\gamma^2 = \| \vec{v}_0 \|_\gamma^2$$

or equivalently,

$$a_0^2 + b_0^2 + (\frac{1}{\gamma^2})c_0^2 = a_1^2 + b_1^2 + (\frac{1}{\gamma^2})c_1^2.$$

However, we are really more interested in the "x and y part" of the energy. Its change is given by

$$2\Delta E = (a_1^2 + b_1^2) - (a_0^2 + b_0^2) = (\frac{1}{\gamma^2})c_0^2 - (\frac{1}{\gamma^2})c_1^2$$

Taking the limit as $\gamma \to 0$ one gets

Proposition 2 *The energy change at each collision is given by*

$$\Delta E = 2(\frac{\partial f}{\partial z})(\frac{\partial f/\partial x\, a_0 + \partial f/\partial y\, b_0 + \partial f/\partial z}{(\partial f/\partial x)^2 + (\partial f/\partial y)^2})$$

or, equivalently,

$$E_1 = E_0 + 2(f_t/|\nabla f|)(f_t/|\nabla f| + \vec{v}_0 \cdot \nabla f/|\nabla f|)$$

(as before, the partials in the equations below are calculated at the point (q_1, t_1))

3. Hamiltonian structure

Under Legendre's transformation $p_x = x'$, $p_y = y'$, $p_z = (1/\gamma^2)z'$, an autonomous Lagrangian with metric (1) becomes the Hamiltonian

$$H = \frac{1}{2}[p_x^2 + p_y^2 + \gamma^2 p_z^2] + V(x, y, z). \tag{9}$$

To circumnvent the singular behaviour as $\gamma \to 0$, rewrite (9) as

$$\begin{aligned} H &= E + \frac{1}{2\gamma^2} z'^2, \\ E &= \frac{1}{2}[p_x^2 + p_y^2] + V(x, y, z), \end{aligned} \tag{10}$$

with E being the (more relevant) finite part of the energy when γ is infinitesimal. Differentiating, we get

$$dH = dE + z'(1/\gamma^2)dz' = dE + z'dp_z \ , \qquad z' = c + O(\gamma^2),$$

where c is a constant.

On an energy manifold $H = const.$, $dH = 0$ and

$$dp_z = -dE/z' = -dE/c + O(\gamma^2). \tag{11}$$

As $\gamma \to 0$ we fix the constant $c = 1$ $(= z')$ by the equivalent assumption that the constant energy shell $H = const.$ is asymptotically of the form $H = O(1/\gamma^2)$.

The moving billiard mapping iterates consecutive phase space points immediately after collision. The domain of this mapping is *four* dimensional: fixing the total (extended) energy H in (9), one dimension is cut down, so the flow is restricted to a 5-dimensional manifold; the representative points (considered immediately after collision) belong to a 4-dimensional submanifold transversal to the flow in this extended energy hypersurface.

We now change the "cartesian" coordinates x, y, z to generalized coordinates ζ, μ, ξ. Let $x_0 = \alpha(z, \mu)$, $y_0 = \beta(z, \mu)$ represent parametrically the moving boundary $f(x, y, z) = 1$. μ is any convenient parameter for the frozen boundaries (we do not need to assume that the z-curves are orthogonal to the μ-curves). Let ξ describe outward motion along normals (with frozen z). The coordinate transformation $X : (\zeta, \mu, \xi) \longrightarrow (x, y, z)$ is

$$\begin{aligned} x = \ & x(z, \mu, \xi) &= \alpha(z, \mu) + \xi \frac{\partial\beta/\partial\mu}{[(\partial\alpha/\partial\mu)^2 + (\partial\beta/\partial\mu)^2]^{1/2}} \\ y = \ & y(z, \mu, \xi) &= \beta(z, \mu) - \xi \frac{\partial\alpha/\partial\mu}{[(\partial\alpha/\partial\mu)^2 + (\partial\beta/\partial\mu)^2]^{1/2}} \\ z = \ & \gamma\zeta & \end{aligned} \tag{12}$$

Note that $\xi = 0$ means that the particle is instantaneously at the moving boundary. The kinetic energy T is now given by

$$T = (1/2)(\zeta', \mu', \xi')\mathbf{G}(\zeta', \mu', \xi')^{tr} \tag{13}$$

with $\mathbf{G} = A^{tr}A$, where A^{tr} is the matrix whose rows are X_ζ, X_μ, X_ξ. Moreover, Legendre's transformation yields

$$(p_\zeta, p_\mu, p_\xi) = \mathbf{G}\,(\zeta', \mu', \xi')^{tr} = A^{tr}AA^{-1}(x', y', z')^{tr} = A^{tr}(x', y', z')^{tr}$$

Thus for instance, p_μ is equal to the scalar product of X_μ with (x', y', z'). In our case, since $z_\mu \equiv 0$, we have

$$p_\mu = x'x_\mu + y'y_\mu \tag{14}$$

Poincaré's 1-form $\theta = p_x dx + p_y dy + p_z dz$ (see [1] §38) writes, via pullback to the new coordinates, as

$$\theta = x'(x_\mu d\mu + x_\xi d\xi + x_z dz) + y'(y_\mu d\mu + y_\xi d\xi + y_z dz) + p_z dz$$

A 4-dimensional Poincaré section is obtained when we fix the total energy in (9) and take $\xi = 0$. Then θ simplifies into

$$\lambda = \theta_{section} = (x'x_\mu + y'y_\mu)d\mu + (x'x_z + y'y_z + p_z)dz$$

or, in view of (14),

$$\lambda = p_\mu d\mu + (F + p_z)dz$$

where we define the quantity F by

$$F = x'x_z + y'y_z. \tag{15}$$

Finally, from (11) and letting $\gamma \to 0$, we get:

Proposition 3 *The conserved 2-form for the Poincaré section is*

$$\omega = dp_\mu \wedge d\mu + dt \wedge d(E - F)\ . \tag{16}$$

Although the pairs (μ, p_μ), $(E - F, t)$ are canonically conjugate, physically it is more natural to map over $((\mu, p_\mu), (E, t))$.

4. The billiard mapping $\mathcal{T}$

Although not canonically conjugate, we use the variables μ, α (which we call "geometric pair", because they characterize a rigid billiard mapping) and E, t

(the "Fermi pair", as in the 1-dimensional Fermi accelerator problem). α is the angle between the the tangent to the frozen boundary and the billiard ball trajectory immediately after a collision. Denote by α^* the angle between the billiard ball velocity and the instantaneous tangent immediately *before* collision. Let u be the velocity of the boundary in the direction of the exterior normal. In contradistinction with rigid billiards, if the normal velocity u is nonzero at a collision, then $\alpha^* \neq \alpha$. Also, the sines of the angles α and α^* may be either positive or *negative*. There are *four* possibilities at any collision of the ball with the boundary. In case a.2, the ball keeps moving outwardly (no problem, since the boundary moves even faster, outwardly). In case b.2, the billiard ball is struck by the boundary "from behind". These features represent a new twist, in comparison with rigid billiards.

a. $u_1 > 0$ (then necessarily $0 < \alpha_1^* < \pi$).

a.1) $\sqrt{2E_0}sin(\alpha_1^*) > 2u_1$;

a.2) $2u_1 \geq \sqrt{2E_0}sin(\alpha^*) > u_1$.

b. $u_1 \leq 0$. b.1) $sin(\alpha_1^*) \geq 0$;

b.2 $sin(\alpha_1^*) < 0$ (then necessarily $|u_1| > \sqrt{2E_0}|sin(\alpha_1^*)|$).

Proposition 4 *Consider a time-dependent billiard with a smooth moving boundary. For any (x_1, y_1, t_1) in the boundary, and such that $f_t(x_1, y_1, t_1) \neq 0$, assume there is a time interval t_{12} for which $(x_1, y_1, t_1 + t_{12})$ also belongs to the boundary and for which, again, $f_t(x_1, y_1, t_1 + t_{12}) \neq 0$ (this assumption is satisfied, in particular by billiards whose boundaries vary periodically). Consider billiard coordinates (μ, α, E, t) where μ is any parameter for the frozen boundaries. Then the billiard mapping is smooth a.e. in the domain*

$$D = \{(\mu, \alpha, E, t) \in [0, 2\pi) \times [0, 2\pi) \times R^+ \times R \mid (2E)^{1/2}sin(\alpha) + u(\mu, t) > 0\}$$

Moreover, this domain is invariant under the mapping.

Physically, the condition $(2E)^{1/2}sin(\alpha) + u(\mu, t) > 0$ assures that the ball will not cross the moving boundary from inside to outside. The words "a.e" are necessary even for boundaries which are smooth with respect to x, y *and* t. In fact, consider the set C of initial conditions such that, for some iteration n of the mapping, $\vec{v_n} = 2u_{n+1}\vec{\eta}$. C has zero measure (it is given by iteration of a codimension two condition). More interestingly, observe that $\vec{v_0} = \lambda\vec{\eta}$, $\lambda > 2u_1 > 0$, implies $sin(\alpha_1) = 1$, whereas for $2u_1 > \lambda > u_1 > 0$, $sin(\alpha_1) = -1$. Hence, there is an unavoidable discontinuity at $\lambda = 2u_1$.

Heretofore we consider only time dependent billiards such that each frozen boundary $C(t)$ is *convex*. Following Berry [2], we use here as parameter μ the angle φ between the tangent vector and a fixed direction, $\varphi \in [0, 2\pi)$. Thus we have

$$x(\varphi,t) = x_0(t) + \int_0^{\varphi} R(\varphi,t)cos(\varphi)d\varphi$$

$$y(\varphi,t) = y_0(t) + \int_0^{\varphi} R(\varphi,t)sin(\varphi)d\varphi \tag{17}$$

where $R(\varphi, t)$ is the radius of curvature of the instantaneous boundary at the point whose tangent has slope φ. Here $(x_0(t), y_0(t))$ gives the location of a reference point with horizontal tangent.

Elementary geometry yields, in all cases (observe that no harm is done from C_0 and C_1 being different curves):

$$\alpha_0 + \alpha_1^* = \varphi_1 - \varphi_0 \pmod{2\pi} \qquad (\text{hence } d\alpha_1^* = -d\alpha_0 - d\varphi_0 + d\varphi_1) \tag{18}$$

Proposition 5 *Let u_1 be the velocity of the boundary at (q_1, t_1) in the direction of the exterior normal and R the radius of curvature of the instantaneus boundary at the corresponding impact point. Then*

i)

$$u(\varphi,t) = (\frac{\partial x}{\partial t}, \frac{\partial y}{\partial t}).(sin\varphi, -cos\varphi) \tag{19}$$

ii)

$$E_1 = E_0 + 2u_1[u_1 - (2E_0)^{1/2}\sin\alpha_1^*] \tag{20}$$

iii)

$$\cos\alpha_1 = (E_0/E_1)^{1/2}\cos\alpha_1^* \tag{21}$$

iv) *In Poincare's invariant 1-form* (16) $\lambda = p_\varphi d\varphi + [F(\varphi,\alpha,E,t) - E]dt$ *we have:*

$$p_\varphi = (2E)^{1/2}R\cos\alpha, \quad F = (\frac{\partial x}{\partial t}, \frac{\partial y}{\partial t}).\vec{v}, \tag{22}$$

where

$$\vec{v} = \sqrt{2E}(cos(\varphi+\alpha), sin(\varphi+\alpha)) \tag{23}$$

v) *The symplectic form ω (see* (16) *) writes as*

$$\begin{aligned}
\omega &= \sqrt{2E}\,R\,sin(\alpha)\,d\varphi\wedge d\alpha \\
&+ [1 - \frac{1}{\sqrt{2E}}(x_t, y_t).e^{i(\varphi+\alpha)}]\,dt\wedge dE \\
&+ \sqrt{2E}\,[cos(\alpha)\frac{\partial R}{\partial t} - (x_{t\varphi}, y_{t\varphi}).e^{i(\varphi+\alpha)}]\,dt\wedge d\varphi \\
&+ \frac{R\cos(\alpha)}{\sqrt{2E}}\,dE\wedge d\varphi \\
&+ \sqrt{2E}\,(x_t, y_t).(sin(\varphi+\alpha), -cos(\varphi+\alpha))\,dt\wedge d(\varphi+\alpha).
\end{aligned} \tag{24}$$

This 2-form in the extended phase space, in the case of a rigid billiard becomes, interestingly enough,

$$\omega = \sqrt{2E}\, R\, sin(\alpha)\; d\varphi \wedge d\alpha + dt \wedge dE + \frac{R\, cos(\alpha)}{\sqrt{2E}}\, dE \wedge d\varphi. \tag{25}$$

There is an extra term $\frac{R\cos(\alpha)}{\sqrt{2E}}\, dE \wedge d\varphi$, not just the expected one $dt \wedge dE$. Notice that $\sqrt{2E}(t_n - t_{n-1})$, the lengths, can be used as one of the coordinates for a rigid billiard.

The volume form $\omega \wedge \omega$ is simpler and has a nice physical meaning:

Proposition 6 *The volume form* $\nu = \omega \wedge \omega$ *is given by*

$$\nu = \rho(\varphi, \alpha, E, t)\; d\varphi \wedge d\alpha \wedge dt \wedge dE, \qquad \rho = R(\sqrt{2E}\, sin(\alpha) + u). \tag{26}$$

The factor $\rho > 0$ *represents the diference between the normal velocity of the billiard ball (just after impact) and the boundary velocity in the normal direction.*

Remark. Notice that $\rho = 0$ represents the boundary of domain D. Raul Donangelo (personal communication), in numerical experiments for the liquid drop model in nuclear physics, observed that for "many" initial conditions ρ eventualy decreases to nearly zero, situation in which the ball moves close to the boundary for a long while. As a plausible explanation, consider a a small nonzero symplectic volume K for which ρ is bounded from below by a positive constant. Iterate $\mathbf{T}^n K$; suppose that ρ becomes small on a subset K' of K. *Conservation of symplectic volume forces the euclidian volume of* K' *to grow.*

We now describe the jacobian $\mathbf{J}(t_k, t_{k+1})$ of $\mathbf{T}$.

Proposition 7 *("bank of differentials" for the* 4×4 *jacobian* $J(t_k, t_{k+1})$ *).*

i) $dt_1 = J_{41} d\varphi_0 + J_{42} d\alpha_0 + J_{43} dE_0 + J_{44} dt_0$

$$\begin{aligned}
J_{41}[-\sin\alpha_1^*(2E_0)^{1/2} + u_1] &= -t_{01}\cos\alpha_1^*(2E_0)^{1/2} - R_0\sin(\varphi_0 - \varphi_1)\\
J_{42}[-\sin\alpha_1^*(2E_0)^{1/2} + u_1] &= -t_{01}(2E_0)^{1/2}\cos\alpha_1^*\\
J_{43}[-\sin\alpha_1^*(2E_0)^{1/2} + u_1] &= t_{01}\sin\alpha_1^*/(2E_0)^{1/2}\\
J_{44}[-\sin\alpha_1^*(2E_0)^{1/2} + u_1] &= -\sin\alpha_1^*(2E_0)^{1/2} + u_0\cos(\varphi_0 - \varphi_1)
\end{aligned}$$

ii) $d\varphi_1 = J_{11} d\varphi_0 + J_{12} d\alpha_0 + J_{13} dE_0 + J_{14} dt_0$

$$\begin{aligned}
R_1 J_{11} &= R_0\cos(\varphi_0 - \varphi_1) + t_{01}\sin\alpha_1^*(2E_0)^{1/2} + (2E_0)^{1/2}\cos\alpha_1^* J_{41}\\
R_1 J_{12} &= t_{01}\sin\alpha_1^*(2E_0)^{1/2} + (2E_0)^{1/2}\cos\alpha_1^* J_{42}\\
R_1 J_{13} &= t_{01}\cos\alpha_1^*/(2E_0)^{1/2} + (2E_0)^{1/2}\cos\alpha_1^* J_{43}\\
R_1 J_{14} &= u_0\sin(\varphi_0 - \varphi_1) + (2E_0)^{1/2}\cos\alpha_1^*(-1 + J_{44})
\end{aligned}$$

iii) $$dE_1 = J_{31}d\varphi_0 + J_{32}d\alpha_0 + J_{33}dE_0 + J_{34}dt_0$$

$$\begin{aligned}
J_{31} = &\quad 2J_{11}\partial u_1/\partial\varphi_1[2u_1 - \sin\alpha_1^*(2E_0)^{1/2}] + \\
&+ 2J_{41}\partial u_1/\partial t_1[2u_1 - \sin\alpha_1^*(2E_0)^{1/2}] + \\
&+ 2u_1\cos\alpha_1^*(1 - J_{11})(2E_0)^{1/2} \\
J_{32} = &\quad 2J_{12}\partial u_1/\partial\varphi_1[2u_1 - \sin\alpha_1^*(2E_0)^{1/2}] + \\
&+ 2J_{42}\partial u_1/\partial t_1[2u_1 - \sin\alpha_1^*(2E_0)^{1/2}] + \\
&+ 2u_1\cos\alpha_1^*(1 - J_{12})(2E_0)^{1/2} \\
J_{33} = &\quad 1 + 2J_{13}\partial u_1/\partial\varphi_1[2u_1 - \sin\alpha_1^*(2E_0)^{1/2}] + \\
&+ 2J_{43}\partial u_1/\partial t_1[2u_1 - \sin\alpha_1^*(2E_0)^{1/2}] - \\
&- 2u_1[\cos\alpha_1^* J_{13}(2E_0)^{1/2} + \sin\alpha_1^*/(2E_0)^{1/2}] \\
J_{34} = &\quad 2J_{14}\partial u_1/\partial\varphi_1[2u_1 - \sin\alpha_1^*(2E_0)^{1/2}] + \\
&+ 2J_{44}\partial u_1/\partial t_1[2u_1 - \sin\alpha_1^*(2E_0)^{1/2}] - \\
&- 2u_1\cos\alpha_1^* J_{14}(2E_0)^{1/2}
\end{aligned}$$

iv) $$d\alpha_1 = J_{21}d\varphi_0 + J_{22}d\alpha_0 + J_{23}dE_0 + J_{24}dt_0$$

$$\begin{aligned}
-\sin\alpha_1 J_{21} &= (E_0/E_1)^{1/2}[\sin\alpha_1^*(1 - J_{11}) - J_{31}\cos\alpha_1^*/(2E_1)] \\
-\sin\alpha_1 J_{22} &= (E_0/E_1)^{1/2}[\sin\alpha_1^*(1 - J_{12}) - J_{32}\cos\alpha_1^*/(2E_1)] \\
-\sin\alpha_1 J_{23} &= -(E_0/E_1)^{1/2}\sin\alpha_1^* J_{13} + \cos\alpha_1^*/(2(E_0E_1)^{1/2}) - \\
&\quad -J_{33}(E_0/E_1)^{1/2}\cos\alpha_1^*/(2E_1) \\
-\sin\alpha_1 J_{24} &= (E_0/E_1)^{1/2}[-\sin\alpha_1^* J_{14} - J_{34}\cos\alpha_1^*/(2E_1)]
\end{aligned}$$

(Observe the "telescopic" structure of these formulas)

Remark. For the implementation, it is also important to have formulas for the factors $u_1 = \partial u_1/\partial t_1$ and $\partial u_1/\partial\varphi_1$ appearing in the Jacobian. From (19) we get

i)

$$\frac{\partial u}{\partial\varphi} = (\frac{\partial x}{\partial t}, \frac{\partial y}{\partial t}).(cos\varphi, sin\varphi) \tag{27}$$

(this is precisely the tangential velocity of the boundary in the (φ, t) parametrization; notice that differentiating u with respect to φ should produce two terms, but $(\frac{\partial^2 x}{\partial t\partial\varphi}, \frac{\partial^2 y}{\partial t\partial\varphi}).(sin\varphi, -cos\varphi)$ vanishes).

ii)

$$\frac{\partial u}{\partial t} = (\frac{\partial^2 x}{\partial t^2}, \frac{\partial^2 y}{\partial t^2}).(sin\varphi, -cos\varphi) \tag{28}$$

5. An example

In this example, we assume that the moving billiard has a fixed diametral line (say, the y-axis), which would be stable for every "frozen" billiard. As it is well known (see e.g. [2]), a diametral line for a rigid planar billiard is stable if the oscullating circles at the impact points intersect. We will show that *due to the periodic time-dependence of the boundary, the trajectory can become unstable spatially.*

At the points of contact q_0, q_1, q_2, ... in the y-axis,

$$\alpha_0 = \alpha_1 = ... = \pi/2 \quad , \quad \varphi_1 - \varphi_0 = \pi \quad , \text{etc.}$$

the boundary may have nonzero velocities. Nonetheless, from the remark above it follows that $\frac{\partial u}{\partial \varphi} = 0$ so the Jacobian matrices $J(t_k, t_{k+1})$ are block diagonal. The "Fermi part" B (i.e, the E, t block) depends only on the restricted dynamics to the 1-dimensional Fermi oscillator, and is given by

$$J_{43} = \frac{(-1)^{\rho+1} t_{01}}{2E_0[(-1)^\rho - u_1/\sqrt{2E_0}]}$$
$$J_{44} = \frac{(-1)^\rho + u_0/\sqrt{2E_0}}{(-1)^\rho - u_1/\sqrt{2E_0}}$$
$$J_{33} = 1 + 2J_{43}(\partial u_1/\partial t_1)[2u_1 + (-1)^{\rho+1}\sqrt{2E_0}] + 2(-1)^{\rho+1} u_1/\sqrt{2E_0}$$
$$J_{34} = 2J_{44}(\partial u_1/\partial t_1)[2u_1 + (-1)^{\rho+1}\sqrt{2E_0}]$$

A novel feature is the contribution of the Fermi oscillator dynamics to the "geometrical block" A, corresponding to the variables φ, α. Namely, factors of the form $\kappa = \sqrt{E_0/E_1} = |v_0/v_1|$ (v is the billiard ball velocity) are present:

$$J_{11} = -R_0/R_1 + (-1)^\rho\, l_{01}/R_1$$
$$J_{12} = (-1)^\rho\, l_{01}/R_1$$
$$J_{21} = (-1)^\sigma \sqrt{E_0/E_1}(-1 + J_{11})$$
$$J_{22} = (-1)^\sigma \sqrt{E_0/E_1}(-1 + J_{12})$$

Sign conventions are as follows, according to the four possible cases (there is an undefined case, $\sqrt{2E_0} = 2u_1$, when the billiard mapping is discontinuous):

a) $u_1 > 0$: Then $\rho = 0$ and there are two subcases,

a.1 $\sqrt{2E_0} > 2u_1$. Here $\sigma = 0$;

a.2 $2u_1 > \sqrt{2E_0} \geq u_1$. Here $\sigma = -1$.

b) $u_1 < 0$:

b.1 $sin(\alpha_1^*) = 1$. Then $\rho = \sigma = 0$.

b.2 $sin(\alpha_1^*) = -1$. Then $\rho = \sigma = -1$.

In the latter, the boundary hits the ball from behind, with $|u_1| \geq \sqrt{2E_0}$.

In case a) ($u_1 > 0$), the ball velocity decreases at collision, i.e., $\kappa^{-1} \in [0,1)$. Indeed, $v_0 > u_1, v_1 = v_0 - 2u_1$ imply $-v_0 < v_1 < v_0$. In case b) $(u_1 < 0)$, the velocity increases in absolute value: $\kappa^{-1} \in (1, \infty)$. Indeed, since $u_1 < v_0 < \infty$, then $|1 - 2u_1/v_0| > 1$.

In the sequel we assume, for concreteness, that at the collision points only cases a.1 and b.1 will occur. To simplify the analysis, we assume that all curvatures are equal, $R_0 = R_1 = R_2 = ... = R$; we assume also that the distances are all equal, $l_{01} = l_{12} = l_{23} = ... = l$. Under those asumptions, the quantity of interest is the ratio $p = l/R$. We denote by A_j the geometrical block corresponding to the linearization of the mapping from (q_j, t_j) to (q_{j+1}, t_{j+1}). Let $\kappa_j = \sqrt{E_j/E_{j+1}}$. A short computation gives

$$A_1 A_0 = \begin{pmatrix} (p-1)^2 + \kappa_0\, p(p-2) & (\kappa_0 + 1)\, p(p-1) \\ \kappa_1(\kappa_0+1)\,(p-1)(p-2) & \kappa_1\,[p(p-2) + \kappa_0(p-1)^2\,] \end{pmatrix} \tag{29}$$

A 2-bounce periodic orbit happens when $\kappa_0\kappa_1 = 1$. The stability index $\chi = \frac{1}{2}\, tr(A_1 A_0)$ is equal to

$$\chi(p;\kappa_0) = (p-1)^2 \; + \; \frac{1}{2}\,(\kappa_0 + \frac{1}{\kappa_0})\; p(p-2) \tag{30}$$

For $\kappa_0 = 1$ (which corresponds to the rigid billiard) we have $\chi = (p-1)^2 + p(p-2)$, and the stability region $|\chi| \leq 1$ is the interval $0 \leq p = l/R \leq 2$, which is the well known result. When $\kappa_0 \neq 1$ (without loss of generality, we may assume $\kappa_0 > 1$), the stability region shrinks to

$$l/R \in [\,0\,,\, \frac{2}{\kappa_0+1}] \cup [\frac{2\kappa_0}{\kappa_0+1}\,,\, 2\,]. \tag{31}$$

This set empties as $\kappa_0 \to \infty$. Hence, through the effect of boundary vibrations, it is possible to unstabilize a diametral orbit which would stable for every one of the underlying rigid billiards.

6. References

1. V.I.Arnold: *Mathematical Methods of Classical Mechanics* (Springer-Verlag, Berlin, Heidelberg, New York, 1978).

2. M.V.Berry, Regularity and chaos in classical mechanics, illustrated by three deformations of a circular billiard, *Eur.J.Phys.* **2** (1981) 91-102.
3. R.Douady, *Applications du théorème des tores invariants* (Thèse de 3ème Cycle, Univ. Paris VII,1982).
4. E.Fermi, On the origin of the cosmic radiation, *Phys.Rev.* **75** (1949), 1169-1174
   (see also S.Ulam, in *Proc. 4th Berkeley Symp. on Math.Stat. and Prob., vol 3,* (University of California Press, Berkeley, 1961), p. 315.)
5. J.Glanz, Time-reversed sound waves resonate among physicists. *Science* **265** (1994), 474-475.
6. S.Laederich and M. Levi, Invariant curves and time dependent potentials, *Erg.th.& Dyn.Syst.* **11** (1991), 365-378 .
7. M.Levi, Quasiperiodic motions in superquadratic time-periodic potentials, *Commun.Math.Phys* **143** (1991), 43-83 .
8. A.J.Lichtenberg and M.A. Liebermann: *Regular and Stochastic Motion,* (Springer-Verlag, Berlin, Heidelberg, New York, 1983).
9. R.Markarian, Introduction to the ergodic theory of plane billiards, in *Dynamical Systems, Santiago de Chile, 1990* (Harlow, Longman, 1992).
10. A.Pais, Niels Bohr's Times, in *Physics, Philosophy and Policy* (Oxford University Press, New York, 1991)
    (see also the review by F.Wilczek, *Science* **255** (1992), p. 345-347.
11. L.D. Pustyl'nikov, New mechanism of particle acceleration and relativistic analog of the Fermi–Ulam model, *Th.Math.Phys.* **77**:1 (1988), 1110-1115.
12. L.D. Pustyl'nikov, A new mechanism of particle acceleration and rotation numbers, *Th.Math.Phys.* **82**:2 (1990), 180-187.

ANNULUS AND BILLIARD MAPS WITH ASYMPTOTIC ENTROPY

JAREK KWAPISZ*
Mathematics Department, SUNY at Stony Brook
NY 11794-3651, USA
E-mail: jarek@sunysb.math.edu

and

RICHARD SWANSON†
Mathematical Sciences, Montana State University
Bozeman, MT 59717-0240, USA
E-mail: dswanson@math.montana.edu

ABSTRACT

In this paper we derive some properties of a new geometric-topological invariant defined in Swanson[14] and Kwapisz-Swanson[11] that measures the rotational complexity of annulus maps, called asymptotic or rotational entropy. We completely characterize this concept for circle maps of degree one as a linear function of the diameter of the rotation interval. For annulus maps, the authors have proved in prior work[11] that Hölder C^1 diffeomorphisms with nonvanishing asymptotic entropy are isotopic rel a finite set to pseudo-Anosov maps. We provide a brief description of the proof of that result. Finally, in this paper, we prove that nonvanishing asymptotic entropy implies the existence of infinitely many periodic orbits and a nontrivial rotation set. We apply the concept to billiard maps to prove integrable billiard problems have continuous rotation numbers, and, more significantly, rotation discontinuities force generalized rotary horseshoes in billiard maps.

1. Introduction, definitions and main results

Smale horseshoes on the annulus may have trivial rotation sets, despite their complicated dynamics. Horseshoes which give rise to rotation intervals of positive length are called rotary horseshoes. When rotation numbers of periodic orbits vary discontinuously in a precise sense, one can infer the existence of pseudo-Anosov orbits: the original annulus map is pseudo-Anosov rel a finite invariant set (Handel[9]). Asymptotic entropy, defined below, measures this kind of rotational complexity very well. In the second section, devoted to circle maps, we derive a formula for asymptotic entropy in terms of the diameter of the rotation set. This is as it should be. For circle maps, the rotation interval

*Supported in part by Polish Academy of Sciences grant #210469101 "Iteracje i Fraktale"
†Supported in part by National Science Foundation (EPSCoR-DMS): #OSR 93-50-546

completely describes the rotational behavior and is decisive for the existence of rotary horseshoes.

On the annulus, the authors have proved[11] that positive asymptotic entropy implies rotational discontinuities for sufficiently smooth diffeomorphisms. Homeomorphisms of the annulus that are pseudo-Anosov rel a finite set have a nontrivial rotation interval relative to any boundary component. That result is an important motivation for the study of this kind of entropy. It follows that smooth annulus maps with positive asymptotic entropy have infinitely many periodic orbits corresponding to an interval of rotation numbers. The assumption of smoothness may be unnecessary, but we cannot prove this yet.

However, for the homeomorphism case, we can prove in Section 3 that nonvanishing rotational entropy implies a nontrivial rotation interval and, hence, infinitely many periodic points, as in Section 3.

In the last section, we relate these results to the problem of characterizing nonintegrability in billiard maps. We show that in the integrable case, evidently, rotation numbers are well defined and vary continuously. If the rotation is ill-defined or discontinuous — under the generic expectation that periodic points are dense — then we can conclude the existence of pseudo-Anosov orbits.

1.1. The Definition of Asymptotic Entropy

Let $\mathbf{A}$ be the annulus $S \times [0,1]$ i.e. the quotient of $\tilde{\mathbf{A}} = \mathbf{R} \times [0,1]$ under the action of $\mathbf{Z}$ by integer translation along $\mathbf{R}$. We will refer to the quotient map as $\pi : \tilde{\mathbf{A}} \to \mathbf{A}$. On the strip $\tilde{\mathbf{A}}$ we have the Euclidean metric $\tilde{d}$. There is a unique corresponding metric d on $\mathbf{A}$.

From now on it will be assumed that we are given a homeomorphism $f : \mathbf{A} \to \mathbf{A}$, with f isotopic to the identity transformation. We also fix a lift $F : \tilde{\mathbf{A}} \to \tilde{\mathbf{A}}$ of f. This lift determines the *displacement function* of F, $\phi_F : \mathbf{A} \to \mathbf{R}^2$, obtained as the factor of $(j \circ F - j \circ id) : \tilde{\mathbf{A}} \to \mathbf{R}^2$ where j is the embedding of $\tilde{\mathbf{A}}$ into $\mathbf{R}^2$.

Definition 1.1 *For $n \in \mathrm{N}$. Define $\tilde{d}_n$ a metric on $\tilde{\mathbf{A}}$ by the formula:*

$$\tilde{d}_n(x,y) := \max\{\tilde{d}(F^i(x), F^i(y)) : 0 \le i \le n-1\}.$$

We will say that a set $E \subset \tilde{\mathbf{A}}$ is $(\tilde{d}_n, R)$*-separated* if the distance between any two points in E is at least R with respect to $\tilde{d}_n$ metric.

Definition 1.2 *For a subset X of $\mathbf{A}$ and $R > 0$ we define :*

$$\tilde{s}_X(d_n, R) = max\ \#\ of\ (\tilde{d}_n, R) - separated\ subset\ of\ \pi^{-1}(X) \cap [0,1]^2;$$

and the 'R-entropy':

$$\tilde{h}_X(R) := \limsup_{n\to\infty} \frac{1}{n} \log(\tilde{s}_X(d_n, R));$$

Definition 1.3 *For a subset X of $\mathbb{A}$ topological entropy h_X^{top} is defined as*

$$h_X^{top} = \lim_{R\to 0} h_X(R)$$

and asymptotic entropy (called also rotational entropy) as

$$h_X^{as} = \limsup_{R\to\infty} R \cdot \tilde{h}_X(R).$$

Remarks 1.4 *The reader may be accustomed to defining h_X^{top} directly in the base space $\mathbb{A}$, but the two formulations are easily seen to be equivalent.*

Asymptotic entropy is independent of the choice of lift F and of fundamental domain. The reader may prefer to define either kind of entropy using minimal spanning sets, which is equivalent for topological entropy but may differ for asymptotic entropy. However, the spanning version of asymptotic entropy is positive if and only if the separated version is positive, the exact value being of less importance.

What kind of invariant is asymptotic entropy? Technically, it is a geometric invariant, as it depends on the choice of lift metric. If we fix the metric in the lift, however, then rotational entropy is a true topological invariant like topological entropy.

In the next section, we discuss some easy results about asymptotic entropy for circle maps.

1.2. Asymptotic Entropy for Circle Maps

We want to prove the following:

Proposition 1.5 *Let $f : \mathbb{S}^1 \to \mathbb{S}^1$ denote a circle map of degree 1. Fix a lift F of f. The asymptotic entropy of f is exactly*

$$h^{as}(F) = Diam\,\rho(F) \cdot \log 2.$$

This means that for circle maps asymptotic entropy is indistinguishable from knowledge of the rotation set diameter. Rotation sets for circle maps are always closed intervals[2] I. Thus, asymptotic entropy reduces to a well understood concept for circles. In particular, positive asymptotic entropy implies the existence of "rotary horseshoes".

We will need the additivity property, which holds for circle or annulus maps:

Lemma 1.6 $h_X^{as}(F^m) = m \cdot h_X^{as}(F)$.

Proof:

Clearly any (F^m, n, R)-separated pair x, y is (F, mn, R)-separated. So it is easily verified that the "$\le$" part of the inequality is established. By uniform continuity, there exists a constant C_m depending only on m and the choice of metric such that for sufficiently large R,

$$\tilde{d}(x,y) \le R \Rightarrow \tilde{d}_m(x,y) \le R + C_m.$$

Hence, if $E(F,n,R)$ denotes a set of maximal cardinality comprised of (n,R)-separated points in the same fundamental domain, we have

$$\text{card}\, E(F, mn, R + C_m) \le \text{card}\, E(F^m, n, R).$$

Therefore,

$$\tilde{h}(F, R + C_m) \le \frac{1}{m}\tilde{h}(F^m, R),$$

which completes the proof of the lemma. □

Now we return to the proof of the proposition. Since the subject is extremely well known, we provide only a sketch of the ideas involved.

Proof: Suppose $\rho(F) = [\alpha, \beta] \supset [p/q, r/q]$. There exist disjoint intervals $I_0 = [x_0, y_0]$ and $I_1 = [x_1, y_1]$ in $[0,1]$ such that the lift $G := T^{-p}F^q$ of f^q fixes x_0 in I_0 and $F(y_1) = y_1 + (r - p)$. In fact, one can choose I_0 and I_1 so that their images under G cover $I_0 \cup I_1 + (r - p)$. Points of I_0 do not "go around the circle". Points of I_1 go around essentially $r - p$ times. This defines a rotary horseshoe $f^q|_\Lambda$.

It is not hard to compute the asymptotic entropy of such a rotary horseshoe:

$$h_\Lambda^{as} = h_\Lambda^{as}(G) = (r - p) \cdot \log 2$$

We can illustrate the former in the special case, $r = 1, p = 0, q = 1$. Points in I_0 advance 0 units in $\mathbb{R}$, while points of I_1 move 1 unit. The shift based on the Markov rectangles consisting of $\pi(I_i)$, $i = 0, 1$ is the full two-sided two shift. Fix a value of $R > 0$, and write $n = kR + m$, $0 \le m < R$. Then a maximal (n, R) separated set E is given by lifts of words of length kR which are a union of words of length R consisting entirely of 1's or 0's. The remainder m is bounded and makes no contribution to the exponential growth rate. The cardinality of E is 2^k. The R-asymptotic entropy is $\tilde{h}(R) = \limsup_{n\to\infty}(k/n) \cdot \log 2 = (1/R) \cdot \log 2$. The additivity lemma above allows us to divide by q, if that is the common denominator of the endpoints. The reverse inequality is established below in Lemma 2.1. □

2. Prior Results: Hölder diffeomorphisms and Pseudo-Anosov Behavior

Let $F : \widetilde{\mathbf{A}} \to \widetilde{\mathbf{A}}$ be a lift of an orientation preserving and boundary preserving homeomorphism $f : \mathbf{A} \to \mathbf{A}$. Let X denote a compact subset of $\mathbf{A}$. We begin by finding an upper bound for $h_X^{as}(f)$ in terms of the rotation set $\rho_X(F)$. This bound is proved in Kwapisz-Swanson[11], but it will convenient to give the proof here also.

Lemma 2.1 $h_X^{as}(f) \leq DIAM(\rho_X(F)) \cdot \log 2$.

Proof: Let $\{F^n(x) = x(n)\}$ denote a positive trajectory of the the lift F, and pr_1 is projection onto the first coordinate. Choose $\rho^+, \rho^- \in \mathbb{R}$ so that $\rho(f, X) \subset (\rho^-, \rho^+)$. Note that we get, for sufficiently high n :

$$(*) \quad \rho^- \leq pr_1(x(n) - x(0))/n \leq \rho^+.$$

Therefore, there is a constant $C > 0$ such that for all $n \geq 1$

$$(**) \quad \rho^- - C/n \leq pr_1(x(n) - x(0))/n \leq \rho^+ + C/n.$$

In fact, by the compactness of the fundamental domain and periodicity of $F^n - Id$, C can be chosen independently of $x(0)$.

Let $R > 0$ be as in the definition of asymptotic entropy.

Define

$$m = R/(\rho^+ - \rho^-).$$

If m is not already an integer, increase ρ^- and decrease ρ^+ so that m is now an integer and $(*)$ is true for $x(m)$. We may have to increase R, but R is independent of $\rho^\pm$ except for having to be sufficiently large. This may also increase C in $(**)$. Now fix those values of m, R, and C.

Let D be an R-box; i.e., $D := pr_1^{-1}(I)$ where I is an arbitrary interval of length R. Let $x(i), y(i)$ be two trajectories in the universal cover both originating in D.

Claim 2.2 *For the above choice of m, the following holds:*

(1) both $x(m)$ and $y(m)$ are in one of the two boxes $D + m\rho^+$ or $D + m\rho^-$;

(2) For $n = 0, 1, 2, ..., m$, $d(x(n), y(n)) \leq R + 2C + 2$.

Proof of Claim. (Part 1) From $(*)$ we get $\inf I + m\rho^- \leq pr_1(x(m)) \leq \sup I + m\rho^+$. That is, $x(m)$ lies in a region swept out as we slide $D + m\rho^-$ to $D + m\rho^+$. But by the choice of m there is no gap between the two. □

(Part 2) We consider the case when $x(m) \in D+m\rho^+$. The other case is similar. Apply (**) to the trajectory $x(i)$ twice — once starting from $x(0)$ and the second time starting from $x(m)$ and running back in time. We get:

$$pr_1(x(n)) \leq pr_1(x(0)) + n\rho^+ + C;$$

$$pr_1(x(n)) \geq pr_1(x(m)) - (m-n)\rho^+ - C.$$

Since $x(0) \in D$ and $x(m) \in D + m\rho^+$ we see that:

$$\inf I + n\rho^+ - C \leq pr_1(x(n)) \leq \sup I + n\rho^+ + C.$$

Of course we have an analogous inequality for $y(n)$. "Subtracting" the two, yields $pr_1(x(n) - y(n)) \leq R + 2C$. Our claim follows. □

The claim insures that we can match to each trajectory $x(n)$ a sequence of signs $\sigma_i = \pm$ so that $x(i \cdot m) \in D + m\rho^{\sigma_1} + ... + m\rho^{\sigma_i}$. Also this correspondence is injective on $(R+2C+2)$-separated trajectories by (2). Trivial counting implies that $\tilde{h}(R+2C+2) \leq \log 2/m$. The Lemma follows.

□

Let $cl(Per(f))$ denote the closure in **A** of the set of periodic points of f. Handel has provided a somewhat technical criterion[9], called the *pA-Hypothesis*, for an annulus homeomorphism to be pseudo-Anosov rel a finite set. We have proved elsewhere[11] that an equivalent criterion is for the rotation number assignment $\pi(x) \mapsto \rho(F, x)$ to be either not well-defined or not continuous on the set $cl(Per(f))$.

Theorem 2.3 (Handel's Lemma[9]) *Suppose that $f : \mathbf{A} \to \mathbf{A}$ satisfies the pA-hypothesis. The f is isotopic rel a finite invariant set to a pseudo-Anosov map ϕ. Moreover, the rotation set of f contains the rotation set of ϕ, with respect to the original annulus covering space.*

Theorem 2.4 (Theorem A.) *Suppose that f is an annulus homeomorphism. If we have $h^{as}_{cl(Per)} > 0$ then f satisfies the pA-hypothesis.*

Theorem 2.5 (Theorem B.) *If f is a $C^{1+\epsilon}$-smooth annulus diffeomorphism then*

$$h^{as}_{cl(Per)} \geq \frac{1}{2} \cdot h^{as}.$$

If we assemble these latter three results, we get

Theorem 2.6 (Kwapisz-Swanson[11]) *If f is a Hölder smooth annulus diffeomorphism isotopic to the identity and the asymptotic entropy of f is nonvanishing, then f is isotopic rel a finite invariant set to a pseudo-Anosov map.*

Remarks 2.7 *The proof is too long and technical to be included here. The proof proceeds by establishing that at least half the asymptotic entropy can be realized on the supports of invariant measures with nonzero Lyapunov exponents. Such measures are called hyperbolic. When f is a Hölder diffeomorphism hyperbolic measures are supported in the closure of the periodic points by a result due to Katok*[10].

3. The C^0 case: asymptotic entropy implies periodic orbits

In this section we will prove that positive asymptotic entropy implies the existence of a chain transitive set having a nontrivial rotation set. This, in turn, by a result of J. Franks[7], implies the existence of an interval of rotation numbers and infinitely many periodic orbits in the annulus. This correlates well with the Hölder smooth case and suggests that one might be able to remove the smoothness requirement in Theorem A and obtain the pA hypothesis simply given nonvanishing rotational entropy .

3.1. Asymptotic entropy and chain transitive components

There are many good references for chain recurrence, and we refer the reader to any of those (e.g. Conley[4]) for detailed background information.

Definition 3.1 *An f-periodic ϵ-chain is a sequence $\{x_0, \ldots, x_n = x_0\}$ with $d(f(x_i), x_{i+1}) < \epsilon$ for $0 \leq i < n$ and $n > 2$. The point x is chain recurrent if for every $\epsilon > 0$, there exists a periodic ϵ chain containing x. Two points x and y lie in the same chain transitive component C, if for all $\epsilon > 0$, there exists a periodic ϵ chain containing x and y.*

As in Remark 2.7, at least one half of the total asymptotic entropy is attained on the closure of the union of supports of the invariant measures of f. But every invariant measure is supported in the closure of the set of f-recurrent points by the Poicar/'e recurrence theorem (e.g. Walters[18]).

Therefore, we can conclude

Proposition 3.2 *If asymptotic entropy is positive, then the asymptotic entropy is positive on the chain recurrent set CR.*

However, the following conjecture remains open:

Conjecture 3.3 *Let f denote an annulus homeomorphism isotopic to the identity with chain recurrent set CR. If $h^{as}_{CR}(f) > 0$, then there exists a chain transitive component C of CR such that $h^{as}_{C}(f) \geq (1/2)h^{as}_{CR}(f)$.*

Such a result is not difficult for topological entropy , but scaling differences between topological and asymptotic entropy require a new argument in the case of asymptotic entropy. We will prove, however, that for arbitrary homeomorphisms isotopic to the identity, positive asymptotic entropy forces infinitely many periodic orbits. To do this, we will need the following lemma, due to J. Franks[7]:

Lemma 3.4 *If a chain transitive set contains points x and y with rotation numbers $\rho(f,x) = r$ and $\rho(f,y) = s$, $r < s$, then*

a) the full rotation set of f contains the interval $[r,s]$, and

b) for each reduced rational $p/q \in [r,s]$ there exists a periodic point of period q and rotation number p/q.

We need two additional lemmas:

Lemma 3.5 *Let X denote a chain transitive subset of $f : \mathbf{A} \to \mathbf{A}$ such that $\rho(f,X) \subset (a,b)$. then there exists a number $\delta > 0$ such that if $d(x,X) < \delta$, then $\rho(f,x) \subset (a,b)$.*

Lemma 3.6 *If every chain transitive set admits at most a single rotation number, then the rotation number mapping is continuous on the set of chain recurrent points.*

For a proof of the first lemma, see Swanson[15]. where a slightly stronger result is proved. The second lemma is a corollary of the first.

Theorem 3.7 *If asymptotic entropy is nonvanishing, then there are infinitely many periodic orbits corresponding to a nontrivial rotation interval.*

Proof: If some chain component admits more than one rotation number then the theorem follows from the result of Franks cited above. Otherwise, the rotation number mapping is continuous on the chain recurrent set by the last lemma. If so, for each $\delta > 0$ there is finite cover of the chain recurrent set by compact subsets X_i, $i = 1,\ldots,N$ whose rotation sets each have diameters less than δ. However, one of X_i must support full asymptotic entropy $h > 0$. Since by Lemma 2.1, $h \leq (\log 2)\delta$, we arrive at a contradiction, since δ was arbitrary, unless $h^{as} = 0$. □

4. Area Preserving and Billiard Maps of the Annulus

Consider an area preserving diffeomorphism $f : \mathbf{A} \to \mathbf{A}$. Such maps arise naturally for billiard problems[16] and in celestial mechanics. As J. Franks[7] has noted elsewhere, the area preserving case enjoys many advantages in rotation number theory. The recurrent set is now the entire annulus. If the annulus map admits two distinct rotation numbers, then the map realizes the full interval of rotation numbers between those extremes, including infinitely many reduced rationals corresponding to periodic orbits[7].

From Section 2, we can easily infer the following theorem.

Theorem 4.1 *Suppose that* $f : \mathbf{A} \to \mathbf{A}$ *is a* C^2 *area preserving annulus diffeomorphism, whose periodic points are dense. If this map is integrable, then the rotation number map* $\rho : \mathbf{A} \to \mathbb{R}$ *is well-defined and continuous.*

Proof: If the map ρ is not well-defined or is discontinuous, Handel's criterion is met for the pA-hypothesis (as in Theorem 2.1). As a consequence, there exist finite pseudo-Anosov orbits and a compact set K such that $f|_K$ is semiconjugate to a rotary horseshoe. No such map can be integrable. □

If the conclusions of the last theorem hold, then we shall say that f admits a *generalized rotary horseshoe.*

4.1. Billiard Maps

I would like to extend my thanks to the references[8,19] for helpful conversations concerning the material of this section. We remind the reader of the reduction of billiard orbits to the study of area preserving annulus maps. Parameterize the convex table T by a circle coordinate $p \in \mathbb{S}^1$. Trajectories are geodesics or lightrays which meet the table and exit at a common angle α. Consider the space $K := (0, \pi) \times \mathbb{S}^1$. The map $f : K \to K$ is such that $f(\alpha, p) = (\beta, q)$, where β is the incident angle at point q in the elastic collision immediately following the one at p. Now K is an open annulus. However, for *convex* billiards we may compactify to a compact annulus and we shall assume that this has been carried out, replacing K with $\mathbf{A}$, the usual annulus.

The Aubry-Mather theorem asserts that the set of rotation numbers is precisely the interval whose endpoints are attained on the boundary circles. More significantly, for each number in this interval there corresponds a periodic orbit or, in the irrational case, a Denjoy minimal set or (minimal) invariant circle.

Sometimes called *the Birkhoff conjecture*, the assertion that a billiard map is integrable iff the billiard table is an ellipse has long defied the most powerful methods of analysis and topology.

For our purposes we will use the version due to S. Tabachnikov[16]:

Conjecture 4.2 (Birkhoff's Billiards' Conjecture) *If a neighborhood of a strictly convex smooth billiard curve is foliated by caustics, then the curve is an ellipse.*

The result then implies that the entire billiard disc is foliated by confocal ellipses[16].

We certainly will not resolve this problem in the present discussion! However, one can at least sharpen the problem a little by looking for pseudo-Anosov orbits as in the previous theorem.

We make the following standing assumption:

(*) Suppose every nonempty open set of nonperiodic points of the billiard transformation has a constant (irrational) rotation number.

In fact, it seems to be open whether there *can* be an open set of nonperiodic points.

Theorem 4.3 *For integrable billiard problems the rotation number is continuous in a neighborhood of the annulus boundary curve corresponding to the billiard table curve. If the rotation mapping is ill defined or discontinuous at some phase cylinder point then, in fact, there is a generalized rotary horseshoe in the phase space cylinder.*

Proof: In the integrable case the phase cylinder is foliated into invariant curves. The curves correspond to caustics in the billiard table. Such an invariant curve has a constant rotation number, since that is true of homeomorphisms on circles. The rotation number varies continuously, for that is true of one-parameter families of circle maps.

For area preserving twist maps an open set of nonperiodic points must have a common irrational rotation number ω. It follows that if ρ is ill-defined or discontinuous on the annulus, then the rotation map is ill-defined or discontinuous on the closure of the periodic orbits. □

Conjecture 4.4 *Suppose the billiard curve is smooth and strictly convex. The asymptotic entropy of the billiard map is positive iff the billiard table is not an ellipse. In other words either the system is completely integrable or it admits a generalized rotary horseshoe.*

5. References

1. R.L. Adler, A.G. Konheim, and M.H. McAndrew, Topological entropy , *Trans. Amer. Math. Soc.* **114**, 1965, 309–319.
2. R. Bamon, E. Malta, J. Palis and F. Takens, Rotation intervals of endomorphisms of the circle, *Ergod. Thy. and Dynam. Sys.*, **4** (1984), 493-499.
3. R. Bowen, Entropy for group endomorphisms and homogeneous spaces, *Trans. Amer. Math. Soc.* **153** 1971, 401–414.

4. C. Conley, Isolated invariant sets and the Morse Index *CBMS Conference Series*, No. 38, 1978.
5. M. Denker, C. Grillenberger, and K. Sigmund, Ergodic theory on compact spaces: *Lect. Notes in Math.* **527** Springer-Verlag, New York, 1976.
6. D. Fried, Flow equivalence, hyperbolic systems, and a new zeta function for flows. *Comment. Math. Helv.*, **57** 1982, 237-259.
7. J. Franks, Periodic points and rotation numbers for area preserving diffeomorphisms of the plane, *Publications Mathématiques I.H.E.S.*, no. 71, 1990, 105-120.
8. E. Gutkin, personal communication, Cocoyoc, Morelos, Mexico, 1994.
9. M. Handel, The rotation set of a homeomorphism of the annulus is closed, *Commun. Math. Phys.*, 127, 1990, 339–349.
10. Katok, A., "Lyapunov exponents, entropy , and periodic orbits for diffeomorphisms," *Publications Mathematiques* 51 (1980), Institut des Hautes Etudes Scientifiques, Paris, 137–173.
11. J. Kwapisz and R. Swanson, Rotational Entropy , Periodic orbits, and Pseudo-Anosov Maps, (preprint), 1994.
12. Misiurewicz, M. and Ziemian, K., Rotation sets of toral maps. *Jour. of London Math. Soc.* (2), **40**, 1989, 490–506.
13. M. Rees, "A minimal positive entropy homeomorphism of the 2-torus", *J. London Math. Soc.* (2), 23 (1981), 537–550.
14. R. Swanson, The Rotational Component of Entropy, *Proceedings of Workshop in Dynamical Systems*, Barcelona, Spain, 1990.
15. ______________ Periodic Orbits and the Continuity of Rotation Numbers, *Proc. of the AMS*, Vol. 117, No. 1, 269-273.
16. S. Tabachnikov, "Billiards", preprint, University of Arkansas (Little Rock), 1994.
17. W. Thurston, "On the geometry and dynamics of diffeomorphisms of surfaces" *Bull. Amer. Math. Soc.* 19 (1988), 417-431.
18. P. Walters, *An introduction to Ergodic Theory*, Springer-Verlag, New York, 1982.
19. M. Wojtkowski, personal communication, Cocoyoc, Morelos, Mexico, 1994.

POLYGONAL VORTEX CONFIGURATIONS

DEBRA LEWIS*
E-mail: lewis@slinky.ucsc.edu

and

TUDOR RATIU†
E-mail: ratiu@cats.ucsc.edu
Board of Mathematics, University of California at Santa Cruz,
Santa Cruz, California, 95064

ABSTRACT

A brief account of recent work of the authors in point vortex dynamics is given. The point vortex approximation to the planar ideal incompressible Euler equations turns out to have some very symmetric stationary point configurations consisting of k regular vortex n-gons and a regular vortex kn-gon oriented such that it is fixed by any reflection fixing one of the n-gons. For small k, the n-gons must be oriented in such a way that the entire configuration shares the symmetries of any of the n-gons. These configurations exist only for specific values of the relative vortex strengths, which can be uniquely expressed as functions of the radii of the polygons. Necessary conditions for the formal stability of general stationary point vortex configurations with respect to arbitrary perturbations are derived; these stability conditions set restrictions on the vortex strengths. An efficient strategy for carrying out a complete linear and formal stability analysis is developed and is implemented for $k = n = 2$. abstract

1. Introduction

There are strong formal analogies between the n-body problem in celestial mechanics and the point vortex problem arising in the study of ideal incompressible two-dimensional fluid flow. Often information about one system suggests new results about the other, even though the physical interpretations may differ significantly. In this paper we present in an informal way some very recent results which we have obtained in the study of relative equilibria of the point vortex problem in the hope that it may lead to similar results in the n-body problem and related systems. We shall not present here the technical details of the proofs and the numerical computations. We refer the interested reader to our recent paper (Lewis and Ratiu [1994]), on which the present account is based, and the references given therein.

The equations of motion for N planar point vortices $z_1, \ldots, z_N \in \mathbb{C}$ with

*Research supported by NSF grant DMS-9122708 and UCSC FRC grant 503128-09523
†Research supported by NSF grant DMS-9122708, DOE grant DE-FG03-92ER25129, ONR grant N00014-92-J-1545, and UCSC FRC grant 503185-19900

vortex strengths $\Gamma_1, \ldots, \Gamma_N$ are

$$\dot{\bar{z}}_\alpha = \frac{1}{2\pi i} \sum_{\beta=1, \beta \neq \alpha}^{N} \frac{\Gamma_\beta}{z_\alpha - z_\beta}. \tag{1}$$

We are interested in a very special type of solutions of this system: rings of regular concentric polygons all rotating with the same constant angular velocity; the point vortices making up each ring are of equal strength. Most studies of vortex polygons to date have treated only collections of rings of equal numbers of vortices, that is, configurations consisting of k collections of n vortices each, n vortices in each collection being of equal strength. See Aref [1983], Campbell and Ziff [1979], or Koiller et al. [1985] and references therein for detailed discussions of such configurations. These collections of n-gons maintain their polygonal character under the dynamics (1). Thus the motion of each n-gon is determined by the motion of a single vortex in the polygon.

We consider a situation which is in some sense a generalization of the class of configurations considered above. If the configuration contains vortex polygons consisting of unequal numbers of vertices, then an initially polygonal configuration typically will not retain its polygonal character. However, the polygonal structure of some initial conditions is preserved for all time. In particular, there are families of polygonal configurations which evolve by a steady rotation, that is, they are relative equilibria of the system.

We consider the nonlinear orbital stability of vortex configurations within the class of all possible planar vortex motions. In particular, we do *not* assume that perturbations lie within the class of polygonal vortex configurations. We focus on necessary conditions for formal stability, that is, the existence of a local extremal modulo symmetries of an energy-momentum functional, for general planar polygonal vortex configurations; if certain technical conditions are satisfied, formal stability implies nonlinear orbital stability. For relatively high dimensional systems, the utilization of easily checked instability criteria can greatly increase the efficiency of the formal stability analysis. As an example, we consider the case of two vortex pairs and a vortex quartet. Simple conditions on the relative signs of the vortex strengths and the relative sizes and orientations of the polygons suffice to show the formal instability of members of two of the three classes of relative equilibria; a numerical eigenvalue analysis shows that the remaining class contains both stable and unstable configurations.

2. Concentric polygonal configurations

A family of vortices $z_{\iota_0}, \ldots, z_{\iota_{n-1}}$ with vortex strengths $\Gamma_{\iota_0} = \cdots = \Gamma_{\iota_{n-1}}$ is said to form a regular vortex n-gon if there exists $r \in \mathbb{C}$ such that $\sum_{\ell=0, \ell \neq j}^{n-1} z_{\iota_\ell} =$

$re^{2\pi j/n}$ for $j = 0, \ldots, n-1$. The case of two groups of vortices with strengths of equal magnitude and opposite orientation is the classical Havelock [1931] problem. The situation in which a collection of kn vortices can be partitioned into k vortex n-gons of equal strength has been studied by several authors, including Aref [1983] and Koiller et al. [1985].

We consider here point vortex configurations consisting of one kn-gon, all of whose vortices are of strength Γ, and k vortex n-gons, the i-th n-gon consisting of vortices of strength $\Gamma\,\gamma_i$. The polygons are assumed to be regular, distinct, and have a common center. As we shall see below, such configurations persist under the dynamics (1) only for very specific combinations of ring radii and vortex strengths. The relative orientations of the polygons are also highly restricted: the kn-gon must be fixed by reflections fixing any one of the n-gons; in addition, we have shown for small values of k that all of the n-gons must be fixed by reflections fixing any one of the n-gons.

We denote the positions of the kn vortices of strength Γ by z_α, $\alpha = 0, \ldots, kn-1$, while those of strength $\Gamma\,\gamma_i$ are denoted by $\zeta_{j\beta}$, $\beta = 0, \ldots, n-1$, $j = 0, \ldots, k-1$. We shall search for configurations consisting of k vortex n-gons and one vortex kn-gon, all centered at the origin. Such an arrangement corresponds to the assumption that

$$\begin{aligned} z_\alpha(t) &= \Phi(\alpha)\, z_0(t) & \alpha &= 1, \ldots, kn-1 \\ \zeta_{j\beta}(t) &= \Phi(k\beta)\, \zeta_{j0}(t) & \beta &= 1, \ldots, n-1, \quad j = 0, \ldots, k-1, \end{aligned} \tag{2}$$

where $\Phi(x) := \exp\left(\frac{2\pi i x}{kn}\right)$, for all times t. In this situation, there are $k+1$ polygons: one kn-gon with 'leader' z_0 and vortex strengths Γ and k n-gons, each with 'leader' ζ_{j0} and vortex strengths $\Gamma\,\gamma_j$.

If we define $\rho_j := (\zeta_{j0}/z_0)^n$ for $j = 0, \ldots, k-1$ and assume that $\rho_j \neq \rho_\ell$ if $j \neq \ell$, then (1) and (2) imply that

$$\begin{aligned} \frac{2\pi i\, z_\alpha \dot{\bar{z}}_\alpha}{\Gamma} &= \sum_{\ell=0,\ell\neq\alpha}^{kn-1} (1 - \Phi(\ell-\alpha))^{-1} + \sum_{j=0}^{k-1} \gamma_j \sum_{\ell=0}^{n-1} (1 - \Phi(k\ell-\alpha)\,\zeta_{j0}/z_0)^{-1} \\ &= \frac{kn-1}{2} + n \sum_{j=0}^{k-1} \frac{\gamma_j}{1 - \Phi(-n\,\alpha)\,\rho_j} \end{aligned} \tag{3}$$

for $\alpha = 0, \ldots, kn-1$ and

$$\begin{aligned} \frac{2\pi i\, \zeta_{j\beta} \dot{\bar{\zeta}}_{j\beta}}{\Gamma} &= \sum_{\ell=0}^{kn-1} (1 - \Phi(\ell - k\beta)\, z_0/\zeta_{j0})^{-1} + \gamma_j \sum_{\ell=0,\ell\neq j}^{n-1} (1 - \Phi(k(\ell-\beta)))^{-1} \\ &\quad + \sum_{m=0}^{k-1} \gamma_m \sum_{\ell=0}^{n-1} (1 - \Phi(k(\ell-\beta))\,\zeta_{m0}/\zeta_{j0})^{-1} \end{aligned}$$

$$= \frac{kn}{1-\rho_j^{-k}} + \frac{\gamma_j(n-1)}{2} + n\sum_{m=0}^{k-1} \frac{\gamma_m}{1-\rho_m/\rho_j}. \tag{4}$$

for all $j = 0, \ldots, k-1$ and $\beta = 0, \ldots, n-1$. If (2) holds, then $z_0 \dot{\bar{z}}_0 = z_\alpha \dot{\bar{z}}_\alpha$ for all $\alpha = 1, \ldots, kn-1$ and $\zeta_{j0}\dot{\bar{\zeta}}_{j0} = \zeta_{j\beta}\dot{\bar{\zeta}}_{j\beta}$ for all $j = 1, \ldots, k-1$ and $\beta = 0, \ldots, n-1$. It turns out (after a reasonably involved direct verification) that the equations for the z_α's have a convenient reformulation which specifies the relative vortex strengths $\gamma_0, \ldots, \gamma_{k-1}$ in terms of the quantities ρ_j determined by the relative position of the ring leaders, namely,

$$\gamma_j = \frac{\sigma\left(1-\rho_j^k\right)}{\prod_{\ell=0,\ell\neq j}^{k-1}(1-\rho_j/\rho_\ell)}, \qquad \text{where} \qquad \sigma = \frac{2\pi i z_0 \dot{\bar{z}}_0}{n\Gamma} + \frac{1}{2n} - \frac{k}{2}. \tag{5}$$

Equations (5) form a system of nonlinear equations for the ρ_j's, parametrized in part by the constant vortex strengths $\gamma_0, \ldots, \gamma_{k-1}$ and the time-dependent scalar σ. Given the γ_j's, one can search for ρ_j's satisfying both (4) and (5). However, neither of these systems of equations appears to possess closed form solutions. Thus, in the interest of obtaining analytic information about the behavior of such systems, we shall reverse the natural sense of the equations and interpret (5) as determining the relative vortex strengths $\gamma_0, \ldots, \gamma_{k-1}$ as functions of σ, $\rho_0, \ldots, \rho_{k-1}$. Clearly this approach is valid only if the evolution of the radii is consistent with preservation of the vortex strengths. In general, if the radii of the rings change, the relationships (5) between the radii and the vortex strengths will fail and the individual vortices will evolve in such a way that the polygonal structure is destroyed.

Rather than search for the most general class of motions that 'preserve' the vortex strengths, we shall consider only rigidly rotating configurations, which clearly yield constant values of the individual components σ, $\rho_0, \ldots, \rho_{k-1}$ of the right-hand sides of equations (5). (Recent studies of stationary vortex configurations using geometric methods have been carried out by Kirwan [1988], Koiller et al. [1985], and Palmore [1982], among others.) The ring configuration determined by (2–4) evolves by a steady rotation with angular velocity ξ if and only if

$$\sigma = \frac{2\pi\xi\,|z_0|^2}{n\,\Gamma} + \frac{1}{2n} - \frac{k}{2} \tag{6}$$

and

$$\left(\sigma - \frac{1}{2n} + \frac{k}{2}\right)|\rho_j|^{2/n} = \frac{k}{1-\rho_j^{-k}} + \frac{\gamma_j(n-1)}{2n} + \sum_{\ell=0,\ell\neq j}^{k-1} \frac{\gamma_j}{1-\rho_\ell/\rho_j} \tag{7}$$

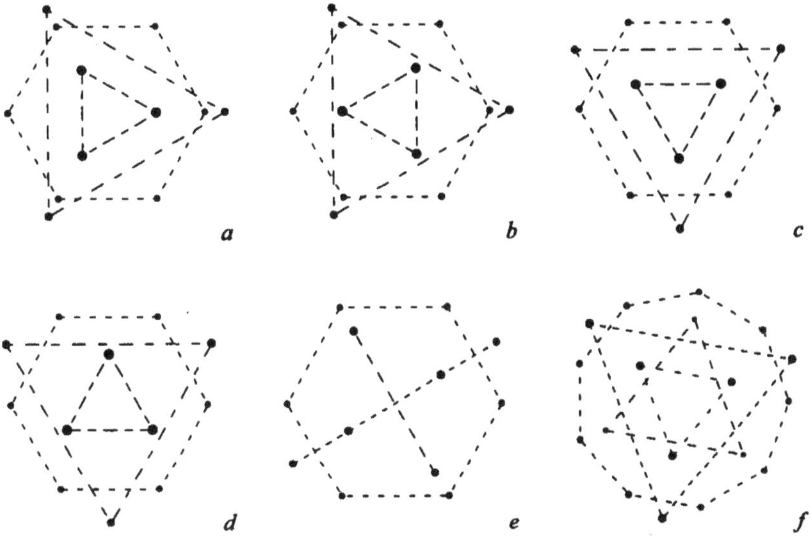

Figure 1: Representative admissible orientations. a) $(k,n) = (2,3)$, m and $\phi_0 - \phi_1$ even; b) $(k,n) = (2,3)$, m even, $\phi_0 - \phi_1$ odd; c) $(k,n) = (2,3)$, m odd, $\phi_0 - \phi_1$ even; d) $(k,n) = (2,3)$, m and $\phi_0 - \phi_1$ odd; e) $(k,n) = (3,2)$, $\phi_0 - \phi_1$ even, $\phi_0 - \phi_2$ odd; f) $(k,n) = (3,3)$, $\phi_0 - \phi_1$ even, $\phi_0 - \phi_2$ odd.

for $j = 0, \ldots, k-1$, where the γ_j's are given by (5). The assumption that the vortices all rotate at equal velocity about the origin implies that σ and the ρ_j's are constant; thus if the relations (5) are initially satisfied, they hold for all time. The remaining k equations (7) can be solved for the ρ_j's.

The $O(2)$ symmetry of the system leads to conditions on the relative orientations of the polygons making up a rigidly rotating relative equilibrium. The kn-gon must be fixed by the reflections across the lines through the vertices of any of the n-gons; this occurs if and only if the lines passing from the origin to the vertices of the n-gons either coincide with one of the lines from the origin to a vertex of the kn-gon or lie precisely halfway between two such lines. This condition can be expressed algebraically as $\zeta_j/z = r_j \Phi(m_j/2)$, where $r_j \in \mathbb{R}^+$ and $m_j \in \{0, \ldots, 2k-1\}$, and hence $\rho_j^k \in \mathbb{R}$, for $j = 0, \ldots, k-1$. In addition, the ratios $\rho_0, \ldots, \rho_{k-1}$ satisfy $\prod_{\ell=0,\ell\neq j}^{k-1}(1 - \rho_j/\rho_\ell) \in \mathbb{R}$ for $j = 0, \ldots, k-1$ if the configuration is in steady rotation. For (at least) $k = 2, 3, 4$ this implies that the ratios ρ_ℓ/ρ_j are themselves real.

If the ratios ρ_ℓ/ρ_j are all real, then $\zeta_j/z = r_j \Phi((k\phi_j + m)/2)$ for some $r_j \in \mathbb{R}^+$ and integers $\phi_0, \ldots, \phi_{k-1}$ and m. In this case *all* of the polygons are fixed by the reflections determined by any of the polygons. Figure 1 illustrates

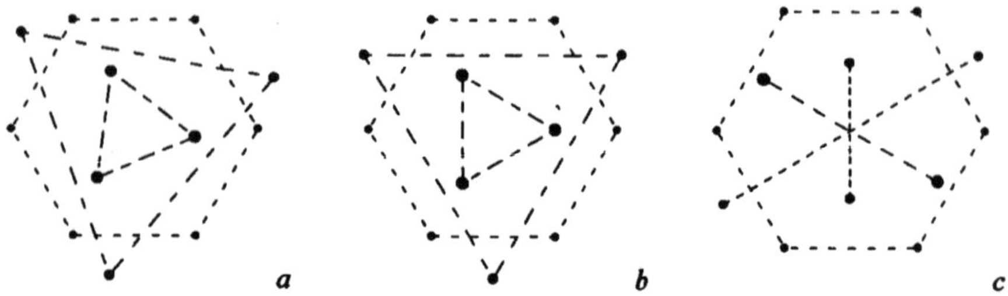

Figure 2: Sample inadmissible orientations: a) $(k,n) = (2,3)$, $\arg\zeta_0 = \frac{\pi}{9}$, $\arg\zeta_1 = \frac{5\pi}{8}$; b) $(k,n) = (2,3)$, $\arg\zeta_0 = 0$, $\arg\zeta_1 = \frac{\pi}{6}$; c) $(k,n) = (3,2)$, $\arg\zeta_0 = \frac{\pi}{6}$, $\arg\zeta_1 = \frac{\pi}{2}$, $\arg\zeta_2 = \frac{5\pi}{6}$.

some representative admissible n-gon orientations. Figure 2 illustrates some *inadmissible* orientations: figure 2.a shows a 'random' orientation of two triangles and a hexagon ($k = 2$, $n = 3$) which violates the condition that $\rho_j^k \in \mathbb{R}$; reflections preserving the triangles do not fix the hexagon. Figures 2.b and 2.c violate the condition that the ratios ρ_j/ρ_ℓ be real for $k = 2, 3, 4$; note that in these two examples reflections preserving one of the n-gons fix the hexagon, but fail to preserve the other n-gons.

If the ρ_j's are known to be real, then the equilibrium conditions (7) simplify to a system of k rational equations

$$\left(\sigma + \frac{k}{2} - \frac{1}{2n}\right) r_j^2 = \frac{k}{1 - (-1)^{m+k\phi_j} r_j^{-kn}} + \sigma \left(\frac{(n-1)\tilde{\gamma}_j}{2n} + \sum_{\ell=0,\ell\neq j}^{k-1} \frac{\tilde{\gamma}_\ell}{1 - (-1)^{(\phi_\ell - \phi_j)} (r_\ell/r_j)^n} \right) \tag{8}$$

in the real variables r_j, where

$$\tilde{\gamma}_j := \frac{1 - (-1)^{m+k\phi_j} r_j^{kn}}{\prod_{\ell=0,\ell\neq j}^{k-1} \left(1 - (-1)^{(\phi_j - \phi_\ell)} (r_\ell/r_j)^n\right)}. \tag{9}$$

In comparison to the general system of equations (7), this system is relatively easy to solve numerically. As an example, we present the numerically computed solutions of (8) for $k = 2$. There are four possible orientations for the two vortex pairs relative to the vortex 'square': the lines through the vortices in each pair are either parallel or perpendicular to one another and either intersect the vertices of the square or pass through the midpoints of the faces of the square. By analyzing each possible orientation separately, we can solve for the distance from the vortices to the origin (two real parameters), rather than the position of the vortices in each pair (two complex parameters).

We introduce a change of variables that simplifies our search for solutions of (8). If we define $p := r_0 r_1$, $q := \frac{r_0}{r_1}$, $\phi := \phi_0 - \phi_1$,

$$P := (-1)^{m+\phi} p^n, \quad \text{and} \quad Q := (-1)^{\phi} q^{-n}, \tag{10}$$

then the equation (8) for $j = 0$ is equivalent to

$$\sigma = f(p,q,n,m,\phi) := \frac{\dfrac{2}{Q/P - 1} + \left(1 - \dfrac{1}{2n}\right) pq}{\dfrac{(n-1)(P-Q)}{2n(1-Q)} + \dfrac{1-PQ}{(1-Q)^2} - pq}, \tag{11}$$

where P and Q are given by (10). The equation (8) for $j = 1$ is obtained by replacing q by $1/q$ (and hence Q by $1/Q$) in (11). Thus (8) is satisfied for both $j = 0$ and $j = 1$ only if $f(p,q,n,m,\phi) = f(p,1/q,n,m,\phi)$. We solve this equation numerically for p, regarding the remaining variables as parameters. Note that we need only consider values of q in the finite range $0 < q < 1$. Solutions satisfying $p = q$ or $p = 1/q$ must be discarded, since they violate the condition that $r_0 \neq r_1$, i.e. that the radii of the rings are distinct. Only the parity of m and ϕ effects the equations: in the case $k = 2$, even values of m lead to 'vertex-vertex' configurations, as in figures 1.a and 1.b, while odd values lead to 'vertex-face' configurations, as in figures 1.c and 1.d.

The results for some sample values of n are given in figure 3. Figure 3 contains plots of the pairs $\left((-1)^{\phi} q, (-1)^{m+\phi} p\right)$. We have parameterized the problem in such a way that q and p are always positive, but we find it instructive to group the solution branches into a single plot.

3. Stability

We now turn to the issue of stability. We shall first present some stability results for general point vortex configurations, and then apply these results to the ring configurations discussed above. Due to the relatively high dimensions of point vortex systems, virtually all stability analyses of these systems to date have considered only stability relative to perturbations preserving the polygonal structure. We do not impose that restriction here.

Consider a general collection of N point vortices $z_0, \ldots, z_N$ in the plane. It was shown by Kirchhoff that the equations (1) are Hamiltonian, with Hamiltonian

$$H(z_1, \ldots, z_N) = -\frac{1}{8\pi} \sum_{\alpha,\beta=1,\alpha\neq\beta}^{N} \Gamma_\alpha \Gamma_\beta \log |z_\alpha - z_\beta|^2 \tag{12}$$

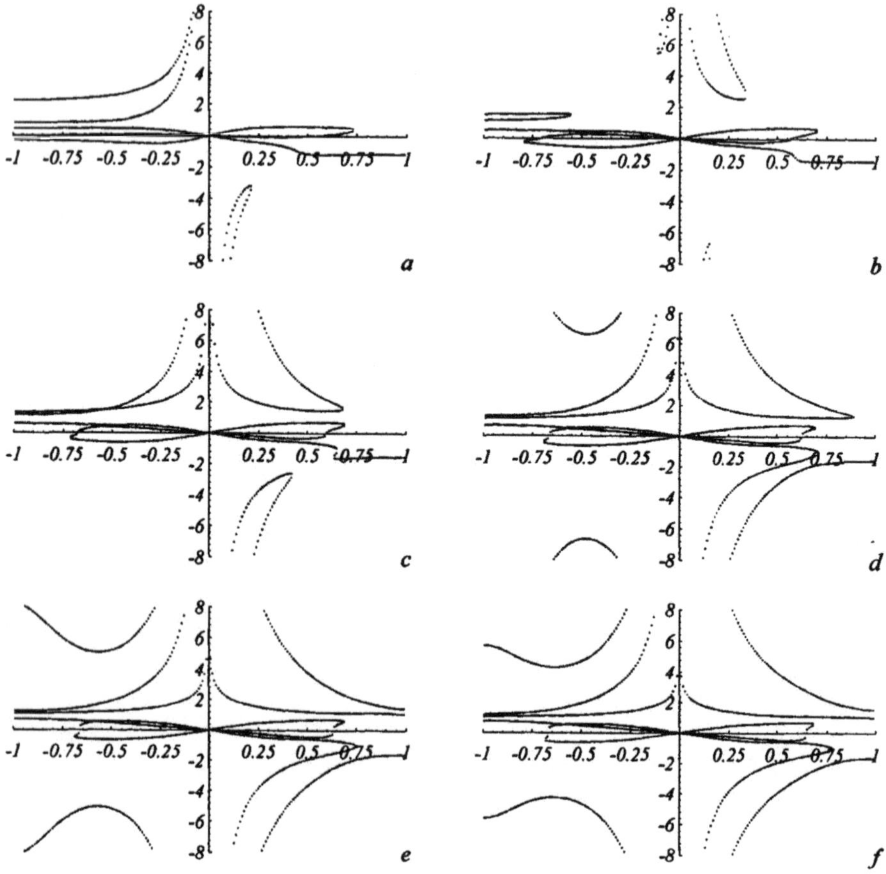

Figure 3: Plots of relative equilibria for $k = 2$, $m = 2, \ldots, 7$. Horizontal axis: $(-1)^{\phi}q$; vertical axis: $(-1)^{m+\phi}p$.

and constant non-canonical symplectic structure

$$\omega((v_1,\ldots,v_N),(w_1,\ldots,w_N))=\sum_{\alpha=1}^{N}\Gamma_\alpha \mathrm{Im}\,[v_\alpha \bar{w}_\alpha], \tag{13}$$

both of which are invariant under Euclidean motions. If the configuration $\boldsymbol{z} := (z_0,\ldots,z_N) \in \mathbb{C}^N$ is in steady rotation with angular velocity ξ, that is, it is a *relative equilibrium* with generator $(\xi, 0) \in e(2) \approx \mathbb{R}\times\mathbb{C}$, then the Hamiltonian vector field X_H satisfies $X_H(\boldsymbol{z}) = i\xi\,\boldsymbol{z}$. We can use a modification of the Hamiltonian, known as the energy-momentum function, to characterize relative equilibria as critical points.

The vector field $(\xi,\tau)_{\mathbb{C}^N}(\boldsymbol{z}) = i\xi\,\boldsymbol{z} + \tau\,\boldsymbol{t}$, where $\xi \in \mathbb{R}$, $\tau \in \mathbb{C}$ and $\boldsymbol{t} := (1,\ldots,1)$, is the infinitesimal generator of the action of the Euclidean group $E(2)$ on the plane. The (non–equivariant) momentum map $\boldsymbol{J} : \mathbb{C}^N \to e(2)^* \approx \mathbb{R}\times\mathbb{C}$ associated to the action of the Euclidean group is given by

$$\boldsymbol{J}(\boldsymbol{z}) = -\sum_{\alpha=1}^{N}\Gamma_\alpha\left(\tfrac{1}{2}|z_\alpha|^2, iz_\alpha\right). \tag{14}$$

The momentum map is determined by the condition that for every algebra element $(\xi,\tau) \in e(2)$ the infinitesimal generator $(\xi,\tau)_{\mathbb{C}^N}$ is the Hamiltonian vector field of the function $J_{(\xi,\tau)} : \mathbb{C}^N \to \mathbb{R}$ defined by $J_{(\xi,\tau)}(\boldsymbol{z}) = \boldsymbol{J}(\boldsymbol{z})\cdot(\xi,\tau)$. Thus non-degeneracy of the symplectic structure implies that $X_H(\boldsymbol{z}) = i\xi\,\boldsymbol{z}$ if and only if $\boldsymbol{z}$ is a critical point of the *energy–momentum function* $H_\xi := H - J_{(\xi,0)}$.

The energy-momentum function H_ξ is conserved by the flow of X_H. Hence if a relative equilibrium $\boldsymbol{z}$ is a local extremal of H_ξ (modulo appropriate symmetries), then H_ξ can be used to demonstrate Lyapunov stability of $\boldsymbol{z}$ (again, modulo appropriate symmetries). A relative equilibrium is said to be *formally stable* if the restriction of the second variation of the energy-momentum function to the kernel of the linearized momentum map is semi-definite, with kernel equal to the space of infinitesimal group motions preserving the momentum constraint. In the case at hand, the energy-momentum function is H_ξ and the momentum preserving group motions are rotations about the origin. A modification of the Lagrange multiplier criterion for constrained extrema that takes into account the symmetry of the system shows that a rigidly rotating relative equilibrium is nonlinearly $SO(2)$-orbitally stable if it is formally stable. Specifically, the equivalence class of the relative equilibrium in the quotient manifold $\mathbb{C}^N/SO(2)$ is a Lyapunov stable fixed point of the dynamics on $\mathbb{C}^N/SO(2)$ induced by the dynamics (1) on $\mathbb{C}^N$. It should be noted that formal *instability*, i.e. indefiniteness of the appropriate restriction of $D^2H_\xi(\boldsymbol{z})$, is not known to imply nonlinear instability and is known *not* to imply linear instability.

We implement the formal stability test as follows: Let $\boldsymbol{z}$ be a critical point of H_ξ. The second variation of the energy-momentum function H_ξ satisfies

$$D^2H_\xi(\boldsymbol{z})(\boldsymbol{\delta z},\boldsymbol{\Delta z}) = \overline{\boldsymbol{\Delta z}}\cdot \boldsymbol{\Gamma}\mathcal{M}(\boldsymbol{z})\boldsymbol{\Gamma}\boldsymbol{\delta z} + \xi\boldsymbol{\Delta z}\cdot\boldsymbol{\Gamma}\boldsymbol{\delta z} \tag{15}$$

where $\boldsymbol{\Gamma}$ is the diagonal matrix of vortex strengths and $\mathcal{M}(\boldsymbol{z})$ is the symmetric $N\times N$ complex matrix $\mathcal{M}(\boldsymbol{z})$ with entries

$$\mathcal{M}(\boldsymbol{z})_{\alpha\beta} = \frac{1}{4\pi}\begin{cases} \dfrac{1}{\Gamma_\alpha}\displaystyle\sum_{\ell\neq\alpha}\Gamma_\ell(z_\alpha - z_\ell)^{-2} & \alpha=\beta \\ -(z_\alpha - z_\beta)^{-2} & \alpha\neq\beta \end{cases}. \tag{16}$$

Note that $D^2H_\xi(\boldsymbol{z})$ is *not* a complex bilinear form. The kernel of the linearized momentum map satisfies

$$\ker[D\boldsymbol{J}(\boldsymbol{z})] = \{\boldsymbol{\Gamma}^{-1}\boldsymbol{\delta z}\in\mathbb{C}^N : \boldsymbol{\delta z}\cdot\boldsymbol{z} = 0 = \boldsymbol{\delta z}\cdot\tau\}. \tag{17}$$

We denote the orthogonal complement to $\mathrm{span}_C[\boldsymbol{t},\boldsymbol{z}]$ by $\mathcal{P}_{\boldsymbol{z}}$. Select an orthonormal basis $\{\boldsymbol{\delta z}_1,\ldots,\boldsymbol{\delta z}_{N-2}\}$ for $\mathcal{P}_{\boldsymbol{z}}$ and let $G_{\boldsymbol{z}}$ denote the (real) matrix with jk-th entry $g_{jk} := \boldsymbol{\delta z}_j\cdot\boldsymbol{\Gamma}^{-1}\boldsymbol{\delta z}_k$. If $G_{\boldsymbol{z}}$ is indefinite, then $\boldsymbol{z}$ is formally unstable. If $G_{\boldsymbol{z}}$ is definite, then let $M_{\boldsymbol{z}}$ denote the (complex) matrix with jk-th entry $m_{jk} := \overline{\boldsymbol{\delta z}_j}\cdot\mathcal{M}(\boldsymbol{z})\boldsymbol{\delta z}_k$. The relative equilibrium $\boldsymbol{z}$ is formally stable if the $2(N-2)$ by $2(N-2)$ real matrix with block structure

$$\begin{pmatrix} \xi G_{\boldsymbol{z}} + \mathrm{Re}[M_{\boldsymbol{z}}] & -\mathrm{Im}[M_{\boldsymbol{z}}] \\ -\mathrm{Im}[M_{\boldsymbol{z}}] & \xi G_{\boldsymbol{z}} - \mathrm{Re}[M_{\boldsymbol{z}}] \end{pmatrix} \tag{18}$$

is definite.

In the interest of efficiency, it is desirable to identify simple, easily tested necessary conditions for formal stability. Such conditions can be derived from the requirement that the restriction of the diagonal matrix $\boldsymbol{\Gamma}^{-1}$ to $\mathcal{P}_{\boldsymbol{z}}$ be definite. For example, let $\boldsymbol{z}$ be a steadily rotating vortex configuration containing at least two vortex polygons centered at the origin, that is, assume that there are pairs of natural numbers $\{\iota_1,n_1\},\ldots,\{\iota_p,n_p\}$, $p\geq 2$, such that

$$\Gamma_k = \Gamma_{\iota_j} \qquad \text{and} \qquad z_k = z_{\iota_j}e^{2\pi i(k-\iota_j)/n_j} \qquad \text{if} \qquad 1\leq k-\iota_j\leq n_j \tag{19}$$

for $j=1,\ldots,p$. If the restriction of $\boldsymbol{\Gamma}^{-1}$ to $\mathcal{P}_{\boldsymbol{z}}$ is positive-definite (respectively negative-definite), then at most one of the vortex strengths Γ_{ι_j}, $j=1,\ldots,p$, is negative (positive). In this case, if $\Gamma_{\iota_\ell}\Gamma_{\iota_j}<0$ for all $j\neq\ell$, then:

1. The number of vortices in the polygon with vortex strength Γ_{ι_ℓ} is prime.

2. $|\Gamma_{\iota_\ell}| > |\Gamma_{\iota_j}|$ and $|\Gamma_{\iota_\ell}|\,|z_{\iota_\ell}|^2 > |\Gamma_{\iota_j}|\,|z_{\iota_j}|^2$ for all $j \in \{1,\ldots,p\}$, $j \neq \ell$.

The formal instability of the Havelock rings, consisting of two rings with vortex strengths of equal magnitude and opposite sign, is an immediate consequence of the second criterion.

Let us now apply this criterion to the case of the polygonal point vortex configurations discussed in §2. For example, if n is even, the necessary conditions for formal stability given above imply that positivity of the relative vortex strengths $\gamma_0, \ldots, \gamma_{k-1}$ is a prerequisite of formal stability. In the case $k = 2$ the equations (5) for the relative vortex strengths γ_j are

$$\gamma_0 = \frac{\sigma\,(1-\rho_0^2)}{1-\rho_0/\rho_1} \qquad \text{and} \qquad \gamma_1 = \frac{\sigma\,(1-\rho_1^2)}{1-\rho_1/\rho_0}. \tag{20}$$

As discussed in §2, ρ_0^2, ρ_1^2, and ρ_0/ρ_1 are all real. Equation (20) implies that γ_0 and γ_1 are both positive if and only if $(\rho_0 - 1/\rho_0)\,(\rho_1 - 1/\rho_1)$ is negative. Thus for even values of n all formally stable configurations consisting of two n-gons and an $2n$-gon belong to one of the following classes:

1. *Vertex-vertex, parallel.* Both the ζ_{0i} and ζ_{1j} lie on the same line through a vertex of the $2n$-gon, with $\sigma > 0$ and either $|\zeta_{0i}| > 1 > |\zeta_{1j}|$ or $|\zeta_{1j}| > 1 > |\zeta_{0i}|$.

2. *Vertex-vertex, perpendicular.* The ζ_{0i} and ζ_{1j} lie on perpendicular lines passing through vertices of the $2n$-gon, with $\sigma(1 - |\zeta_{\ell i}|) > 0$, $\ell = 0, 1$.

3. *Vertex-face, perpendicular.* The ζ_{0i} and ζ_{1j} lie on perpendicular lines passing through midpoints of the faces of the $2n$-gon, with $\sigma > 0$.

These criteria allow us to immediately recognize that several of the numerically determined solution branches consist of unstable configurations. Analogous criteria can be developed for more complicated vortex configurations.

To determine which, if any, of the relative equilibria in the classes described above are actually stable, we must consider the restriction of $D^2H_\xi(\boldsymbol{z})$ to $\mathcal{P}_{\boldsymbol{z}}$ in more detail. Numerical computations show that in the case $k = n = 2$ some vertex–face, perpendicular configurations are formally and hence nonlinearly orbitally stable. The data for these configurations is displayed in Figure 4.

4. Conclusion

The point vortex systems are an example of systems containing relatively large number of relatively simple, nearly identical components. Our goal in

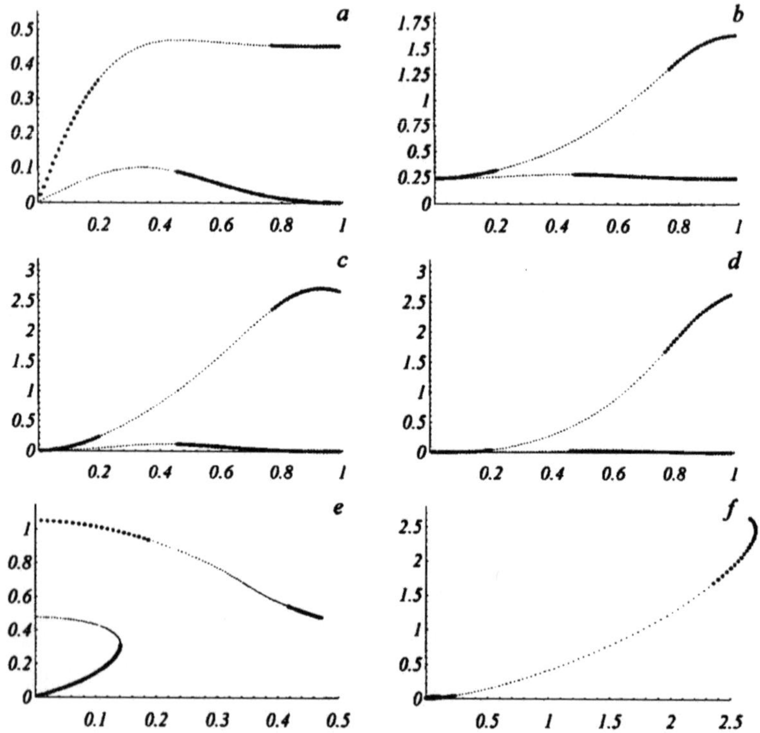

Figure 4: Stable unaligned vertex-face equilibria (m and ϕ odd) for $k = n = 2$, $|z_0| = \Gamma = 1$. a) $p = |\zeta_0\zeta_1|$ vs. $q = |\zeta_0/\zeta_1|$; b) ξ vs. q; c) γ_0 vs. q; d) γ_1 vs. q; e) $|\zeta_1|$ $vs.$ $|\zeta_0|$; f) γ_1 vs. γ_0. Large points indicate stable equilibria, while small points indicate unstable equilibria on the same solution curve.

the study of many-component systems is not, in general, a complete analytic treatment of the problem. Rather, we hope to develop techniques for efficiently obtaining partial results analytically, with the expectation that such an analysis will be supplemented by a numerical study. Symmetries appear to play an important role in the analysis of such systems. For example, we can effectively halve the number of equations to be solved numerically for the polygonal kn-gon/n-gon point vortex configurations by first identifying the configurations satisfying some simple equilibrium conditions determined by symmetry considerations.

One motivation for studying the kn-gon/n-gon configurations is the development of a collection of examples exhibiting varying degrees of discrete symmetry. Once we have assumed a certain degree of symmetry in our configurations, namely a collection of vortex polygons undergoing a steady rotation about a common center, these configurations may automatically possess additional symmetries. For example, the kn-gon must be fixed by reflections fixing any one of the n-gons. In addition, we have shown for $k = 2, 3, 4$ that all of the polygons must be fixed by such reflections. We would like to know if this is true for all values of k and if analogous results hold for related systems, such as the n-body problem.

Simple necessary conditions for formal stability can be obtained by selecting appropriate variations; many informative variations are determined by the symmetries of the system. While the stability of individual states of large systems can be determined relatively easily numerically (unless the system is *extremely* large), it is difficult to develop a reliable qualitative understanding of the system from a few isolated examples. Identifying and analyzing entire families of steady motions can be prohibitively time consuming and meaningful interpretation of numerical results is often difficult. Correlations between various parameters of the system can easily remain hidden unless the choice of variables used to display the results is guided by the expectation of such correlations.

We are studying the relationships between restricted tests for stability based on symmetry considerations and actual nonlinear stability. Both discrete and continuous symmetries set restrictions on the structure of various matrices associated to the analysis of steady motions, determining an easily specified block structure that does not depend on detailed information about the equilibria. For many systems, some of the subblocks can be treated analytically and yield simple, physically meaningful relationships between some of the parameters. We intend to further explore this block structure for point vortex systems and related models.

Acknowledgements. We thank Martin Golubitsky and Jair Koiller for many helpful discussions.

5. References

Adams M. and Ratiu T. [1988] "The three point vortex problem: commutative and non-commutative integrability", *Cont. Math.*, **81**, 245–257.

Abraham R. and Marsden J. [1978] *Foundations of Mechanics*, Benjamin/ Cummings.

Aref H. [1983] "Integrable, chaotic, and turbulent vortex motion in two-dimensional flows", *Ann. Rev. Fluid Mech.*, **15**, 345–389.

Campbell L.J. and Ziff R. [1979] "A catalog of two-dimensional vortex patterns", LA-7384-MS informal report.

Havelock T.H. [1931] The stability of motion of rectilinear vortices in ring formation. *Philos. Mag.* (7), **11**, 617–633.

Kirwan F. [1988] "The topology of reduced phase spaces of the motion of vortices on a sphere", *Physica D*, **30**, 99–123.

Koiller J., Carvalho S., Rodriques da Silva R., and Conçalves de Oliveira L.C.[1985] "On Aref's vortex motions with a symmetry center", *Physica D*, **16**, 27–61.

Lewis D. and Ratiu T. [1994] "Rotating n-gon/kn-gon vortex configurations. (preprint).

Palmore J.I. [1982] "Bifurcation of stationary vortex configurations" in *Nonlinear Problems: Present and Future*, A.R. Bishop, D.K. Campbell, and B. Nicolaenko editors. North-Holland.

PARALLEL ESCAPES IN THE RESTRICTED COULOMBIAN ISOSCELES 3-BODY PROBLEM

ERNESTO A. LACOMBA
E-mail: lace@xanum.uam.mx
and
J. GUADALUPE REYES
E-mail: grv@xanum.uam.mx
Departamento de Matemáticas, Universidad Autónoma Metropolitana-Iztapalapa
Av. Michoacán y la Purísima, Apdo. Postal 55-534, CP. 09340,
México, D.F.
E-mail: lace@xanum.uam.mx

ABSTRACT

We consider here escape motions in the restricted repulsive Coulombian 3-body problem, by studying analytic solutions of a complex differential equation about a singularity in $\Omega \times \mathbb{C}P^1$, where Ω is a region in the complex plane and $\mathbb{C}P^1$ is the projective line. The corresponding solutions projected in the Hill's region of the problem, turn out to be parallel to the homothetic solutions and are analytic at infinity.

1. Introduction

A repulsive n-body problem is obtained from the Newtonian attractive problem when the sign is changed in the potential of the equations of motion. That is, if n punctual bodies of masses $m_1, m_2, \ldots, m_n$ are moving in the Euclidean space $\mathbb{R}^N$, under these repulsive forces, the equations for the system are given by

$$\ddot{r}_i(t) = \frac{1}{m_i} \sum_{1 \le j \le n_{j \neq i}} \frac{\gamma m_i m_j}{|r_i - r_j|^3}(r_i - r_j) \qquad i, j = 1, 2, \ldots, n$$

where $r_k(t)$ is the position at physical time t of the particle with mass m_k in the space $\mathbb{R}^N$ and $\gamma > 0$ is the (Coulombian) repulsion constant between the masses.

Using the corresponding Hamiltonian, it is easy to see that the energy is always positive in this system. Even more, there are no singularity (collision) solutions in this problem (See [6]), and all the velocities are bounded for arbitrary time.

From the fact that energy is always positive, it follows that most of the particles must escape [6], and the question is about the behavior of those which do not do that, having instead an asymptotic motion to an equilibrium position close to the center of mass. The conjeture is, that at most one particle can have such behavior.

Other studies have been made on particular cases of this problem.

Gal'perin [4] and Moser [5] have considered the case where the system has a central repulsion, that is, there exists a function $V : \mathbb{R} \to \mathbb{R}$ known as the effective potential such that the potential $U = U(r)$ of the system satisfies

a) $U(r) = V(|r|)$ (U is a central potential).

b) $V(x)$ is a monotone non increasing function for $x > 0$.

They have found that with some boundedness and integrability conditions at infinity for V, one has the following asymptotic behavior at ∞:

$$r_i(t) = c_i + v_{i,\infty} \cdot t + 0(1), \qquad \text{as} \qquad t \to \infty$$

where c_i are vectors determined uniquely by the initial conditions and $v_{i,\infty}$ the velocity at ∞ of the particle with mass m_i.

Peredo [6] proved that in the collinear n-body problem with repulsive forces the two exterior bodies escape to ∞ and $-\infty$ respectively.

We study here the repulsive isosceles planar 3-body problem in the case where two particles have equal masses and the other one is zero. This is done by using an inversion-projective change of coordinates. Then one looks for analytic solutions of a second order differential equation about the singularity $v = 0$.

Going back to the original coordinates, we get in the Hill's region of the problem solutions parallel to the homothetic solutions: Euler solution $y \equiv 0$ and Lagrange solutions $y = \pm\sqrt{3}x$.

Solutions parallel to Euler were sketched by Peredo [6], and we refine the result computing solutions at infinity.

Our method was motivated by the conclusion of Proposition B in the next section, asserting that the cartesian slope of a solution $(x(t), y(t))$ has a limit at infinity and is bounded in Hill's region on a fixed energy level h. In this case,. it is possible to compute limiting escape velocities for $t \to \infty$.

2. The Isosceles problem

Using Jacobi coordinates for the planar isosceles problem [1], with center of mass at the origin, and considering two masses equal to $m > 0$ and the third mass equal to $\mu \geq 0$, as the Figure 1 shows, then the equations of motion are given by

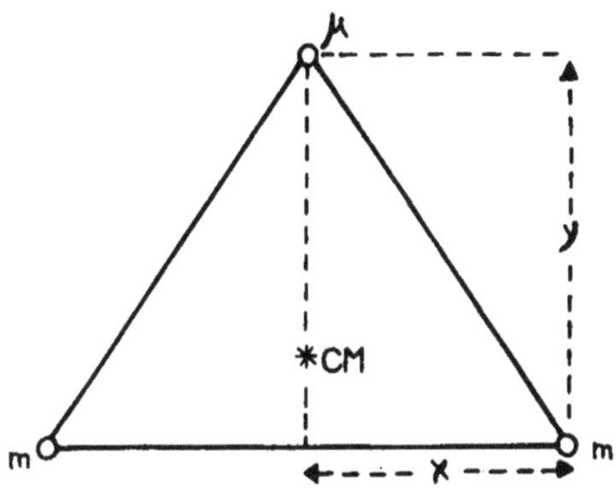

Figure 1: The Coulombian isosceles 3-body problem in Jacobi coordinates.

$$\begin{cases} \ddot{x} = \dfrac{\gamma m}{4x^2} + \dfrac{\gamma \mu x}{(x^2+y^2)^{3/2}} \\ \\ \ddot{y} = \dfrac{\gamma(\mu+2m)y}{(x^2+y^2)^{3/2}} \end{cases} \tag{1}$$

Here, $\gamma > 0$, x is the distance from each particle of mass m to their center of mass and y is the distance from the particle of mass μ to the same point. CM is the center of mass of the whole system.

For this problem there are three simple solutions, the so-called homothetic solutions

$y \equiv 0$ (Euler's solution or collinear).

$y \equiv \pm\sqrt{3}x$ (Lagrange's solutions or equilateral).

When the third mass is equal to zero, we obtain the restricted problem, which has equations of motion

$$\begin{cases} \ddot{x} = \dfrac{m}{4x^2}, \\ \\ \ddot{y} = \dfrac{2my}{(x^2+y^2)^{3/2}}, \end{cases} \tag{2}$$

where $m > 0$; supposing without loss of generality, that the constant is $\gamma = 1$.

Peredo has obtained the following results on the solutions of the system for an energy level $h > 0$.

Proposition A *(Particles of mass $m > 0$). If we introduce the uniformizing variable τ such that $dt/d\tau = x$, then the parametric equations for the movement* *of the particles with mass m are given by the pair*

$$\begin{cases} x(\tau) & = m(1+\cosh 2\tau) \\ t(\tau) & = \frac{m}{2}(2\tau+\sinh 2\tau) \end{cases} \tag{3}$$

We observe that the velocities at infinity exist and are bounded, in fact $\lim_{t\to\infty} \dot{x}(t) = \sqrt{2h}$ while $\lim_{t\to\infty} \dot{y}(t) = v_\infty$ can be any real number.

Proposition B. *(Angle at ∞). Let $x = x(t)$, $y = y(t)$ be solutions of the differential equation (2) and let us define $\theta(t) = \arctan \frac{y(t)}{x(t)}$. Then $\lim_{t\to\infty} \theta(t)$ exists and $\lim_{t\to\infty} \theta(t) = \arctan\left(\frac{v_\infty}{\sqrt{2h}}\right)$, with*

$$-\frac{\pi}{2} < \arctan\left(\frac{v_\infty}{\sqrt{2h}}\right) < \frac{\pi}{2}.$$

That is, the angle of the solution $(x(t), y(t))$ at infinity has a limit and is bounded.

BLOW UP AT INFINITY. Using a spherical blow up at infinity, Peredo F. [6] has shown that for a fixed energy level $h > 0$, the isoenergetic surface E_h of the problem can be provided with a boundary at infinity, the so called manifold at infinity.

This manifold is formed by two 2-dimensional connected components, the flow extends to them (so called a fictitious flow), and has critical points only on such components.

These critical points form two curves, each one contained in one of the components and they are normally hyperbolic.

The flow is gradient-like in the direction of a component of the velocity, and this fact implies that all solution leaves a critical point in one component as $t \to -\infty$ and arrives to another critical point in the other component as $t \to -\infty$. The flow is invariant on the manifold at infinity.

Althought this blow up describes topologically the behavior of escapes and the asymptotic motions, it does not distinguish the solutions becoming parallel to straight lines by the origin, in particular to the central configuration $y = 0$ y $y = \pm\sqrt{3}x$.

For this purpose, we are going to do a transformation of coordinates which carries the ∞ of the x-axis to zero and distinguishes escapes parallel to straight lines passing by the origin of coordinates.

3. Transformation of Coordinates

By Proposition A in Section 1, we can use the variable x as independent variable, because $x \sim 2|t|$ as $t \to \infty$.

From the first equation of (2), we obtain that for a fixed energy level $h > 0$

$$\dot{x}^2 = 2\left(h - \frac{m}{4x}\right)$$

Since $\dot{x}^2 \geq 0$ then $M_h = \{(x, y) | x \geq \frac{m}{4h}\}$ is the Hill's region of permitted motions in the plane.

By chain's rule, for $x >> 0$, we have

$$\dot{y} = \frac{dy}{dx}\dot{x} \qquad \text{and} \qquad \ddot{y} = \frac{d^2y}{dx^2}\dot{x}^2 + \frac{dy}{dx}\ddot{x}.$$

Using the energy relation and substituting in the second equation of (2) we obtain

$$2\left(h - \frac{m}{4x}\right) \cdot \frac{d^2y}{dx^2} + \frac{dy}{dx}\left(\frac{m}{4x^2}\right) = \frac{2my}{(x^2 + y^2)^{3/2}}. \tag{4}$$

In this way, for physical time t big enough, we can replace such independent variable by the variable x, so that we have $y = y(x)$.

Since $\lim_{t\to\infty} \theta(t) = \lim_{x\to\infty} \theta(x) = \lim_{x\to\infty} \arctan\left(\frac{y(x)}{x}\right)$ exists by proposition B, then one looks for asymptotic solutions $y = y(x)$ of the differential equation (4), which have a given slope at ∞. They will be detected by the projective transformation $M_h \to H \times \mathbb{R}P^1$,given by $(x, y) \to (x, u) = (x, y/x)$, where $H = \{x > 0 | x \geq \frac{m}{4h}\} \subset \mathbb{R}$.

By a straightforward computation, we obtain from the relation $y = ux$:

$$\frac{dy}{dx} = u + x\frac{du}{dx} \quad \text{and} \quad \frac{d^2y}{dx^2} = x\frac{d^2u}{dx^2} + 2\frac{du}{dx}.$$

And therefore equation (4) becomes

$$2\left(h - \frac{m}{4x}\right)x\frac{d^2u}{dx} + \left(4h - \frac{3m}{4x}\right)\frac{du}{dx} + \frac{mu}{4x^2} = \frac{2mu}{x^2(1 + u^2)^{3/2}}, \tag{5}$$

defined in $H \times \mathbb{R}P^1$. We now take the inversion transformation $v = \frac{1}{x}$ and using again the chain's rule, we obtain the second order differential equation

$$v\frac{d^2u}{dv^2} = \frac{4mu}{(4h - mv)(1 + u^2)^{3/2}} + \frac{m\left(v\frac{du}{dv} - u\right)}{2(4h - mv)} \tag{6}$$

defined in $V \times \mathbb{R}P^1$, where $V = \{v \epsilon \mathbb{R} | 0 \leq v \leq \frac{4h}{m}\} \subset \mathbb{R}$.

That is, the transformation $(x, y) \to (1/x, y/x) = (v, u)$ will detect any solution of (4) with slope α at infinity in M_h, as a solution of equation (6) which cross the line $\{0\} \times \mathbb{R}P^1$ at the point $(0, \alpha)$. This is because such transformation sends the straight line $y = \alpha x + \beta$ into the straight line $u = \beta v + \alpha$.

On other hand, $v = 0$ is a singularity of the differential equation (6). Then for finding analytic solutions with initial conditions $v = 0$, $u(0) = \alpha$, $u'(0) = \beta$, we use analytic continuation complexifying the variables (v, u):

$$\begin{array}{ll} V \times \mathbb{R}P^1 & \rightarrow \Omega \times \mathbb{C}P^1 \\ (v, u) & \rightarrow (z, \omega) \end{array}$$

We obtain the differential equation in $\Omega \times \mathbb{C}P^1$

$$z\omega'' = \frac{4m\omega}{(4h - mz)(1 + \omega^2)^{3/2}} + \frac{m(z\omega' - \omega)}{2(4h - mz)} \tag{7}$$

where $'$ denotes the derivative respect to z.

The equation (7) is of the form $z\omega'' = F(\omega', \omega, z)$, where F is analytic in its arguments.

A necessary condition for (7) to have analytic solutions at initial conditions $z = 0$, $\omega(0) = \alpha$, $\omega'(0) = \beta$ is that the right hand side of (7) vanishes when we evaluate at those initial conditions: $F(\beta, \alpha, 0) = 0$ (See [3], pags 193-194).

Such condition is equivalent to $0 = \frac{4m\alpha}{4h(1+\alpha^2)^{3/2}} - \frac{m\alpha}{8h}$, that is, $\alpha = 0, \pm\sqrt{3}$ and β arbitrary.

On the other hand, a sufficient condition for the existence of analytic solutions for those initial conditions is that the critical quadratic associated to the differential equation (7) has no positive integer roots.

The critical quadratic in this case is $(\lambda - f_1)\lambda = 0$, where $f_1 = \frac{\partial F}{\partial \omega'}(\beta, \alpha, 0)$. A calculation shows $f_1 \equiv 0$ at the given initial conditions.

Then, the differential equation (7) has analytic solutions which are unique $\omega = \omega(z)$ for the initial conditions $z = 0$, $\omega(0) = 0, \pm\sqrt{3}$, $\omega'(0) = \beta$ arbitrary. (See [3]).

We summarize in the

Theorem 1. *The differential equation (7) defined in $\Omega \times \mathbb{C}P^1$ has unique analytic solutions $\omega = \omega(z)$ on the line $\{0\} \times \mathbb{C}P^1$, if and only if the initial conditions are $z = 0$, $\omega = 0, \pm\sqrt{3}$, and $\omega' = \beta$ arbitrary.*

Doing the correspondence with $V \times \mathbb{R}P^1$ we obtain the

Proposition 1. *The differential equation (6) has analytic solutions on the line $\{0\} \times \mathbb{R}P^1$ if and only if $\alpha \in \mathbb{R}P^1$ satisfies that $\alpha = 0$ or $\alpha = \pm\sqrt{3}$, $z = 0$ and $u'(0) = \beta$ is arbitrary.*

We observe that $u = 0$ and $u = \pm\sqrt{3}$ does correspond in the real space to the central configurations of the initial repulsive problem.

Moreover, the central configurations $u = 0, \pm\sqrt{3}$ are surrounded around $v = 0$ by curves in all directions β, which cross $0 \times \mathbb{R}P^1$ at the same points. Thus, for every curve in M_h corresponding to any of those solutions, the cartesian slope at ∞ is precisely α.

In the other words

Proposition 2. *For every central configuration, there exists an invariant 2-dimensional submanifold projecting to the Hill's region, formed by solution curves of (3) all parallel to such central configurations as $t \to \infty$.*

4. Eulerian Solutions

The differential equation (7) has $\omega \equiv 0$, as solution. It corresponds to the Euler solution in the initial problem.

Theorem 1 of the Section 2 indicates us that for the given initial conditions, there exist a unique solution, and we proceed to find it locally. For this, we expand the factor $\frac{\omega}{(1+\omega^2)^{3/2}}$ around $\omega = 0$:

$$\frac{\omega}{(1+\omega^2)^{3/2}} = \omega - \frac{3}{2}\omega^3 + \frac{15}{8}\omega^5 + \dots \quad \text{with} \quad |\omega| < 1 \tag{8}$$

To analyze the solutions of the Equation (7) with such initial conditions, we make a truncation of series (8), by using the theory of holomorphic singularities, and particularly the classification of germs of conformal maps $(\mathbb{C}, 0) \to (\mathbb{C}, 0)$, with identity linear part [2],[7].

Let A the set of germs $(\mathbb{C}, 0) \to (\mathbb{C}, 0)$ of conformal mappings of the form $f(u) = u + au^2 + bu^3 + \dots$

Definition. *Two germs $f, g \in A$ are said formally equivalent if there exists a formal change of coordinates $\omega = H(u)$ such that the following diagram commutes*

$$\begin{array}{ccc} (\mathbb{C},0) & \xrightarrow{f} & (\mathbb{C},0) \\ H\downarrow & & \downarrow H \\ (\mathbb{C},0) & \xrightarrow{g} & (\mathbb{C},0) \end{array}$$

Let $A_n = \{f \in A | f(u) = u + au^n + \dots,\}$. If $f \in A_n$ we can suppose without loss of generality (by a linear change of coordinates) that $a \equiv 1$.

Theorem. (Venkov [7]). *Let $f(u) = u + u^n + \ldots$ be an element of A_n, then there exists an unique $\lambda \in \mathbb{C}$, such that the mapping f is formally equivalent to the mapping $f_{\lambda,n}(u) = u + u^n + \lambda u^{2n-1}$.*

If $f_{\lambda,n}$ and $f_{\lambda',n'}$ are formally equivalent, then $\lambda = \lambda'$ and $n = n'$.

Corollary [7]. *If $f(u) = u + u^n + \lambda u^{2n-1} + \ldots$, then $f_{\lambda,n} = u + u^n + \lambda u^{2n-1}$ is in fact the formal normal form of f.*

Coming back to our problem, if we set $\omega = \sqrt{\frac{2}{3}}iu$ in the right hand side of the series (8) we obtain

$$\frac{\omega}{(1+\omega^2)^{3/2}} = \omega - \frac{3}{2}\omega^3 + \frac{15}{8}\omega^5 + \ldots = \sqrt{\frac{2}{3}}i(u + u^3 + \frac{5}{6}u^5 + \ldots). \tag{9}$$

Together with $\omega' = \sqrt{\frac{2}{3}}iu'$, $\omega'' = \sqrt{\frac{2}{3}}u''i$, equation (7) becomes

$$z\sqrt{\frac{2}{3}}iu'' = \frac{4m}{(4h - mz)}\sqrt{\frac{2}{3}}i(u + u^3 + \frac{5}{6}u^5 + \ldots) + \sqrt{\frac{2}{3}}i\left(\frac{m(zu' - u)}{2(4h - mz)}\right).$$

So, finally we obtain

$$zu'' = \frac{4m}{4h - mz}(u + u^3 + \frac{5}{6}u^5 + \ldots) + \frac{m(zu' - u)}{2(4h - mz)}. \tag{10}$$

Now, using the corollary to Venkov's, theorem, with $\lambda = 5/6$ the normal form $u + u^3 + \frac{5}{6}u^5$ substituted in (10), converts it into

$$zu'' = \frac{4m}{(4h - mz)}(u + u^3 + \frac{5}{6}u^5) + \frac{m(zu' - u)}{2(4h - mz)}. \tag{11}$$

From here, for the initial conditions $z = 0, u = 0$, $u' = a_1$ arbitrary, we can use the differential equation (11) to obtain analytic solutions.

We remark that the truncated equation (11) for (10) is of type (P) of Painlevé-Gambier [2], which has non-movable singular points.

By calculating the coefficients, the solution of equation (11) with such initial conditions is

$$u(z) = a_1 z + a_1\frac{m}{2h}z^2 + a_1\frac{39m^2}{288h^2}z^3 + \cdots \tag{12}$$

which converges around $z = 0$ by Theorem 1.

When we substitute $\omega(z) = \sqrt{\frac{2}{3}}iu(z)$ for passing to $\Omega \times \mathbb{C}P^1$, we have the solutions of the form

$$\omega(z) = \beta z + \beta\frac{m}{2h}z^2 + \beta\frac{39m^2}{288h^2}z^3 + \cdots. \tag{13}$$

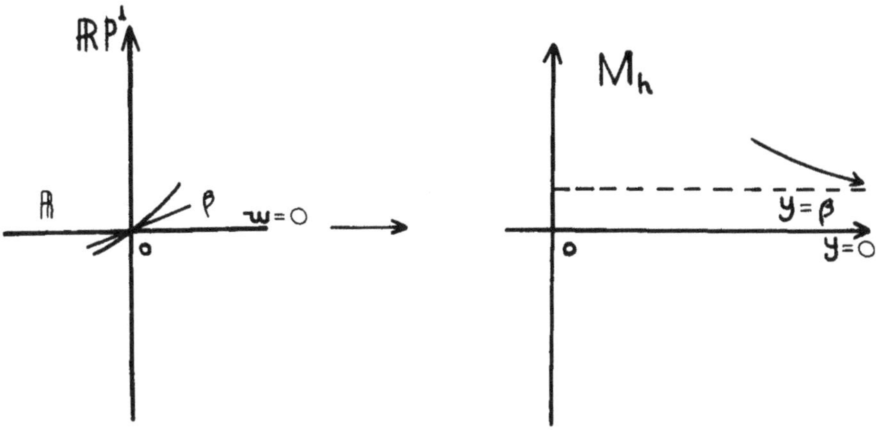

Figure 2: Asymptotic behaviour of Eulerian solutions at infinity.

By restriction to the reals, the solutions in $V \times \mathbb{R}P^1$ for the initial conditions $v = 0$, $u = 0$, $u' = \beta$ arbitrary are

$$u(v) = \beta v + \beta \frac{m}{2h} v^2 + \beta \frac{39m^2}{288h^2} v^3 + \cdots$$

And finally, in the Hill's region, we have that at ∞, the asymptotic parallel solutions to Euler solution are given by

$$y(x) = \beta + \beta \frac{m}{2h} \cdot \frac{1}{x} + \beta \frac{39m^2}{288h^2} \frac{1}{x^2} + \cdots \tag{14}$$

We summarize in the following result

Theorem 2. *(Asymptotic parallel motions to Euler solution). Let (x_0, y_0) be a point in the Hill's region for a fixed energy level $h > 0$, and let $y = y(x)$ be the solution curve of (4) through this point such that $\lim_{x\to\infty} y(x)$ exists and is finite. Then $\lim_{x\to\infty} y(x) = \beta$ where β is the slope of the corresponding solution curve for the differential equation (7) in $\Omega \times \mathbb{C}P^1$ at $z = 0$, $\omega = 0$.*

See Figure 2.

Corollary. *If $(x_0, y_0, \dot{x}_0, \dot{y}_0)$ is any initial condition for the differential equation (2) such that $\lim_{t\to\infty} y(t) = 0$, then the solution for such initial configuration is the Euler solution $y \equiv 0$. For any other solution with initial configuration $y_0 \neq 0$ the asymptotic value of the coordinate y is either infinity or a non-zero real number.*

5. Lagrangian Solutions

If we consider the translation $v = \omega - \sqrt{3}$ along the line $\mathbb{C}P^1$ and proceed

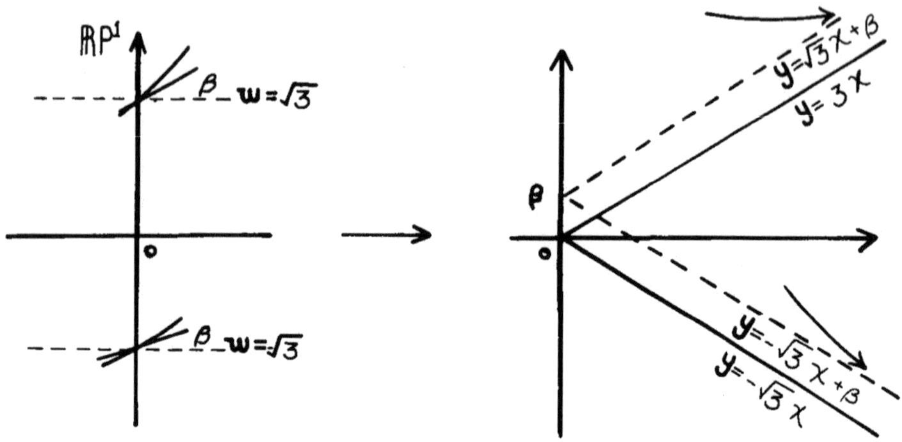

Figure 3: Asymptotic behaviour of Lagrangian solutions at infinity.

as in Section 3, it is easy to see that the solutions of (7) are the form

$$\omega(z) = \sqrt{3} + \beta z + \beta \frac{m}{2h} z^2 + \beta \frac{39m^2}{288h^2} z^3 + \cdots \tag{15}$$

around $z = 0$, $\alpha = \sqrt{3}$, $\omega'(0) = \beta$ arbitrary.

From this, in the Hill's region we obtain the solutions at ∞:

$$y(x) = \beta + \sqrt{3}x + \beta \frac{m}{2h}\frac{1}{x} + \beta \frac{39m^2}{288h^2}\frac{1}{x^2} + \cdots \tag{16}$$

and we have the

Theorem 3. *(Asymptotic parallel motions to the Lagrange solution $y = \sqrt{3}x$). Let (x_0, y_0) be a point in M_h and consider the solution curve of (4) through this point such that $\lim_{x\to\infty} y(x) = \infty$ and $\lim_{x\to\infty} \frac{dy}{dx}(x) = \sqrt{3}$. Then, if (z_0, ω_0) is the corresponding point in $\Omega \times \mathbb{C}P^1$ and the corresponding solution to the differential equation (7) passes through $(0, \sqrt{3})$ with slope β at this point, then $y(t)$ converges asymptotically to the straight line $y = \sqrt{3}x + \beta$ as $t \to \infty$.*

See Figure 3.

In a similar way, for the case $\lim_{t\to\infty} y(t) = -\infty$, with the above conditions $y(x)$ have asymptotic behavior to the straight line $y = -\sqrt{3}x + \beta$ as $t \to \infty$. (See Figure 3).

Corollary. *(Velocities at infinity). For a fixed energy level $h > 0$ and for the asymptotic solutions to the central configurations we have*

i) If $\lim_{x\to\infty} y(x) = \infty$ *then* $v_\infty = \sqrt{6h}$.
ii) If $\lim_{x\to\infty} y(x) = \beta$ *then* $v_\infty = 0$.
iii) If $\lim_{x\to\infty} y(x) = -\infty$ *then* $v_\infty = -\sqrt{6h}$.

Proof. It follows from Proposition B, since the values of the tangent function in the central configurations are $\sqrt{3}$,0, $-\sqrt{3}$ in each case.

Remarks

1) In the coordinates (z, ω) using the different methods of [3] with $z = 0$ and other initial conditions for the differential equation (7), one gets ramification points of different types.

2) It is now possible to compute the asymptotic numbers by numerical methods. The range of velocity by the last corollary is $(0, -\sqrt{6h})$ for the fixed level of energy $h > 0$.

6. References

1. Arnold V.I., *Mathematical Methods of classical Mechanics.* Springer-Verlag, New-York, Heidelberg-Berlin. 1978.

2. Anosov, D.V., Arnold,V.I., *Dynamical Systems I.* Springer-Verlag. 1988.

3. Forsyth A.R., *Theory of differential equations* Vol.III. Dover 1900.

4. Galperin Yu., *The asymptotic behavior of a system of particles with repulsion.* Russian Math. Surveys, 36 (1981), 147-148.

5. Moser, J., *The scattering problem for some particle systems on the line.* Springer Verlag. Lecture Notes in Math. 597 (1977), 441-463.

6. Peredo, F., *El problema Isósceles Restringido con Fuerzas Repulsivas,* Ph. D. dissertation at the CINVESTAV, Mexico City, 1984.

7. Voronin, S.M., *Analytic classification of germs of conformal mappings* $(\mathbf{C}, 0) \to (\mathbf{C}, 0)$ *with identity linear part.* Funct. Annal. Appl. 15 (1981),1-13.

QUALITATIVE BEHAVIOUR OF THE FLOW OF THE n-BODY PROBLEM IN THE ZERO ENERGY LEVEL

JAUME LLIBRE

Departament de Matemàtiques, Universitat Autònoma de Barcelona
08193, Bellaterra, Barcelona, Spain.
E-mail: jllibre@mat.uab.es

and

JÖRG WALDVOGEL

Seminar für Angewandte Mathematik, ETH-Zentrum
CH-8092, Zürich, Schweiz
E-mail: walvoge@sam.math.ethz.ch

ABSTRACT

In this note we summarize some recent techniques and results on the singularities in the equations of motion of a class of classical mechanical systems. We use these geometric techniques to give a characterization of the global behaviour of the solutions of the flow on the zero energy level for this class of classical mechanical systems which contain the gravitational n-body problem.

1. Introduction

Newton's equation may be written

$$\dot{q} = M^{-1}p, \quad \dot{p} = -\nabla V(q), \tag{1}$$

with $q = (q_1, \ldots, q_n)^t$ and $p = (p_1, \ldots, p_n)^t$, where q_i, $p_i \in \mathbb{R}^d$ are the position and momentum of the body with mass m_i. Here q is a point in Q, an open subset of $\mathbb{R}^{nd}$, and the *potential energy function* $V : Q \to \mathbb{R}$ is real analytic, homogeneous of degree $-k$ (i.e., $V(rq) = r^{-k}V(q)$ for all $r \in \mathbb{R}$), and has an isolated singularity at the origin. The mass matrix M is a diagonal matrix with positive entries $m_1, \ldots, m_1, \ldots, m_n, \ldots, m_n$. Each mass appears exactly d times. Q is called the *configuration space* and $Q \times \mathbb{R}^{nd}$, the *phase space* of system (1). On the phase space our system is a first order system of differential equations or a vector field.

System (1) may be written in Hamiltonian form by introducing the *Hamiltonian* or *total energy function*

$$H(q,p) = \frac{1}{2}p^t M^{-1} p + V(q). \tag{2}$$

The term $\frac{1}{2}p^tM^{-1}p$ is called the *kinetic energy* of the system. Then (1) becomes

$$\dot{q} = \frac{\partial H}{\partial p}, \quad \dot{p} = -\frac{\partial H}{\partial q}. \tag{3}$$

It is well known that H is a first integral or constant of the motion for the system (3). Hence we can reduce the dimension of the system by one by considering (3) as a vector field on the energy level $H^{-1}(h)$. Note that $H^{-1}(h)$ is a smooth submanifold of $Q \times \mathbb{R}^{nd}$, if h is a regular value of H.

We shall classify the global behaviour of the flow on the zero energy level for the class (1) of classical mechanical systems. This class was studied by McGehee [7] and Devaney [3] and we call it the McGehee-Devaney class.

Studies on the behaviour of the flow on the zero energy level for the gravitational n-body problem (i.e. a particular case for which $k = 1$), but from another point of view, have been made by Saari and Hulkower [8].

Here we shall first follow McGehee, Devaney, Sim and Lacomba in describing the total collision manifold T and the infinity manifold I_0^∞ in the zero energy level, see Section 2. Afterwards we shall study, in Section 3, the flow on the zero energy level.

2. The total collision manifold and the infinity manifold

In this section, first we shall follow Devaney [3] in his description of the total collision manifold using the changes of coordinates due to McGehee [6] and [7]. After, we shall use Devaney's notation to introduce the infinity manifold in the zero energy level defined by Lacomba and Sim [4].

Recall that V has an isolated singularity at the origin in the configuration space (by a *singularity* we understand a point or a set of points where the system is undefined, as in the total collision or escape at the infinity in the n-body problem). Certain orbits of the system end at this singularity in finite time, the *(total) collision orbits*; while others begin at the singularity, the *(total) ejection orbits*.

We introduce the McGehee coordinates by using three coordinates changes. The first change $(q, p, t) \mapsto (r, s, y, s, t)$ is defined by

$$r = (q^tMq)^{1/2}, \quad s = r^{-1}q, \quad y = s^tp, \quad x = M^{-1}p - ys.$$

Note that r, $y \in \mathbb{R}$ and s, $x \in \mathbb{R}^{nd}$. It is clear that r^2 is the *moment of inertia* of the system and s is the configuration. Since $s^tMs = 1$, we think of s as a point of the ellipsoid $S = \{s : s^tMs = 1\}$ in $\mathbb{R}^{nd}$. Then y is the radial component of the velocity $\dot{q}$, and x is the component of the velocity $\dot{q}$ tangent to S because $s^tMx = 0$. So we consider the pair (s, x) as defining a point in the tangent bundle TS of S.

The system (1) becomes

$$\begin{aligned} \dot{r} &= y, \\ \dot{y} &= r^{-1}(x^tMx) + r^{-k-1}kV(x), \\ \dot{s} &= r^{-1}x, \\ \dot{x} &= -r^{-1}(yx + (x^tMx)s) - r^{-k-1}(kV(s)s + M^{-1}\nabla V(s)). \end{aligned} \tag{4}$$

This system is a real analytic vector field on the manifold $(0,\infty) \times \mathbb{R} \times TS$, and it is no longer Hamiltonian. Note that the singularity set $\{0\} \times \mathbb{R}^{nd}$ for system (1) now is given by $r = 0$ for system (4). The second change $(r, s, y, x, t) \mapsto (r, s, u, v, t)$ is given by

$$u = r^{k/2}x, \quad v = r^{k/2}y.$$

In these coordinates, system (4) is written as

$$\begin{aligned} \dot{r} &= r^{-k/2}v, \\ \dot{v} &= r^{-(k+2)/2}(u^tMu + (k/2)v^2 + kV(s)), \\ \dot{s} &= r^{-(k+2)/2}u, \\ \dot{r} &= r^{-(k+2)/2}((k-2)vu/2 - (u^tMu)s - kV(s)s - M^{-1}\nabla V(s)). \end{aligned} \tag{5}$$

The set $r = 0$ still is a set of singularities of (5), but now it can be removed by a change of time scale. The last change $(r, s, u, v, t) \mapsto (r, s, u, v, \tau)$ is defined by

$$\frac{dt}{d\tau} = r^{(k+2)/2}.$$

Then (5) becomes

$$\begin{aligned} r' &= rv, \\ v' &= (1 - k/2)u^tMu + kr^kh, \\ s' &= u, \\ u' &= (k/2 - 1)vu - (u^tMu)s - kV(s)s - M^{-1}\nabla V(s), \end{aligned} \tag{6}$$

where the prime means differentiation with respect to τ. In (6) we have used the energy relation (2) that now reads as

$$\frac{1}{2}(u^tMu + v^2) + V(s) = r^kh. \tag{7}$$

System (6) is an analytic vector field on $[0,\infty) \times \mathbb{R} \times TS$ because the singularities at $r = 0$ have been removed. The system has been extended to the boundary $r = 0$. Since $r' = 0$ when $r = 0$, the boundary is invariant by the flow. The energy relation (7) also extends to the boundary, giving

$$\frac{1}{2}(u^t M u + v^2) + V(s) = 0. \tag{8}$$

We denote by T the boundary of $H^{-1}(h)$ determined by (8); T is called the *total collison manifold* of the system. Therefore

$$T = \left\{(r,s,u,v) : r = 0,\ s^t M s = 1,\ s^t M u = 0,\ \frac{1}{2}(u^t M u + v^2) + V(s) = 0\right\}.$$

If h is a regular value of H, then the dimension of T is the dimension of $H^{-1}(h)$ minus one. Note that T is independent of h; that is each energy level has the same boundary. Since $dt/d\tau = 0$ on T, orbits which previously reached or escaped T in finite time t are now asymptotic to T in time τ. As it is well known the study of the flow on T determines the local behaviour of solutions near total collision.

Now, we introduce the infinity manifold only on the zero energy level. In contrast to the total collision manifold the infinity manifold depends on the energy level, for more details see [4]. First, we consider the change $(q,p,t) \mapsto (\overline{r}, \overline{s}, \overline{y}, \overline{x}, t)$ defined by

$$\overline{r} = (q^t M q)^{-1/2}, \quad \overline{s} = \overline{r} q, \quad \overline{y} = \overline{s}^t q, \quad \overline{x} = M^{-1} p - \overline{y}\overline{s}.$$

Note that this change put the infinity at the origin, and that $\overline{s} = s$, $\overline{y} = y$, $\overline{x} = x$. So again $\overline{s}^t M \overline{s} = 1$, $\overline{s}^t M \overline{x} = 0$ and we define the ellipsoid

$$\overline{S} = \{\overline{s} : \overline{s}^t M \overline{s} = 1\}.$$

The system (1) becomes

$$\begin{aligned}
\dot{\overline{r}} &= -\overline{r}^2 \overline{y}, \\
\dot{\overline{y}} &= \overline{r}(\overline{x}^t M \overline{x}) + \overline{r}^{k+1} k V(\overline{s}), \\
\dot{\overline{s}} &= \overline{r}\overline{x}, \\
\dot{\overline{x}} &= -\overline{r}(\overline{y}\overline{x} + (\overline{x}^t M \overline{x})\overline{s}) - \overline{r}^{k+1}(M^{-1}\nabla V(\overline{s}) + k V(\overline{s})\overline{s}).
\end{aligned} \tag{9}$$

The energy relation (2) goes over to

$$\frac{1}{2}(\overline{x}^t M \overline{x} + \overline{y}^2) + \overline{r}^k V(\overline{s}) = h. \tag{10}$$

The system (9) is a real analytic vector field on the manifold $[0,\infty)\times\mathbb{R}\times T\overline{S}$, because the singularities at $\overline{r}=0$ of (1) have been removed. Note that $\overline{r}=0$ means that at least one of the n bodies of the system reaches the infinity. Since $\dot{\overline{r}}=\dot{\overline{y}}=\dot{\overline{s}}=\dot{\overline{x}}=0$ when $\overline{r}=0$, the boundary is invariant by the flow. In fact, it is formed by equilibrium points. The energy relation (10) extends to this boundary if $h\geq 0$, giving

$$\frac{1}{2}(\overline{x}^tM\overline{x}+\overline{y}^2)=h.$$

If $h=0$ or $h>0$ the boundary $\overline{r}=0$ reduces to a point or to an ellipsoid, respectively. Since the boundary is independent of the potential energy V, we lose information about the behaviour near the infinity. Then, this boundary is not the good infinity manifold. Therefore, we make the change $(\overline{r},\overline{s},\overline{y},\overline{x},t)\mapsto(\overline{r},\overline{s},\overline{u},\overline{v},\tau)$ given by

$$\overline{u}=\overline{r}^{-k/2}x,\quad \overline{v}=\overline{r}^{-k/2}\overline{y},\quad \frac{dt}{d\tau}=\overline{r}^{-(k+2)/2},$$

which eliminates the unpleasant $\overline{r}$ in the system (9). Again $\overline{u}=u$ and $\overline{v}=v$.

In these coordinates, system (8) is written

$$\begin{aligned}
\overline{r}' &= -\overline{r}\overline{v},\\
\overline{v}' &= (1-k/2)\overline{u}^tM\overline{u}+k\overline{r}^{-k}h,\\
\overline{s}' &= \overline{u},\\
\overline{u}' &= (k/2-1)\overline{v}\overline{u}-(\overline{u}^tM\overline{u})\overline{s}-kV(\overline{s})\overline{s}-M^{-1}\nabla V(\overline{s}).
\end{aligned}\tag{11}$$

Here we have used the energy relation which now it becomes

$$\frac{1}{2}(\overline{u}^tM\overline{u}+\overline{v}^2)+V(\overline{s})=\overline{r}^{-k}h.\tag{12}$$

Notice that (11) is singular when $\overline{r}=0$ and $h\neq 0$, then to study the infinity we must consider the case $h=0$. We see that the boundary

$$I_0^\infty=\left\{(\overline{r},\overline{s},\overline{v},\overline{u}):\overline{r}=0,\ \overline{s}^tM\overline{s}=1,\ \overline{s}^tM\overline{u}=0,\ \frac{1}{2}(\overline{u}^tM\overline{u}+\overline{v}^2)+V(\overline{s})=0\right\}$$

is invariant under the flow (11). We call this the *infinity manifold* at the zero energy level, and it indeed appears as a boundary submanifold glued to the zero energy level

$$I_0=\left\{(\overline{r},\overline{s},\overline{u},\overline{v}):\overline{r}>0,\ \overline{s}^tM\overline{s}=1,\ \overline{s}^tM\overline{u}=0,\ \frac{1}{2}(\overline{u}^tM\overline{u}+\overline{v}^2)+V(\overline{s})=0\right\}.$$

The examples of the McGehee-Devaney class that we shall discuss are systems with two degrees of freedom. Then, following to Devaney [3], if $n = 2$ the "ellipsoidal" coordinates (s, u) may be replaced by the polar coordinates $(\theta, u = r\dot{\theta})$, in the M-metric. Systems (6) and (11) become in this case

$$\begin{aligned} r' &= rv, \\ v' &= (1 - k/2)u^2 + kr^k h = u^2 + (k/2)v^2 + kV(\theta), \\ \theta' &= u, \\ u' &= (k/2 - 1)vu - \frac{dV(\theta)}{d\theta}, \end{aligned} \tag{13}$$

with the energy relation

$$\frac{1}{2}(u^2 + v^2) + V(\theta) = r^k h,$$

and

$$\begin{aligned} \overline{r}' &= -\overline{r}\overline{v}, \\ \overline{v}' &= (1 - k/2)\overline{u}^2 + k\overline{r}^{-k} h = \overline{u}^2 + (k/2)\overline{v}^2 + kV(\overline{\theta}), \\ \overline{\theta}' &= \overline{u}, \\ \overline{u}' &= (k/2 - 1)\overline{v}\overline{u} - \frac{dV(\overline{\theta})}{d\overline{\theta}}, \end{aligned} \tag{14}$$

with the energy relation

$$\frac{1}{2}(\overline{u}^2 + \overline{v}^2) + V(\overline{\theta}) = \overline{r}^{-k} h,$$

respectively. The variables (r, v, θ, u) or $(\overline{r}, \overline{v}, \overline{\theta}, \overline{u})$ are called the McGehee variables.

3. The flow on the zero energy level

We recall that, from (6), the flow on the zero energy level I_0 assume the form

$$\begin{aligned} r' &= rv, \\ v' &= (1 - k/2)u^t M u, \\ s' &= u, \\ u' &= (k/2 - 1)vu - (u^t M u)s - kV(s)s - M^{-1}\nabla V(s), \end{aligned} \tag{15}$$

with the energy relation

$$\frac{1}{2}(u^tMu + v^2) + V(s) = 0.$$

While the flow on the total collision manifold T and on the infinity manifold I_0^∞ are defined by

$$\begin{aligned} v' &= (1 - k/2)u^tMu, \\ s' &= u, \\ u' &= (k/2 - 1)vu - (u^tMu)s - kV(s)s - M^{-1}\nabla V(s), \end{aligned} \tag{16}$$

and

$$\begin{aligned} \overline{v}' &= (1 - k/2)\overline{u}^tM\overline{u}, \\ \overline{s}' &= \overline{u}, \\ \overline{u}' &= (k/2 - 1)\overline{v}\overline{u} - (\overline{u}^tM\overline{u})\overline{s} - kV(\overline{s})\overline{s} - M^{-1}\nabla V(\overline{s}), \end{aligned} \tag{17}$$

respectively.

We recall that a vector field is called *gradient-like with respect to a function* f if f increases along all non-equilibrium point orbits. Gradient-like vector fields are simple dynamical systems. Now, we give a unified version of Proposition 3 of [3] and of Theorem 1 of [4] that it follows easily from (15), (16) and (17).

Proposition 1 *If $k < 2$ (resp. $k > 2$), then the flow on I_0, T and I_0^∞ is gradient-like with respect to the v or $\overline{v}$ coordinate (resp. $-v$ or $-\overline{v}$). Furthermore, the following three statements are equivalent:*

(1) $(r(\tau), v(\tau), s(\tau), u(\tau))$ is a solution of the flow on I_0 with $r' = rv$ and $r(0) = r_0$.

(2) $(v(\tau), s(\tau), u(\tau))$ is a solution of the flow on T.

(3) $(v(\tau), s(\tau), u(\tau))$ is a solution of the flow on I_0^∞.

From Proposition 1, the flow on T and I_0^∞ is the projection of the flow on I_0. The fact that the flow on the total collision manifold T is equivalent to the flow on the original problem with zero energy was already observed in a particular case by Waldvogel [10].

The critical points s of the restriction of V to S are called central configuration. Then the following proposition is well known, see [3].

Proposition 2 *The flow on T has an equilibrium point at (v, s, u) if and only if s is a central configuration, $u = 0$, and $v = \pm(-2V(s))^{1/2}$.*

From Propositions 1 and 2, it follows that we have exactly four equilibrium points associated with each central configuration s if $V(s) < 0$. Two of them are on T. We denote them by

$$s^+ = (0, (-2V(s))^{1/2}, s, 0),$$
$$s^- = (0, -(-2V(s))^{1/2}, s, 0).$$

The other two are on I_0^∞. In a similar way, we denote them by $\bar{s}^+$ and $\bar{s}^-$. By convenience, we do not consider the case where $V(s) = 0$ for some central configuration s. In this case, we only have two equilibrium points associated with s. Recall that for the gravitational n-body problem $V(s) < 0$.

The following proposition is due to Devaney [3].

Proposition 3 *Suppose $k \neq 2$. If the central configuration s is a non-degenerate critical point of $V(s)$, then the equilibrium points s^+, s^-, $\bar{s}^+$, $\bar{s}^-$ for the flow on I_0' are hyperbolic, where I_0' denotes the manifold $T \cup I_0 \cup I_0^\infty$.*

We denote by $W^s(s^*)$ and $W^u(s^*)$ the stable and unstable invariant manifolds at the equilibrium point s^* in I_0'. Note that

$$(W^s(s^-) \cup W^u(s^-) \cup W^s(s^+) \cup W^u(s^+)) \subset (T \cup I_0),$$
$$(W^s(\bar{s}^-) \cup W^u(\bar{s}^-) \cup W^s(\bar{s}^+) \cup W^u(\bar{s}^+)) \subset (I_0 \cup I_0^\infty).$$

The dimensions of the stable and unstable invariant manifolds at the equilibrium points associated with non-degenerate central configuration for the gravitational n-body problem have been computed in [2] and [1].

For the McGehee-Devaney class of vector fields, there are special collision and ejection orbits which are associated to each central configuration called the *homothetic orbits*. These are given as follows. From (6) we have that if s is a central configuration and $u = 0$, then $s' = 0$ and $u' = 0$, where we have used the Euler's formula $s^t \nabla V(s) = -kV(s)$. Then on the plane (r, v) system (6) reduces to

$$r' = rv,$$
$$v' = kr^k h = (k/2)v^2 + kV(s).$$

We have exactly two homothetic orbits, H_s^+ and H_s^-, associated with each central configuration s such that $V(s) < 0$, whose expression is

$$r(\tau) = r(0)\exp(\pm(-2V(s))^{1/2}\tau),$$
$$v(\tau) = \pm(-2V(s))^{1/2},$$
$$s(\tau) = s,$$
$$u(\tau) = 0.$$

Then the homothetic orbit $H_s^+ \subset I_0$ projects on T and on I_0^∞ onto the equilibrium points s^+ and $\overline{s}^+$, respectively. These homothetic orbits can therefore be interpreted as heteroclinic orbits. The homothetic orbits from this point of view for the gravitational n-body problem have been studied in [2] and [5].

The main result of this note is the following.

Theorem 4 *Assume that the central configurations are non-degenerate critical points of V. If $k < 2$ then the flow on I_0' is gradient-like with respect to the v or $\overline{v}$-coordinate. Furthermore, we can classify the global behaviour of the solutions of the flow on I_0' in twenty-three kinds, see Table 1. The case $k > 2$ is similar.*

In Table 1, s, s_1 and s_2 denote central configurations, with $V(s) \neq V(s_1)$, but $V(s)$ and $V(s_2)$ can be equal. Recall that $\overline{v} = v$.

The proof of Theorem 4 follows, from Propositions 1 and 3, taking into account that $r' = rv$ and $\overline{r}' = -\overline{r}\overline{v}$.

The following proposition follows from Proposition 1 of [3].

Proposition 5 *If 0 is a regular value of $V(s)$, then $T \approx I_0^\infty$ is a compact manifold.*

On compact manifolds gradient vector fields are extremely simple dynamical systems because all non-equilibrium point solutions both begin and end at equilibrium points.

Then the following corollary follows from Theorem 4.

Corollary 6 *If $T \approx I_0^\infty$ is a compact manifold, then only the following fourteen kinds of solutions given in Table 1 are possible on I_0'* : 6, 7, 8, 9, 10, 13, 14, 15, 16, 17, 18, 19, 20 *and* 21.

Now, we give two examples.

Example 1 (For more details see [3]). For $m \geq 1$, the anisotropic Kepler theorem is described by the Hamiltonian equations corresponding to the Hamiltonian

$$H(q,p) = \frac{1}{2}(p_1^2 + p_2^2) - (mq_1^2 + q_2^2)^{-1/2}.$$

When $m = 1$, we have the usual Kepler problem. In McGehee coordinates, the anisotropic Kepler problem goes over to

$$\begin{aligned} r' &= rv, \\ v' &= u^2 + \frac{1}{2}v^2 + V(\theta), \\ \theta' &= u, \\ u' &= -\frac{1}{2}vu - \frac{dV(\theta)}{d\theta}, \end{aligned} \tag{18}$$

where $V(\theta) = -(m\cos^2\theta + \sin^2\theta)^{-1/2}$.

The total collision manifold (resp. the infinity manifold) of (18) is the torus T (resp. I_0^∞) of Figure 1. We can think of the zero energy level I_0 as the space contained between T and I_0^∞. When $m > 1$ there are sixteen equilibrium points for the flow on I_0', we denote them by $s_i^-, s_i^+, \overline{s}_i^-, \overline{s}_i^+$, $i = 1, 2, 3, 4$, see Figure 1. Then we have the eight corresponding homothetic orbits H_i^-, H_i^+. Furthermore, on T and on I_0^∞, s_i^-, $\overline{s}_i^-$ are sources and s_i^+, $\overline{s}_i^+$ are sinks for $i = 2, 4$ (if $m > 9/8$ the orbits spiral into and away from the sinks and sources); the other eight equilibrium points are saddles. For all $m > 1$, on T and on I_0^∞ each branch of one of the invariant manifolds of the saddles is asymptotic to a distinct sink or source, after θ increases or decreases by $3\pi/2$ or $\pi/2$. Hence, since $T \approx I_0^\infty$ is a compact manifold for the anisotropic Kepler problem when $m > 1$, the fourteen possible kinds of solutions on I_0', given in Corollary 6, exist.

Example 2 (For more details see [3], [9] and [4]). For $m > 0$, the planar isosceles three body problem may written in hamiltonian form with the Hamiltonian

$$H = \frac{1}{2}(2p_1^2 + (2+m)(2mp_2^2)^{-1}) - q_1^{-1} - 4m(q_1^2 + 4q_2^2)^{-1/2}.$$

In McGehee coordinates the equations become equations (18) with $V(\theta) = -(2^{1/2}\cos\theta)^{-1} - 4m^{3/2}(2m + 4\sin^2\theta)^{-1/2}$, where $-\pi/2 < \theta < \pi/2$. If we regularize the double collision we can extend the potential on $[-\pi/2, \pi/2]$. In a similar way to Example 1, Figure 2 gives us T, I_0^∞, I_0, the twelve equilibrium points and the six homothetic orbits for the isosceles problem. Then, on T and on I_0^∞, s_2^-, $\overline{s}_2^-$ are sources and s_2^+, $\overline{s}_2^+$ are sinks; the other eight equilibrium points are saddles. Now, to study the qualitative picture of the flow on T (resp. I_0^∞) it is neccesary to consider five cases according the value of m. From [9] and since $T \approx I_0^\infty$ is no compact for the isosceles problem, we have that the twenty-three possible kinds of solutions on I_0' described in Theorem 4, exist.

4. Acknowledgements

The first author is partially supported by a DGICYT grant n PB93-0860.

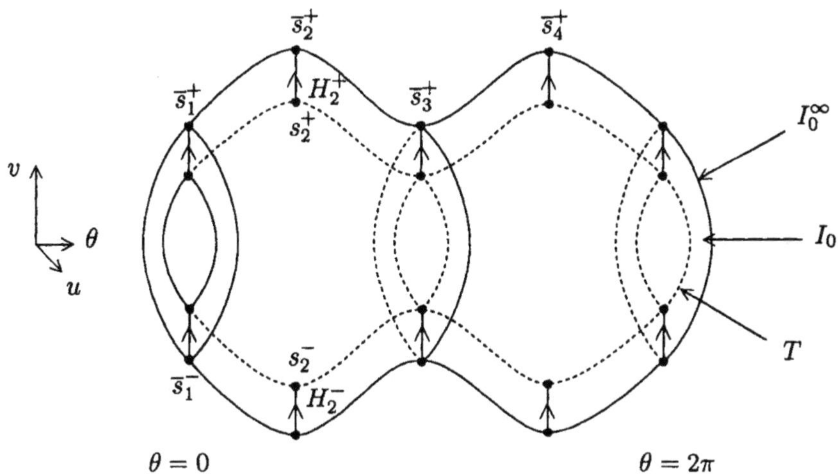

Figure 1: The zero energy level I_0' for the anisotropic Kepler problem

The set of solutions on I_0' such that the value of v on them satisfies:	is formed by
1. $-\infty < v = \overline{v} < +\infty$	I_0'-{the union of the stable and unstable manifolds associated with all the equilibrium points on I_0'}
2. $-\infty < v < -\sqrt{-2V(s)}$	$W^s(s^-)$
3. $-\infty < \overline{v} < -\sqrt{-2V(s)}$	$W^s(\overline{s}^-) \cap I_0$
4. $-\infty < v < +\sqrt{-2V(s)}$	$W^s(s^+) \cap T$
5. $-\infty < \overline{v} < +\sqrt{-2V(s)}$	$W^s(\overline{s}^+)$
6. $-\sqrt{-2V(s)} < v < -\sqrt{-2V(s_1)}$	$W^u(s^-) \cap W^s(s_1^-) \cap T$
7. $-\sqrt{-2V(s)} < v = \overline{v} < -\sqrt{-2V(s_1)}$	$W^u(\overline{s}^-) \cap W^s(s_1^-) \cap I_0$
8. $-\sqrt{-2V(s)} < \overline{v} < -\sqrt{-2V(s_1)}$	$W^u(\overline{s}^-) \cap W^s(\overline{s}_1^-) \cap I_0^\infty$
9. $-\sqrt{-2V(s)} < v < +\sqrt{-2V(s_2)}$	$W^u(s^-) \cap W^s(s_2^+) \cap T$
10. $-\sqrt{-2V(s)} < \overline{v} < +\sqrt{-2V(s_2)}$	$W^u(\overline{s}^-) \cap W^s(\overline{s}_2^+)$
11. $-\sqrt{-2V(s)} < v < +\infty$	$W^u(s^-) \cap T$
12. $-\sqrt{-2V(s)} < \overline{v} < +\infty$	$W^u(\overline{s}^-)$
13. $v = -\sqrt{-2V(s)}$	s^-
14. $v = \overline{v} = -\sqrt{-2V(s)}$	H_s^-
15. $\overline{v} = -\sqrt{-2V(s)}$	$\overline{s}^-$
16. $v = +\sqrt{-2V(s)}$	s^+
17. $v = \overline{v} = +\sqrt{-2V(s)}$	H_s^+
18. $\overline{v} = +\sqrt{-2V(s)}$	$\overline{s}^+$
19. $+\sqrt{-2V(s_1)} < v < +\sqrt{-2V(s)}$	$W^u(s_1^+) \cap W^s(s^+) \cap T$
20. $+\sqrt{-2V(s_1)} < v = \overline{v} < +\sqrt{-2V(s)}$	$W^u(s_1^+) \cap W^s(\overline{s}^+) \cap I_0$
21. $+\sqrt{-2V(s_1)} < \overline{v} < +\sqrt{-2V(s)}$	$W^u(\overline{s}_1^+) \cap W^s(\overline{s}^+) \cap I_0^\infty$
22. $+\sqrt{-2V(s)} < v < +\infty$	$W^u(s^+)$
23. $+\sqrt{-2V(s)} < \overline{v} < +\infty$	$W^u(\overline{s}^+) \cap I_0^\infty$

Table 1.

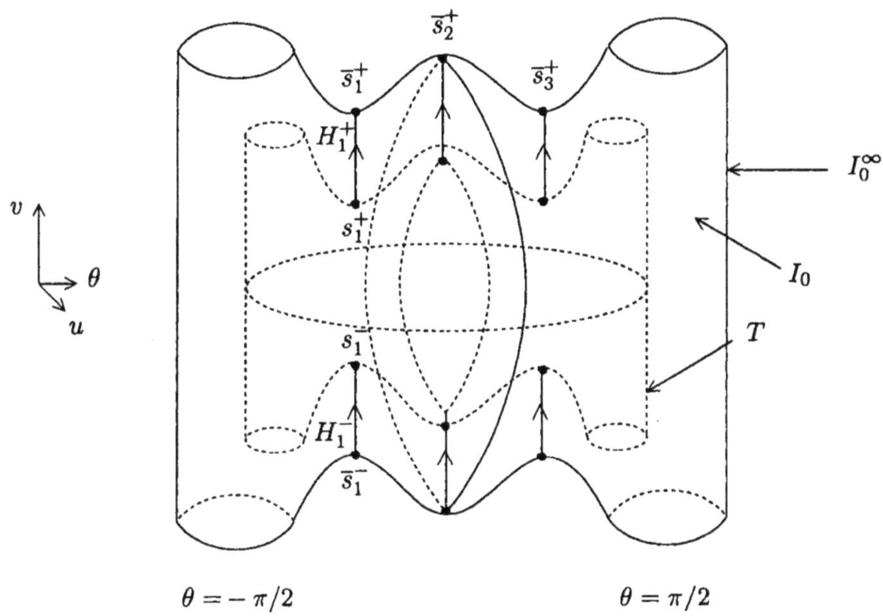

Figure 2: The zero energy level I_0' for the isosceles problem after the regularization of double collisions

5. References

1. J. Casasayas and J. Llibre. Invariant manifolds associated to homothetic orbits in the n-body problem. *Indiana Univ. Math. J.* **31**, 1982, pp. 463–470.
2. R.L. Devaney. Structural stability of homothetic solutions of the collinear n-body problem. *Celestial Mech.* **19**, 1979, pp. 391–404.
3. R.L. Devaney. Singularities in Classical Mechanical Systems, in *Ergodic Theory and Dynamical Systems I*, ed. A. Katok (Birkhäuser, 1981, pp. 211–333).
4. E.A. Lacomba and C. Simó. Boundary manifolds for energy surfaces in Celestial Mechanics. *Celestial Mech.* **28**, 1982, pp. 37–48.
5. J. Llibre and C. Simó. Characterization of transversal homothetic solutions in the n-body problem. *Archive for Rat. Mech. and Anal.* **77**, 1981, pp. 189–198.

6. R. McGehee. Triple collision in the collinear three-body problem. *Invent. Math.* **27**, 1974, pp. 191–227.
7. R. McGehee. Singularities in Classical Celestial Mechanics, in *Proceeding of the International Congress of Mathematics* ed. Helsinki, 1978, pp. 827–833.
8. D.G. Saari and N.D. Hulkower. On the manifolds of total collapse orbits and of completely parabolic orbits for the n-body problem. *J. of Diff. Equations* **41**, 1981, pp. 27–43.
9. C. Simó. Analysis of triple collision in the isosceles problem, in *Classical Mechanics and Dynamical Systems*, ed. Marcel Dekker, 1981, pp. 203–224.
10. J. Waldvogel. Stable and unstable manifolds in planar triple collision, in *Instabilities in Dynamical Systems*, ed. V. Szebehely (Reidel, 1979, pp. 263–271).

EQUIVARIANT GENERATING FUNCTIONS AND PERIODIC POINTS

KENNETH R. MEYER*
Department of Mathematical Sciences, University of Cincinnati
Cincinnati, OH 45221-0025, U.S.A.
E-mail: ken.meyer@uc.edu

ABSTRACT

This note shows that to an area preserving map with a fixed point at the origin whose multipliers are n-th roots of unity there is associated a generating function which has an n-fold symmetry. The critical points of the function other than the one at the origin correspond to periodic points of the map of period n. A maximum or minimum of the generating function corresponds to an elliptic periodic point of the map and a saddle point of the generating function corresponds to hyperbolic periodic point of the map. Thus the study of periodic points is reduced to catastrophe theory with an n-fold symmetry.

1. Introduction.

Poincaré [12] associated a generating function to an area preserving map of a subset of the plane with the property that fixed points of the map correspond to critical points of the generating function. In particular a maximum or minimum of the generating function corresponds to an elliptic fixed point of the area preserving map and a saddle point of the generating function corresponds to hyperbolic fixed point of the map. Thus a rough idea of the dynamics of the area preserving map can be obtained by plotting the level curves of the generating function. Of course Poincaré's construction of the generating function is local and there are some mild eigenvalue conditions imposed. See Meyer [9] for details.

In my 1970 study of the generic bifurcation of periodic points of area preserving mappings I use Poincaré's generating function to analyze the bifurcations of fixed points of a map since the bifurcation of critical points was well studied in catastrophe theory. Catastrophe theory gave a nice clean solution to that part of the problem. For the bifurcation of periodic points of higher period I used normal form methods to study the bifurcations. As is well known the normalizing transformation does not converge in general, but it does provide enough information to eke out the nature of the bifurcations – an adequate but not aesthetic solution. In the intervening years normal form methods have been used by many authors on a variety of problems. Also the methods of catastrophe and singularity theory have been adapted to bifurcation problems [4,6,7].

*This research partially supported by grants from the National Science Foundation and the Taft Foundation.

The methods are distinct, but yield the same results in general. I discuss the interconnections between the methods below.

In this note I will show that to an area preserving map with a fixed point at the origin whose multipliers are n-th roots of unity there is associated a generating function which has an n-fold symmetry. The critical points of the function other than the one at the origin correspond to periodic points of the map of period n. As with Poincaré's generating function a maximum or minimum of the generating function corresponds to an elliptic periodic point of the map and a saddle point of the generating function corresponds to hyperbolic periodic point of the map. Thus the study of periodic points is again reduced to catastrophe theory but now the functions must have an n-fold symmetry.

This result is carried out in detail in Bridges and Furter [4] using a Lagrangian variational formulation pioneered by Aubry [3]. Those schooled in singularity theory may find their presentation easier, but I think a differential equationists will find the proof given here closer to their heritage. The construction of the generating function uses the classical technique called the Liapunov-Schmidt method or the alternative method by Hale – see Chapter IX of Hale [8] for details and historical remarks. Here I will modify the presentation of this method as given in Moser [11] to the present problem. Moser makes it very clear that his proof carries over to the n-degrees of freedom problem with the only addition expence being some more notation. Considering the 2-dimensional problem simplifies the presentation without loss of essential features.

In the next section, Section 2 , the relation between periodic systems and Poincaré maps for Hamiltonian systems is summarized and some standard results about normal forms for periodic Hamiltonian systems are given. This gives some motivation for the results presented later. Section 3 sets the notation and states the main result of the paper about the existence of the generating function. In Section 4 only a sketch is given of the proof because it is only a minor modification of the proof given in Moser [11] and his proof is straight forward. Section 5 indicates some extensions and applications of the main result.

2. Periodic Systems, Mappings, and Normal Forms.

Most of the results of this section are well known and can be found in Meyer and Hall [10]. The page numbers refer to that book. Bifurcations occur in systems which depend on various parameters, but for the time being the dependence on parameters of the various maps and equations will be suppressed.

Let $\mathcal{O}$ be an open neighborhood of the origin in $\mathbb{R}^2$ and $P : \mathcal{O} \rightarrow \mathbb{R}^2$ be a smooth area preserving mapping which has the origin as a fixed point. Smooth will mean $C^k, k \geq 1$, or real analytic and area preserving means that the determinant of the Jacobian matrix of P is identically equal to $+1$, i.e. $\det DP(\xi) \equiv +1$ for $\xi \in \mathcal{O}$. Let $B = DP(0)$ so $P(\xi) = B\xi + \cdots$. Assume that the eigenvalues of B are n-th roots of unity, $n > 2$, so $B^n = I$. (The case when $n = 2$ will be discussed in the last section.) The symplectic matrix B has a Hamiltonian logarithm A, i.e. $e^A = B$ and $\operatorname{trace} A = 0$ - see (page 54). (Remark: The statement of the lemma on page 67 is incorrect and should be replaced by the statement of Theorem II.E.2.)

P is the period map of a 1-periodic Hamiltonian system of the form

$$\dot{x} = F(x,t) = Ax + f(x,t) = J\nabla H(x,t) = JSx + J\nabla K(x,t) \tag{1}$$

where

$$H(x,t) = \frac{1}{2}x^T Sx + K(x,t) \text{ and } A = \text{ JS}. \tag{2}$$

Here F, f, H and K are as smooth as P in $x \in \mathcal{O} \subset \mathbb{R}^2$ and are 1-periodic and C^∞ in $t \in \mathbb{R}$. Also $f(0,t) \equiv 0, Df(0,t) \equiv 0, \nabla K(x,0) \equiv 0, D^2K(x,0) \equiv 0$ where the derivatives are with respect to x only (page $116ff$).(The neighborhood $\mathcal{O}$ may have to be shrunk a bit.)

For the rest of this section only assume that F, f, H and K are formal power series in x with coefficients that are smooth, periodic functions in t. Furthermore assume that they are in normal form, i.e. assume that

$$H(e^{At}x,t) \equiv H(x,0) \tag{3}$$

and

$$f(e^{At}x,t) \equiv e^{At}f(x,0) \tag{4}$$

(page 193). Make the time-periodic, symplectic change of variables $x = e^{At}y$ so the equations of motion 1 become

$$\dot{y} = V(y) = J\nabla L(y) \tag{5}$$

where

$$L(y) = K(y,0), V(y) = f(y,0) \tag{6}$$

Equations (3) and (4) imply that

$$L(By) = L(y), V(By) = BV(y). \tag{7}$$

Thus if the system (1) with Hamiltonian (2) is in normal form then a linear, symplectic, periodic change of variable reduces the system to an autonomous system (5) with the n-fold symmetry expressed by (7).

Let $x(t,\xi)$ and $y(t,\xi)$ be the solutions of (1) and (5) respectively through ξ at $t=0$. So $P(\xi)=x(1,\xi)=e^{A}y(1,\xi)=By(1,\xi)$. Thus the normal form for P is $By(1,\xi)$ where y is the time one map of an autonomous system with the n-fold symmetry (7).

We can think of L as a formal analog of Poincaré's generating function. If ξ_0 is a critical point of L then by (7) so are $B\xi_0, B^2\xi_0, \ldots, B^n\xi_0=\xi_0$. Since ξ_0 is a critical point of L it is a fixed point of $y(1,\xi)$ and hence a periodic point of period n of P. So the critical points of L give rise to periodic points of P of period n.

Let the Hessian of L at the critical point ξ_0 be R so the linearization of (5) about the critical point is $\dot{y}=\text{JRy}=\text{Qy}$. Therefore the linearization of P^n at ξ_0 is $e^{nQ}\xi$. If ξ_0 is a nondegenerate maximum or a minimum of L then the eigenvalues of Q are of the form $\pm\omega i, \omega>0$, and the eigenvalues of e^{nQ} are $e^{\pm n\omega i}$. Thus the periodic point ξ_0 of P is elliptic (provided $n\omega$ is not a multiple of π). Similarly, a nondegenerate saddle point of L corresponds to a hyperbolic periodic point of P.

Thus critical points of L other than the origin correspond to periodic points of P of period n. Saddle points of L correspond to hyperbolic points of P and most maxima and minima of L correspond to elliptic periodic points of P.

To illustrate this result use complex coordinates in the plane, i.e. change coordinates from real coordinates $\xi=(\xi_1,\xi_2)$ to complex coordinates $(z,\bar{z})$ by $z=\xi_1+i\xi_2$, $\bar{z}=\xi_1-i\xi_2$. If the matrix B is in normal form then

$$B=\begin{pmatrix}\cos\omega & \sin\omega\\ -\sin\omega & \cos\omega\end{pmatrix}$$

where $\omega=2\pi/n$ and the transformation $\xi\to B\xi$ is $z\to e^{\omega i}z$ in complex coordinates. The only monomials invariant under this transformation are of the form

$$(z\bar{z})^{\alpha}z^{n\beta}\bar{z}^{n\gamma} \tag{8}$$

where α,β,γ are non-negative integers. Thus the formal generating function L is a formal series in these terms.

For example when $n=3$ the first few terms are $az\bar{z}+bz^3+\bar{b}\bar{z}^3+\cdots$. The prototype bifurcation which occurs at a fixed point whose multipliers are cube roots of unity with one parameter μ can be modeled with $a=\mu$ and $b\neq 0$. In this case you can rotating coordinates so that b is real and so the model generating function is

$$L=\mu z\bar{z}+b(z^3+\bar{z}^3) \tag{9}$$

in complex coordinates or

$$L=\mu r^2+2br^3\cos 3\theta \tag{10}$$

The critical points other than the one at the origin occur in two sets namely $r = -\mu/3b, \theta = 0, 2\pi/3, 4\pi/3$ and $r = +\mu/3b, \theta = \pi/3, \pi, 5\pi/3$. Both sets are saddle points for L and correspond to a single periodic orbit of period three. This is the standard bifurcation as given in Meyer [9].

3. Main result.

It is well known that it is impossible in general to construct a change of variables which will linearize equation (1) – even a formal change of variables. It is also known that in general you cannot find a smooth transformation which reduces a system of the form (1) to normal form. The alternative method takes another tack. It asks what terms must be subtracted from (1) in order that a linearizing transformation exist.

In particular, we shall show that (1) can be modified to the form

$$\dot{x} = F(x,t) + w(x,t) = Ax + f(x,t) + w(x,t) \tag{11}$$

so that there is a change of variables

$$x = u(y,t) \tag{12}$$

which reduces (1) to the linear equation

$$\dot{y} = Ay. \tag{13}$$

Of course the u and w are not arbitrary. They are as smooth as f and 1-periodic in t. u is invertible with inverse $y = u^{-1}(x,t)$, and most importantly

$$w(x,t) = v(u^{-1}(x,t),t) \tag{14}$$

where $v(y,t)$ satisfies

$$v(e^{At}y,t) = e^{At}v(y,0) \tag{15}$$

for all t and s.

The importance of these properties comes from the following observation. For each y_0 the function $x(t) = u(e^{At}y_0,t)$ is an periodic solution of (11) of period n, so if $w(x(t),t) \equiv v(e^{At}y_0,t) \equiv 0$ then it is a n-periodic solution of the original equation (1). But by (15) $v(e^{At}y_0,t) \equiv 0$ if and only if

$$V(y_0) = v(y_0,0) = 0. \tag{16}$$

Thus the search for n-periodic solutions of (1) is reduced to solving the system of equations (16) which are called the determining equations.

It follows from (15) that

$$V(By) = BV(y) \tag{17}$$

so V has an n-fold symmetry. Also, since the equation (1) is Hamiltonian $V(y)$ is the gradient of a function $L(y)$ and by (17) L satisfies

$$L(By) = L(y). \tag{18}$$

This L serves as the analog of the Poincaré generating function for the study of the bifurcation of periodic points of period n. In contrast to the normal form approach outlined in the previous section this is not a formal procedure since we will establish the existence of smooth u and v satisfying the relations above. Let $U(x) = u(x,0) = u(x,1)$; U has a local inverse U^{-1}.

In summary: *Let $P : \mathcal{O} \longrightarrow \mathbb{R}^2 : x \longrightarrow Bx + \cdots$ ($\mathcal{O}$ a neighborhood of the origin in $\mathbb{R}^2$) be a smooth area preserving mapping which has the origin as a fixed point. Assume that the eigenvalues of B are n^{th} roots of unity, $n > 2$, so $B^n = I$. Then there exist a local diffeomorphism $y = U(x)$ and a smooth $L : \mathcal{O}' \to \mathbb{R}$, ($\mathcal{O}'$ a neighborhood of the origin in $\mathbb{R}^2$) such that $L(By) = L(y)$. If y_0 is a critical point of L other than at the origin then $x_0 = U^{-1}(y_0)$ is a periodic point of P of period n and conversely. Nondegenerate maxima or minima of L correspond to elliptic periodic points of P and nondegenerate saddle points of L correspond to hyperbolic periodic points of P. If P depends smoothly on parameters then so does U and L.*

4. Proof Outline.

Moser [11] gives a variational derivation for the various formulas given below. I shall simple verify the results. Since there are two variables $y \in \mathcal{O} \subset \mathbb{R}^2$ and $t \in \mathbb{R}$, let D denote the derivative with respect to y ($D = \partial/\partial y$) and an overdot the derivative with respect to t ($(\dot{\cdot}) = d/dt$). Let the matrices A and B be in normal form so $\| e^{At} \| = \| B \| = 1$.

The first step is to solve the linearized functional equation for the transformation sought. Here it is given by simple formulas (see (24 and (17). Let $\mathcal{D}$ be the differential operator defined by

$$\mathcal{D}v(y,t) = Dv(y,t)Ay - Av(y,t) + \dot{v}(y,t). \tag{19}$$

Now $\mathcal{D}v = 0$ implies $\frac{d}{ds}e^{-As}v(e^{As}y, t+s) = 0$ or $v(e^{As}y, t+s) = e^{As}v(y,t)$. Define a projection operator P by

$$Ph(y,t) = \int_0^1 e^{-As}h(e^{As}y, t+s)ds. \tag{20}$$

Then

$$Pv(y,t) = \int_0^1 e^{-As} v(e^{As}y, t+s)ds = \int_0^1 v(y,t)ds = v(y,t),$$

so $Pv = v$.

Lemma 1 *Let $g(y,t)$ smooth for $y \in O$, $t \in \mathbb{R}$ and be 1-periodic in t, then there exists functions $u(y,t)$ and $v(y,t)$, smooth for $y \in O$, $t \in \mathbb{R}$ and be 1-periodic in t such that*

$$\mathcal{D}v = 0 \tag{21}$$

$$\mathcal{D}u - v = g \tag{22}$$

$$Pu = 0 \tag{23}$$

Proof. Let

$$v(y,t) = -\int_0^1 e^{-As} g(e^{As}y, t+s)ds, \tag{24}$$

$$u(y,t) = \int_0^1 (\frac{1}{2} + s) e^{-As} g(e^{As}y, t+s)ds \tag{25}$$

To prove (21):

$$\begin{aligned} Dv(y,t)Ay &= -\textstyle\int_0^1 e^{-As} Dg(e^{As}y, t+s) e^{As} Ayds \\ &= -\textstyle\int_0^1 e^{-As} \left(\frac{dg}{ds}(e^{As}y, t+s) - \dot{g}(e^{As}y, t+s)\right) ds \\ &= -\textstyle\int_0^1 Ae^{-As} g(e^{As}y, t+s)ds - \dot{v}(y,t) \\ &= Av(y,t) - \dot{v}(y,t). \end{aligned}$$

The second to last step is an integration by parts – the nonintegral terms drop out by the periodicity of e^{As}.

To prove (22):

$$\begin{aligned} Du(y,t)Ay &= \textstyle\int_0^1 (\frac{1}{2} + s) e^{-As} Dg(e^{As}y, t+s) e^{As} Ayds \\ &= \textstyle\int_0^1 (\frac{1}{2} + s) e^{-As} \left(\frac{dg}{ds}(e^{As}y, t+s) - \dot{g}(e^{As}y, t+s)\right) ds \\ &= (\tfrac{1}{2} + s) e^{-As} g(e^{As}y, t+s) \mid_0^1 \\ &\qquad - \textstyle\int_0^1 \left(-A(\frac{1}{2} + s) + 1\right) e^{-As} g(e^{As}y, t+s)ds - \dot{u}(y,t) \\ &= g(y,t) + Au(y,t) + v(y,t) - \dot{u}(y,t). \end{aligned}$$

To prove (23):

$$\begin{aligned} Pu(y,t) &= \int_0^1 e^{-As}u(e^{As}y,t+s)ds \\ &= \int_0^1 e^{-As}\int_0^1 \left(\tfrac{1}{2}+\tau\right)e^{-A\tau}g(e^{As}e^{A\tau}y,t+s+\tau)d\tau ds \\ &= -\tfrac{1}{2}Pv(y,t)+\int_0^1 \tau e^{-A\tau}\int_0^1 e^{-As}g(e^{As}e^{A\tau}y,t+s+\tau)dsdt \\ &= -\tfrac{1}{2}v(y,t)+\int_0^1 \tau e^{-A\tau}v(e^{A\tau}y,t+\tau)d\tau \\ &= -\tfrac{1}{2}v(y,t)+\int_0^1 \tau v(y,t)d\tau = 0.\square \end{aligned}$$

Thus having solved the linearized functional equation the next step is to solve the nonlinear equation by the contracting mapping principle.

Lemma 2 *Let $F(x,t) = Ax + f(x,t)$ be smooth for $x \in \mathcal{O}$, $t \in \mathbb{R}$, be 1-periodic in t, and $f(0,t) = 0, Df(0,t) \in 0$. Then there is an open neighborhood $\mathcal{Q} \subset \mathcal{O}$ and there are functions $u(y,t)$ and $v(y,t)$, smooth for $y \in \mathcal{Q}$, $t \in \mathbb{R}$ and be 1-periodic in t such that*

$$\mathcal{D}v = 0 \tag{26}$$

$$\mathcal{D}u - v = f \circ u \tag{27}$$

$$(Pu)(y,t) = y \tag{28}$$

$Du(y,t) = I$. (Equation (27) means $(\mathcal{D}u)(y,t) - v(y,t) = f(u(y,t),t)$.).

Proof Outline. To solve (27) let $u(y,y) = y + u'(y,t)$ so equation (27) becomes

$$(Du')(y,t) - v(y,t) = f(y + u'(y,t),t) \tag{29}$$

and equation (28) becomes $Pu' = 0$. The linearization of these equations is precisely the equations given in Lemma 1. The proof from here on is a standard contracting mapping principle argument similar to the argument given in Moser [11]. Since the details are given in detail there I will omit them. □

This lemma reduces the problem of finding n-periodic solutions to solving a system of equations which with an n-fold symmetry. To see this let $x = u(y,t)$, and $y = u^{-1}(x,t)$ be the change of variables given by this lemma. If $y(t)$ satisfies

$$\dot{y} = Ay \tag{30}$$

then $x(t) = u(y(t),t)$ satisfies

$$\dot{x} = Ax + f(x,t) + v(u^{-1}(x,t),t). \tag{31}$$

So if

$$V(y_0) = v(y_0, 0) = 0 \tag{32}$$

then $v(e^{At}, t) = 0$ and $x(t) = u(e^{At}y_0, t)$ satisfies

$$\dot{x} = Ax + f(x, t). \tag{33}$$

Thus solving (32) gives rise to an n-periodic solution of (33). Since $\mathcal{D}v = 0$ it follows as before that

$$BV(y) = V(By).$$

Up to this point I have not used the fact that the equations are Hamiltonian.

Lemma 3 *Let the conditions of Lemma 2 hold. If the equations (1) are Hamiltonian with Hamiltonian (2) then there is a smooth function L: $Q \to \mathbb{R}$ such that $V(y) = v(y, 0) = \nabla L(y)$ and $L(By) = L(y)$.*

Proof. Here we assume that $f(x, t) = J\nabla K(x, t)$. Let

$$S(y,t) = \int_0^1 \left(\frac{1}{2} \frac{du}{ds}(e^{As}y, t+s)^T Ju(e^{As}y, s+t) - K(u(e^{As}y, t+s), t+s) \right) ds.$$

Then

$$\begin{aligned} DS(y,t) &= \textstyle\int_0^1 \left(\frac{du}{ds}(e^s y, t+s) - J\nabla K(u(e^{As}y, t+s)\right)^T Du(e^{As}y, s+t)e^{As}ds \\ &= \textstyle\int_0^1 \left((\mathcal{D}u)(e^{As}y, t+s) - f(u(e^{As}y, t+s)\right) TJDue^{As}y, s+t)e^{As}ds \\ &= \textstyle\int_0^1 \left(v(e^{As}y, t+s)\right)^T JDu(e^{As}y, s+t)e^{As}ds \\ &= v(y,t)^T J \textstyle\int_0^1 e^{As} Du(e^{As}y, s+t)e^{As}ds \\ &= v(y,t)^T J. \end{aligned}$$

Therefore $v(y, t) = J\nabla S(y, t)$ or $V(y) = J\nabla L(y)$ where $L(y) = S(y, 0)$. In the sequence of expressions for DS given above the first inequality requires an integration by parts, the second uses some definitions, the third uses (22), the forth uses (21) and the remark preceding Lemma 1, and the last comes from differentiating (23).

5. Final Remarks.

It remains now to discuss the unfoldings of the critical points of functions which have a n-fold symmetry. Fortunately, this has been done in detail in

Bridges and Furter [4]. For example, the codimension 3 functions with a 3-fold symmetry with there unfoldings are

$$\frac{1}{3}z\bar{z} + \frac{1}{4}\alpha(z^3 + \bar{z}^3)$$

$$\frac{1}{4}(z\bar{z})^2 + \frac{1}{24}(z^3 + \bar{z}^3)^2 + \frac{1}{2}\alpha z\bar{z} + \frac{1}{6}\beta(z^3 + \bar{z}^3) + \frac{i}{4}\gamma(\bar{z}^3 - z^3)$$

where α, β, γ are the unfolding parameters. These functions are easy to analyse, but the analysis is tedious when there are many unfolding parameters. [4] contains many tables which give the complete catalog of functions with a n-fold symmetry with their unfoldings up to codimension 3.

This reduction to a problem in equivariant catastrophy theory is very pleasing, but if you just want to know the bifurcations in a particular situation it seems to me that a straight forward application of normal form methods will yield the answer quicker. I say this for two reasons. First of all, if you are interested in periodic solutions of a Hamiltonian system, you do not need to know the canonical form of the generating function. You really don't care about the function itself, only its critical points - all its other properties have no meaning. Secondly, normal form algorithms are fully developed and implimented a various computer platforms.

6. References

1. R. Abraham and J. Marsden, *Foundations of Mechanics*, Benjamin-Cummings, London, 1978.
2. V. I. Arnold, *Mathematical Methods of Classical Mechanics*, Springer-Verlag, New York, 1978.
3. S. Aubry, The twist map, the extended Rrenkel-Kontorova model and the devil's staircase, *Physica* **7D**, 1983, 240-58.
4. T. J. Bridges and J. E. Furter, Singularity Theory and Equivariant Symplectic Maps, *Lecture Notes in Mathematics* **1558**, Springer-Verlag, 1993.
5. A. Deprit, Canonical transformation depending on a small parameter, *Celestial Mechanics* **72**, 1969, 173-79.
6. M. Golubitsky and D. G. Schaeffer, *Singularities and Groups in Bifurcation Theory I*, Springer-Verlag, 1985.
7. M. Golubitsky, I. Stewart and D. G. Schaeffer, *Singularities and Groups in Bifurcation Theory II*, Springer-Verlag, 1988.
8. J. K. Hale, *Ordinary Differential Equations*, John Wiley, New York, 1972.
9. K. R. Meyer, Generic bifurcation of periodic points, *Transactions of the American Mathematics Society* **149**, 1970, 95-107.

10. K. R. Meyer and G. R. Hall, *Introduction to Hamiltonian Dynamical Systems and the N-Body Problem*, Springer-Verlag, New York, 1992.
11. J. K. Moser, Periodic orbits near an equilibrium and a theorem of Alan Weinstein, *Communications in Pure and Applied Mathematics* **XXIX**, 1979, 727-747.
12. H. Poincaré, *Les methods nouvelles de la mechanique celeste*, Gauthier-Villar, Paris, 1899.
13. H. G. Poinhoff, *An analytic closing lemma, Proceedings of the Midwest Dynamical Systems Seminar*, Northwestern Univ. Press, Evanston, Ill, 1973, 128-256.
14. T. Poston and I. Stewart, *Catastrophe Theory and its Applications*, Pitman, Boston, 1978.
15. R. Rimmer, Symmetry and bifurcation of fixed points of area preserving maps, *Journal of Differential Equations* **29**(3), 1978, 329-44.
16. R. Rimmer, Generic bifurcation for involuntary area preserving maps, *Memoirs of the American Mathematics Society* **41**(272), 1983, 1-165.

THE MOTION OF TWO DIMENSIONAL VORTICES WITH MASS AS A SINGULAR PERTURBATION HAMILTONIAN PROBLEM

WALDYR M. OLIVA*
Departamento de Matemática, Instituto Superior Técnico
Universidade Técnica de Lisboa, Av. Rovisco Pais 1
1096 LISBOA Codex - PORTUGAL
E-mail: wamoliva@sisdin9.ist.utl.pt

1. Introduction

In the paper "On the Motion of two dimensional Vortices with Mass" [Ragazzo, Koiller and Oliva, 1994] it has been considered a mathematical model related to the problem of two-dimensional massive charges with logarithmic potential in a transversal magnetic field; moreover, in the context of hydrodynamics, it provides a model for impurities in thin flow films and is also a toy model for early planetary or galactical models (see [Ragazzo, Koiller and Oliva, 1994] for references). As it was mentioned there, a singular perturbation question has a very special feature when we consider vortices with mass; one begins with a Hamiltonian system such that after the limit as the masses go to zero we obtain another Hamiltonian with a smaller number of degrees of freedom and a symplectic structure independent of the starting one. Our intention is to emphasize the stability questions associated to this limit and relate them with the changes on the sympletic geometry.

2. Vortices with mass and without mass

2.1. Newtonian formulation for the case of vortices with mass

I will choose, for the present paper, the notations considered in [Ragazzo, Koiller and Oliva, 1994]. So, the motion of N vortices with masses $m_1, \ldots, m_N$ is governed by the Newton's equations:

$$m_j \ddot{\vec{q}}_j = \rho \Gamma_j \vec{e}_3 \times (\dot{\vec{q}}_j - \vec{u}_R^{(j)}(\vec{q}_j)), j = 1, \ldots, N, \tag{1}$$

where ρ is the density of the fluid and

$$\Gamma_j \vec{u}_R^{(j)}(\vec{q}_j) = \vec{e}_3 \times \nabla_j W = \vec{e}_3 \times \left(\frac{\partial W}{\partial q_{j_1}} \vec{e}_1 + \frac{\partial W}{\partial q_{j_2}} \vec{e}_2 \right); \tag{2}$$

*This research was supported in part by JNICT Portugal (STRDA/C/CEN/528/92).

here we used an orthonormal basis $(\vec{e}_1, \vec{e}_2, \vec{e}_3)$ as a positively oriented reference frame, the vectors $\vec{e}_1, \vec{e}_2$ being contained in the plane of the fluid and the symbol $\times$ meaning the usual vector product in $\mathbb{R}^3$. $\vec{q}_j, \dot{\vec{q}}_j$ and $\ddot{\vec{q}}_j$ represent, respectively, the position, the velocity and the acceleration of the jth-vortex; the pair (q_{j_1}, q_{j_2}) gives the cartesian coordinates of the jth-vortex, that is,

$$\vec{q}_j = q_{j_1}\,\vec{e}_1 + q_{j_2}\,\vec{e}_2\,. \tag{3}$$

The function $W = W(\vec{q}_1, \ldots, \vec{q}_N)$ is the so-called Kirchhoff-Routh function that governs the motion of N vortices without masses and intensities $\Gamma_j > 0, j = 1, \ldots, N$.
We will be restricted to the case in which there are N vortices in $\mathbb{R}^2$ with no boundaries.

2.2. Hamiltonian formulation for the case of vortices without mass

Since there is no boundary, we have:

$$W = \frac{1}{4\pi} \sum_{\langle j,k\rangle=1}^{N} \Gamma_j \Gamma_k \; \ell n \mid \vec{q}_j - \vec{q}_k \mid \tag{4}$$

where $\langle j, k\rangle$ means that the sum is considered over all pairs $(j,k), j,k = 1, \ldots, N$, $j \neq k$. Let (W, ω_o) in $\mathbb{R}^{2N}$ to be the Hamiltonian structure governing the motion of N vortices with positive intensities, no masses, and symplectic form given by

$$\omega_o = \sum_{j=1}^{N} \Gamma_j dq_{j_1} \wedge dq_{j_2}. \tag{5}$$

The corresponding Hamiltonian vector field has components

$$X_j = -\frac{1}{\Gamma_j}\frac{\partial W}{\partial q_{j_2}} \quad , \quad Y_j = \frac{1}{\Gamma_j}\frac{\partial W}{\partial q_{j_1}} \tag{6}$$

and first integrals W, I_2, I_3 and I_4:

$$I_2 = \sum_{j=1}^{N} \Gamma_j q_{j_1} \quad , \quad I_3 = \sum_{j=1}^{N} \Gamma_j q_{j_2} \quad , \quad I_4 = \sum_{j=1}^{N} \Gamma_j ({q_{j_1}}^2 + {q_{j_2}}^2). \tag{7}$$

Natural canonical coordinates for ω_o are

$$x'_j = \sqrt{\Gamma_j} q_{j_1} \quad \text{and} \quad y'_j = \sqrt{\Gamma_j} q_{j_2} \tag{8}$$

and, with the Poisson bracket defined by ω_o (see [Castilla, Moauro, Negrini and Oliva, 1993]) one has

$$\{I_2{}^2 + I_3{}^2, I_4\} = 0. \tag{9}$$

2.3. Lagrangian and Hamiltonian formulations for the massive case

The Lagrangian approach for the Newton's equations (1) and (2) is given by the Lagrangian function

$$\mathcal{L} : T(\mathbb{R}^{2N}) \sim \mathbb{R}^{4N} \longrightarrow \mathbb{R}$$

defined by

$$\mathcal{L} = \frac{1}{2}\sum_{j=1}^{N} m_j \| \dot{\vec{q_j}} \|^2 - \frac{\rho}{2}\sum_{j=1}^{N} \Gamma_j \dot{\vec{q_j}} \cdot (\vec{e_3} \times \vec{q_j}) + \rho W \tag{10}$$

with $W = W(\vec{q_1}, \ldots, \vec{q_N})$ given by (4). In fact, the corresponding Euler-Lagrange equations are

$$\frac{d}{dt}\frac{\partial \mathcal{L}}{\partial \dot{q}_{j_i}} - \frac{\partial \mathcal{L}}{\partial q_{j_i}} = 0, \tag{11}$$

$j = 1, \ldots, N$, $i = 1, 2$, and they are equivalent to

$$\frac{d}{dt}\left[m_j \dot{\vec{q_j}} - \rho \Gamma_j \vec{e_3} \times \dot{\vec{q_j}} \right] - \rho \nabla_j W = 0, \tag{12}$$

that are, precisely, equations (1) and (2). The corresponding Legendre's transformation $F\mathcal{L}$ associated to $\mathcal{L}$ is the map

$$F\mathcal{L} : T(\mathbb{R}^{2N}) \longrightarrow T^*(\mathbb{R}^{2N})$$

given by

$$(\vec{q_j}, \dot{\vec{q_j}}) \in T(\mathbb{R}^{2N}) \xrightarrow{F\mathcal{L}} (\vec{q_j}, \vec{p_j}) \in T^*(\mathbb{R}^{2N})$$

where

$$\vec{p_j} = \frac{\partial \mathcal{L}}{\partial \dot{\vec{q_j}}} = m_j \dot{\vec{q_j}} - \frac{\rho}{2}\Gamma_j \vec{e_3} \times \vec{q_j} \,. \tag{13}$$

For $m_j > 0$, $j = 1, \ldots, N$, the Legendre's transformation is a bijection between $T(\mathbb{R}^{2N})$ and $T^*(\mathbb{R}^{2N})$. The canonical symplectic form of $T^*(\mathbb{R}^{2N})$ is

$$\omega = \sum_{j=1}^{N} dp_{j_1} \wedge dq_{j_1} + dp_{j_2} \wedge dq_{j_2}, \tag{14}$$

where $((q_{j_1}, q_{j_2}), (p_{j_1}, p_{j_2}))$ are the usual coordinates of the cotangent bundle $T^*(\mathbb{R}^{2N})$. The energy of $\mathcal{L}$ (see [Abraham, Marsden, 1978] for more details) is the function $E : T(\mathbb{R}^{2N}) \longrightarrow \mathbb{R}$ expressed as

$$E = \sum_{j=1}^{N} \vec{p_j} \cdot \dot{\vec{q_j}} - \mathcal{L} \tag{15}$$

and a Hamiltonian function $H : T^*(\mathbb{R}^{2N}) \rightarrow \mathbb{R}$ is defined by $H = E \circ (F\mathcal{L})^{-1}$.

If $\dot{\vec{q_j}}, j = 1, \ldots, N$, given by (13), are now introduced in (10) and (15), we get for $H = E \circ (F\mathcal{L})^{-1}$:

$$\begin{aligned} H &= \sum_{j=1}^{N} \vec{p_j} \cdot \frac{1}{m_j}[\vec{p_j} + \frac{\rho}{2}\Gamma_j\, \vec{e_3} \times \vec{q_j}] - \frac{1}{2}\sum_{j=1}^{N} m_j \|\frac{1}{m_j}(\vec{p_j} + \frac{\rho}{2}\Gamma_j\, \vec{e_3} \times \vec{q_j})\|^2 \\ &+ \frac{\rho}{2}\sum_{j=1}^{N} \Gamma_j \frac{1}{m_j}[\vec{p_j} + \frac{\rho}{2}\Gamma_j\, \vec{e_3} \times \vec{q_j}].(\vec{e_3} \times \vec{q_j}) - \rho W(\vec{q_1}, \ldots, \vec{q_N}), \end{aligned} \tag{16}$$

and finally

$$H = \frac{1}{2}\sum_{j=1}^{N} \frac{1}{m_j} \| \vec{p_j} + \frac{\Gamma_j \rho}{2}\, \vec{e_3} \times \vec{q_j} \|^2 - \rho W(\vec{q_1}, \ldots, \vec{q_N}). \tag{17}$$

The functions $(\vec{q_j}, \dot{\vec{q_j}})$, where the $\dot{\vec{q_j}}$ are given by (13), are canonical coordinates for $T(\mathbb{R}^{2N})$, relative to $(F\mathcal{L})^*\omega$, provided that the masses m_j do not vanish. The function H defines on $T^*(\mathbb{R}^{2N})$ a Hamiltonian vector-field with respect to the symplectic form ω given by (14). The symplectic structure $(T^*(\mathbb{R}^{2N}), \omega)$ is the canonical one, so is fixed on $T^*(\mathbb{R}^{2N})$; on the other hand $(T(\mathbb{R}^{2N}), (F\mathcal{L})^*\omega)$ depends on the masses. Thus, we decided to analyze our problem in the cotangent bundle $T^*(\mathbb{R}^{2N})$ with H given by (17). It is also interesting to remark that the Lagrangian function $\mathcal{L}$ given by (10) can be written

$$\mathcal{L} = \frac{1}{2}\sum_{j=1}^{N} m_j \| \dot{\vec{q_j}} \|^2 - U(\vec{q_1}, \ldots, \vec{q_N}, \dot{\vec{q_1}}, \ldots, \dot{\vec{q_N}}) \tag{18}$$

where the function

$$U = \left[\frac{\rho}{2} \sum_{j=1}^{N} \Gamma_j \, \dot{\vec{q_j}} \, (\vec{e_3} \times \vec{q_j}) - \rho W(\vec{q_1}, \ldots, \vec{q_N}) \right] \tag{19}$$

does not depend on the masses.

2.4. Remarks

2.4.1. Formally, if the $m_j = 0$ $(j = 1, \ldots, N)$, equation (11) becomes

$$\frac{d}{dt} \frac{\partial U}{\partial \dot{\vec{q_j}}} - \frac{\partial U}{\partial \vec{q_j}} = 0 \tag{20}$$

that represents a first order system; and, it is easy to see that system (20) is equivalent to the Hamiltonian vector field on $\mathbb{R}^{2N}$ given by (6). In other words, system (20) gives the Lagrangian approach to the motion of N vortices without masses and positive intensities Γ_j, $j = 1, \ldots, N$.

2.4.2. As was already observed in [Ragazzo, Koiller and Oliva, 1994], the symmetries of H given by (17) imply that the corresponding Hamiltonian system admits the following integrals:

$$L = \sum_{j=1}^{N} (\vec{q_j} \times \vec{p_j}) \cdot \vec{e_3} \quad , \quad \vec{\pi} = \sum_{j=1}^{N} \left(\vec{p_j} + \frac{\rho \Gamma_j}{2} \vec{q_j} \times \vec{e_3} \right); \tag{21}$$

Moreover, $\pi^2 = \vec{\pi} \cdot \vec{\pi}$ and L do depend on the masses and satisfy

$$\{\pi^2, L\} = 0. \tag{22}$$

H, π^2 and L are three first integrals for the Hamiltonian system defined by (17) and are two by two in involution relatively to the canonical symplectic structure $(T^*(\mathbb{R}^{2N}), \omega)$.

3. Stability of the limit system

As we saw, our Hamiltonian system is defined on $T^*(\mathbb{R}^{2N})$ and the Hamiltonian function is given by (17). For simplicity, and in order to study this system as the masses go to zero, we will assume that $m_j/\Gamma_j = \beta$ for all $j = 1, \ldots, N$, with the Γ_j fixed. Given a compact manifold $\tilde{M} \subset \mathbb{R}^{2N}$, invariant under system (6), it defines a submanifold $\tilde{M}_\beta \subset T^*(\mathbb{R}^{2N})$; in fact $\tilde{M}_\beta = (F\mathcal{L})\bar{M}$ where $\bar{M} \subset T(\mathbb{R}^{2N})$ is the image of $\tilde{M}$ by the section defined by the vector field (6). A typical example is the case in which $\tilde{M}$ is an invariant torus for the Hamiltonian system of vortices without mass. Since for $N = 2$ or 3 system (6) is Liouville integrable, one can choose, in this case, 2 or 3-dimensional invariant tori as examples of $\tilde{M}$. For $N \geq 4$ the Hamiltonian system (6) is not necessarily integrable. But, using properly the KAM theory, Khanin ([Khanin, 1982]) proved that in $\mathbb{R}^{2N}(N \geq 4)$ there exists a set of initial conditions for (6), with positive measure, containing invariant tori of dimension $(N-1)$ that are the union of quasi-periodic motions.

For such a given compact invariant manifold $\tilde{M}$, we will try to understand what happens with the solutions of the system with Hamiltonian (17), that start in M_β or in a neighborhood of M_β. The equations of motion defined by (17) and (14) are:

$$
\begin{aligned}
\dot{q}_{j1} &= \frac{\partial H}{\partial p_{j_1}} = \frac{1}{m_j}\left[p_{j1} - \frac{\Gamma_j \rho}{2} q_{j2}\right] \\
\dot{p}_{j1} &= -\frac{\partial H}{\partial q_{j_1}} = \frac{-1}{m_j}\left[\left(\frac{\Gamma_j \rho}{2}\right)^2 q_{j1} + \left(\frac{\Gamma_j \rho}{2}\right) p_{j2}\right] + \rho \frac{\partial W}{\partial q_{j_1}} \\
\dot{q}_{j2} &= \frac{\partial H}{\partial p_{j_2}} = \frac{1}{m_j}\left[p_{j2} + \left(\frac{\Gamma_j \rho}{2}\right) q_{j_1}\right] \\
\dot{p}_{j2} &= -\frac{\partial H}{\partial q_{j_2}} = \frac{-1}{m_j}\left[\left(\frac{\Gamma_j \rho}{2}\right)^2 q_{j2} - \left(\frac{\Gamma_j \rho}{2}\right) p_{j1}\right] + \rho \frac{\partial W}{\partial q_{j_2}}.
\end{aligned}
$$

Using the hypothesis $m_j = \beta \Gamma_j$ we get (23):

$$
\begin{aligned}
\dot{q}_{j1} &= \frac{1}{\beta \Gamma_j}\left[p_{j1} - \left(\frac{\Gamma_j \rho}{2}\right) q_{j2}\right] \\
\dot{p}_{j1} &= -\frac{\rho}{2\beta}\left[\left(\frac{\Gamma_j \rho}{2}\right) q_{j1} + p_{j2}\right] + \rho \frac{\partial W}{\partial q_{j_1}} \\
\dot{q}_{j2} &= \frac{1}{\beta \Gamma_j}\left[\left(\frac{\Gamma_j \rho}{2}\right) q_{j1} + p_{j2}\right] \\
\dot{p}_{j2} &= \frac{\rho}{2\beta}\left[p_{j1} - \left(\frac{\Gamma_j \rho}{2}\right) q_{j2}\right] + \rho \frac{\partial W}{\partial q_{j_2}} \quad , \quad j = 1, \ldots, N.
\end{aligned}
\tag{23}
$$

Let us reescale time in system (23), that is, if τ is a new time given by $\frac{dt}{d\tau} = \beta$ one obtains a reescaled and regularized system:

$$\begin{aligned}
q'_{j_1} &= \frac{1}{\Gamma_j}\left[p_{j1} - \left(\frac{\Gamma_j\rho}{2}\right) q_{j2}\right] \\
p'_{j_1} &= -\frac{\rho}{2}\left[p_{j2} + \left(\frac{\Gamma_j\rho}{2}\right) q_{j1}\right] + \beta\rho\frac{\partial W}{\partial q_{j_1}} \\
q'_{j_2} &= \frac{1}{\Gamma_j}\left[p_{j2} + \left(\frac{\Gamma_j\rho}{2}\right) q_{j1}\right] \\
p'_{j_2} &= \frac{\rho}{2}\left[p_{j1} - \left(\frac{\Gamma_j\rho}{2}\right) q_{j2}\right] + \beta\rho\frac{\partial W}{\partial q_{j_2}},
\end{aligned} \tag{24}$$

where $q'_{j_1} = \frac{dq_{j1}}{d\tau}$, $p'_{j_1} = \frac{dp_{j1}}{d\tau}$, $q'_{j_2} = \frac{dq_{j2}}{d\tau}$ and $p'_{j_2} = \frac{dp_{j2}}{d\tau}$.

System (24) depends analytically on the parameter β, for $\beta \in \mathbb{R}$; and, for $\beta > 0$, it is equivalent to system (23). It is easy to see that (24) is also a Hamiltonian system on $(T^*(\mathbb{R}^{2N}), \omega)$ with Hamiltonian function $H_\beta = \beta H$ (see (26)). The conclusion is that (23) is not defined for $\beta = 0$ but (24) can itself be considered for $\beta = 0$.

System (24) can be also viewed on $\mathbb{R} \times T^*(\mathbb{R}^{2N})$, that is, with one more variable β:

$$\begin{aligned}
\beta' &= 0 \\
q'_{j_1} &= \frac{1}{\Gamma_j}\left[p_{j1} - \left(\frac{\Gamma_j\rho}{2}\right) q_{j2}\right] \\
p'_{j_1} &= -\frac{\rho}{2}\left[p_{j2} + \left(\frac{\Gamma_j\rho}{2}\right) q_{j1}\right] + \beta\rho\frac{\partial W}{\partial q_{j1}} \\
q'_{j_2} &= \frac{1}{\Gamma_j}\left[p_{j2} + \left(\frac{\Gamma_j\rho}{2}\right) q_{j1}\right] \\
p'_{j_2} &= \frac{\rho}{2}\left[p_{j1} - \left(\frac{\Gamma_j\rho}{2}\right) q_{j2}\right] + \beta\rho\frac{\partial W}{\partial q_{j2}}.
\end{aligned} \tag{25}$$

The functions H_β and β are constants of motion for system (25) and

$$H_\beta = \frac{1}{2}\sum_{j=1}^{N}\frac{1}{\Gamma_j}\left\{\left[p_{j1} - \left(\frac{\Gamma_j\rho}{2}\right) q_{j2}\right]^2 + \left[p_{j2} + \left(\frac{\Gamma_j\rho}{2}\right) q_{j1}\right]^2\right\} - \beta\rho W. \tag{26}$$

Equations (6) and (13) imply that the submanifold $\tilde{M}_\beta \subset T^*(\mathbb{R}^{2N})$ can be given by the relations ($\beta > 0$ fixed):

$$
\begin{aligned}
p_{j_1} &= \left(\frac{\Gamma_j\rho}{2}\right) q_{j_2} - \beta \frac{\partial W}{\partial q_{j_2}} \\
p_{j_2} &= -\left(\frac{\Gamma_j\rho}{2}\right) q_{j_1} + \beta \frac{\partial W}{\partial q_{j_1}}, \quad j = 1, \ldots, N
\end{aligned}
\tag{27}
$$

provided that $q = (\vec{q_1}, \ldots, \vec{q_N}) \in \tilde{M}$.

Consider now the equations obtained from (27) making $\beta = 0$:

$$
\begin{aligned}
p_{j_1} &= \left(\frac{\Gamma_j\rho}{2}\right) q_{j_2} \\
p_{j_2} &= -\left(\frac{\Gamma_j\rho}{2}\right) q_{j_1}
\end{aligned}
\tag{28}
$$

These equations (28) define a section $S_o : \mathbb{R}^{2N} \longrightarrow T^*(\mathbb{R}^{2N})$ with respect to the canonical projection τ^* of $T^*(\mathbb{R}^{2N})$ onto $\mathbb{R}^{2N}$, and if (q_{j_1}, q_{j_2}) vary on $\tilde{M}$, one obtains a compact submanifold $S_o(\tilde{M}) \subset T^*(\mathbb{R}^{2N})$. Remark that $S_o(\tilde{M}) = (F\mathcal{L} \circ \sigma)\tilde{M}$, where $\sigma : \mathbb{R}^{2N} \longrightarrow T(\mathbb{R}^{2N})$ is the zero section.

Assume now that $\tilde{M}$ does not have singularities of the vector field (6); so, one can construct a suitable compact tubular neighborhood M of $\tilde{M}$ in $\mathbb{R}^{2N}$ such that the function $A(q) = \sum_{j=1}^N \frac{1}{2\Gamma_j} \left\{ \left[\left(\frac{\partial W}{\partial q_{j_1}}\right)^2 + \left(\frac{\partial W}{\partial q_{j_2}}\right)^2 \right](q) \right\}$ is positive on M. For a fixed $c > 0$, the compactness of M implies that there exists $\beta_o > 0$ such that for each $\beta \in (0, \beta_o]$ we have $c + \beta\rho W > 0$ on M and the inequality

$$
H_\beta = \frac{1}{2}\sum_{j=1}^N \frac{1}{\Gamma_j} \left\{ \left[p_{j_1} - \left(\frac{\Gamma_j\rho}{2}\right) q_{j_2}\right]^2 + \left[p_{j_2} + \left(\frac{\Gamma_j\rho}{2}\right) q_{j_1}\right]^2 \right\} - \beta\rho W(q) \leq c
\tag{29}
$$

defines a bundle $\mathbf{B}_\beta$ over M of solid ellipsoids, that is, for $q = (\vec{q_1}, \ldots, \vec{q_N}) \in M$ the corresponding ellipsoid is centered at $S_o(q) \in S_o(M)$. We clearly have $\tau^*(\mathbf{B}_\beta) \subset M$.

The equations (27) for $q \in M$ define, as before, a section M_β over M and the condition $M_\beta \subset \overset{\circ}{\mathbf{B}}_\beta$ is true for the values of $\beta \in (0, \beta_o]$ satisfying the inequality

$$
\frac{1}{2}\sum_{j=1}^N \frac{\beta^2}{\Gamma_j} \left\{ \left[\left(\frac{\partial W}{\partial q_{j_1}}\right)^2 + \left(\frac{\partial W}{\partial q_{j_2}}\right)^2 \right](q) \right\} - \beta\rho W(q) < c, \quad \text{for all } q \in M,
\tag{30}
$$

or

$$A(q)\beta^2 - \beta\rho W(q) - c < 0 \quad \text{for all } q \in M. \tag{31}$$

The quadratic polynomial in (31) has a continuous (on q) and positive root

$$r(q) = \frac{\rho W(q) + \sqrt{\rho^2[W(q)]^2 + 4cA(q)}}{2A(q)} \tag{32}$$

for each $q \in M$; then $r(q)$ has a minimum $r_o > 0$ on the compact set M. Call $\beta_1 > 0$ the number

$$\beta_1 = min\ \{\frac{r_o}{2}, \beta_o\}.$$

One may conclude that if $\beta \in (0, \beta_1]$, the inequalities (30) and (31) hold, and then, $M_\beta \subset \overset{\circ}{\mathbb{B}}_\beta$. Any solution $(p(\tau), q(\tau))$ of system (24)starting on M_β (and in particular on $\tilde{M}_\beta$) at $\tau = 0$, stays, at least for $N = 2$, in $\overset{\circ}{\mathbb{B}}_\beta$ for all time $\tau \in \mathbb{R}$. In fact $H_\beta(p(\tau), q(\tau))$ is constant and strictly smaller than c, and in the case of two vortices, we were able to show ([Ragazzo, Koiller and Oliva, 1994]) that $(p(\tau), q(\tau))$ cannot cross the boundary $\partial\mathbb{B}_\beta$, where $H_\beta = c$. This way, the orbit defined by $(p(\tau), q(\tau))$ is a subset of $\overset{\circ}{\mathbb{B}}_\beta$, and since $\tau^*(\mathbb{B}_\beta) \subset M$, we have:

Theorem 3.1. *Let $\tilde{M} \subset \mathbb{R}^4$ be a compact submanifold, invariant under the system (6) of two massless vortices of intensities $\Gamma_j > 0$, $j = 1, 2$, and assume that $\tilde{M}$ does not have singularities of (6). Let us denote by M the closure of a suitable small tubular neighborhood of $\tilde{M}$ in $\mathbb{R}^4$. Then there exists $\beta_1 > 0$ such that for any $\beta \in (0, \beta_1]$, the Hamiltonian system of vortices with masses $m_j = \beta\Gamma_j$, $j = 1, 2$, defined on the canonical symplectic manifold $T^*(\mathbb{R}^4)$ by the Hamiltonian function (17) and given by system (23) has the following property: any motion $q(t)$ of (23), on the configuration space $\mathbb{R}^4$, that starts on a point of $\tilde{M}$ with a velocity given by (6), never leaves the given small neighborhood M.*

That result corresponds to Theorem 8 of section 7.1.2. of [Ragazzo, Koiller and Oliva, 1994], that is, the limit of the solutions as the masses tend to zero are orbitally stable with respect to the initial conditions on $\tilde{M}$. But, as it was shown in [Ragazzo, Koiller and Oliva, 1994], there is no Lyapunov stability.

4. Final Remarks

4.1 Is the result above true for the case $N \geq 3$?
4.2 The case $N = 2$, just considered, is an integrable Hamiltonian system in $T^*(\mathbb{R}^4)$, essentially because we imposed the restriction $m_1/\Gamma_1 = m_2/\Gamma_2$, (see

[Ragazzo, Koiller and Oliva, 1994]). If the masses m_1 and m_2 tend to zero in a such way that m_1/Γ_1 remains allways distinct from m_2/Γ_2, do we have an analogous result on orbital stability?

4.3 Does exist orbital stability in the case of two vortices when one vortex does not have mass and the other has a mass that tends to zero?

5. References

1. Abraham R., Marsden J.E., 1978, *Foundations of Mechanics*, Benjamin.
2. Castilla M.S.A.C., Moauro V., Negrini P., Oliva W.M., 1993, "The four positive vortices problem: regions of chaotic behavior and the non-integrability", *Ann. Inst. Henri Poincaré*, vol. 59, no. 1, pp. 99-115
3. Khanin K.M., 1982, "Quasi-Periodic Motions on Vortex Systems", *Physics*, 4D, pp. 261-269
4. Ragazzo C.G., Koiller J., Oliva W.M., 1994, "On the motion of two dimensional vortices with mass", *J. of Nonlinear Science*, vol. 4, pp. 375-418.

POSTULATES OF HAMILTONIAN GEOMETRICAL OPTICS IN ISOTROPIC AND ANISOTROPIC MEDIA *

ANA LEONOR RIVERA
Facultad de Ciencias, UNAM
E-mail: rivera@ce.ifisicam.unam.mx

SERGEY M. CHUMAKOV
E-mail: sergei@ce.ifisicam.unam.mx

and

KURT BERNARDO WOLF
Instituto de Investigaciones en Matemáticas Aplicadas y en Sistemas
Universidad Nacional Autónoma de México
Apartado Postal 139-B, 62191 Cuernavaca, Morelos, México

ABSTRACT

We provide geometric and dynamic postulates that define the Hamiltonian system that models anisotropic geometric optics. We use the ray arc length as evolution parameter for the Hamilton equations, besides time and distance along a chosen optical axis. Arbitrary anisotropy is then restricted according to symmetry and physical considerations.

1. Introduction: Postulates of Isotropic Optics

In reference [1] we studied the Hamiltonian formulation of geometric optics in anisotropic media based upon the extremal principle and within the Lagrangian and Hamiltonian formalism. The results were applied to the characterization of optical systems composed out of refracting surfaces between generally anisotropic elements. We have noticed, however, that a rather direct connection exists between a basic pair of postulates and the Hamiltonian formulation that is particularly transparent when the evolution parameter is arc length along the light rays.

In this Section we recapitulate the postulates that lead to geometric optics in isotropic media [2], and the Section 2 will propose a generalization of the same postulates when the refractive index depends on ray direction, *i.e.*, for anisotropic media, for the arc length parameter s. Section 3 will review the Lagrangian and Hamiltonian formulations based on the extremal principle for the cases when the evolution parameter is length along an optical axis z and time t. Hamilton's equations present three congruent faces when expressed in terms of z, t, or s. Section 4 examines the restrictions of symmetry and physics on the direction dependence of the refractive index.

*Support from Project DGAPA-UNAM IN 104293 "Optica Matemática".

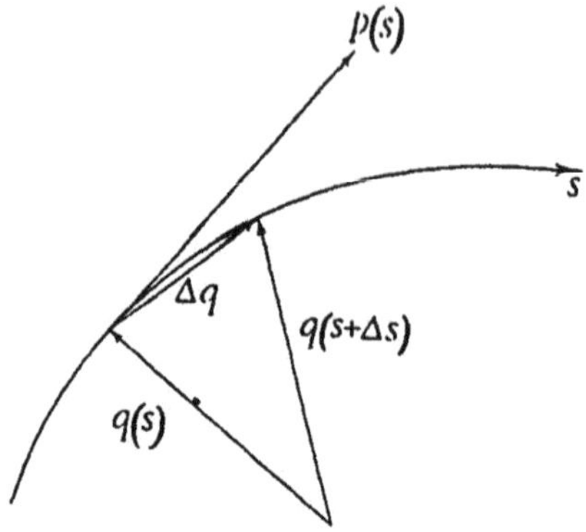

Figure 1: Light rays are modeled as piecewise differentiable lines $q(s)$ with tangent vectors $p(s)$.

Geometric optics of inhomogeneous media models light rays by lines in three-space $q(s) = (q_x(s), q_y(s), q_z(s))^T$, where $s \in \mathbb{R}$ is the parameter of length along the ray and, to save text space, we indicate column vectors by the transpose (^{T}) of row vectors. Now, refering to Figure 1, we state [2]

Geometric postulate: *The rays $q(s)$, $s \in \mathbb{R}$, are continuous and piecewise differentiable.* If we denote by $p(s) = (p_x(s), p_y(s), p_z(s))^T$ its tangent vector (where it is defined, *i.e.*, except possibly at isolated points where $p(s)$ may be discontinuous) then

$$\Delta q \doteq \frac{p}{|p|} \Delta s, \tag{1}$$

We do not (yet) asign any *length* to this tangent vector.

The inhomogeneity of an optical medium is characterized by a single scalar function n of the space points $q \in \mathbb{R}^3$, called the *refractive index* of the medium. A *homogeneous* medium has the same properties everywhere, so n = constant, and light rays are straight lines p = constant vector in the model. A *dynamic* postulate must relate how light rays *bend* in some direction with the non-constancy of $n(q)$. Geometric optics considers refractive indices $n(q)$ that are region-wise smooth, with possible finite discontinuities on smooth 'refracting' surfaces. The gradient $\nabla n(q)$ is the only available *vector* of the medium. This is zero where n is constant; on smooth surfaces of discontinuity, the incremental gradient provides the vector normal to the surface.

Dynamic postulate: *The tangent vector $p(s)$ responds to the space gradient of the function $n(q)$. i.e.,*

$$\Delta p \doteq \nabla n(q) \, \Delta s, \tag{2}$$

See Figure 2. This equation resembles a force acting along the gradient of a

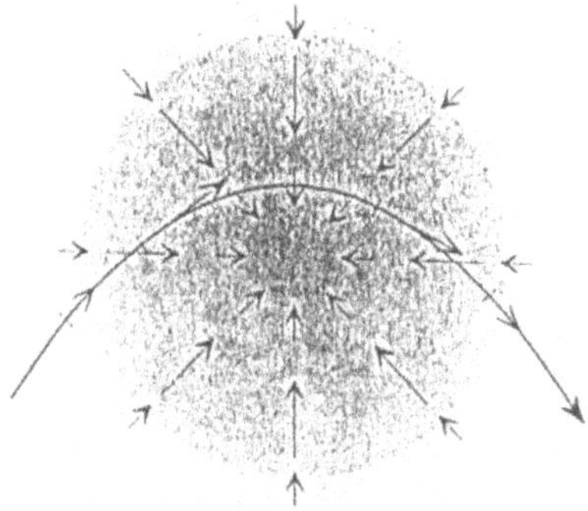

Figure 2: In a smooth inhomogeneous medium, the space gradient of the refractive index bends light rays.

potential; but note that we do not speak of time but of *arc length* s, and not of mechanical potential but *refractive index* $n(q)$ of an optical medium.

In smooth, inhomogeneous regions, the limit $\Delta s \to 0$ leads to the first-degree Hamilton *evolution* equations

$$\begin{aligned} \dot{q}(s) \equiv \frac{dq}{s} &= \frac{p}{|p|} = \frac{\partial \mathcal{H}_{(i)}}{\partial p}, \\ \frac{dp}{s} &= \frac{\partial n(q)}{\partial q} = -\frac{\partial \mathcal{H}_{(i)}}{\partial q} \end{aligned} \tag{3}$$

with the Hamiltonian function

$$\mathcal{H}_{(i)} = |p| - n(q). \tag{4}$$

The three-vector $p(s)$, to which we refered above simply as being along the ray tangent, due to its position in the equation pair (3) qualifies to be the *momentum* vector canonically conjugate [3] to ray *position* vector $q(s)$.

The value of $\mathcal{H}_{(i)}$ is constant along trajectories and we may choose the value $\mathcal{H} = 0$. The momentum vector p is thus constrained to a sphere

$$|p| = n(q), \qquad (isotropic) \tag{5}$$

that we call the *Descartes* sphere. Because by definition $|\dot{q}| = 1$, the momentum vector p may be written in terms of that *unit* ray tangent as

$$p = n(q)\,\dot{q}. \qquad (isotropic) \tag{6}$$

Although this way of obtaining the Hamilton equations is extremely succint, we have not found it in the textbook or research literature. The postulates

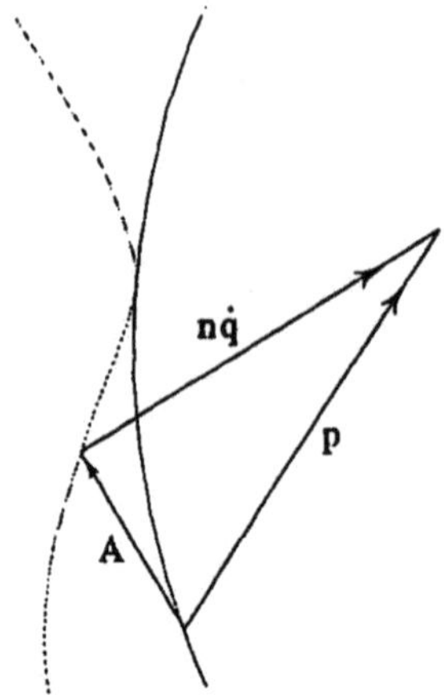

Figure 3: In anisotropic media, the unit tangent vector $\dot{q}$ multiplied by the refractive index and summed to the anisotropy vector $A(q,\dot{q})$, yields the optical momentum vector p; $\dot{q}$ ranges on the unit sphere, $n\dot{q}$ draws the ray surface (dotted line), and p sweeps over the Descartes ovoid (continuous line).

replace the extremal principle and Lagrangian formulation of geometric optics, to arrive directly at its Hamiltonian description; they do so for the arc length evolution parameter s, which is unsuitable for these more common approaches.

2. Postulates of Anisotropic Optics

Anisotropic optical media are those whose refractive index [4] depends on the ray *direction*, $n(q,\dot{q})$. Now we have *two* vectors that characterize the medium: the space gradient $\nabla_q n = \partial n/\partial q$ and the direction gradient $\nabla_{\dot{q}} n = \partial n/\partial \dot{q}$. Since $|\dot{q}| = 1$, it follows that the *anisotropy vector*, that we define as [1]

$$A(q,\dot{q}) = \nabla_{\dot{q}}\, n(q,\dot{q}), \tag{7}$$

is a vector field tangent to the unit sphere, satisfying

$$\dot{q}\cdot A = 0, \tag{8}$$

i.e., this anisotropy vector is perpendicular to the ray tangent $\dot{q}$ and vanishes where the medium is isotropic. See Figure 3.

We propose two postulates for anisotropic optics that generalize the incremental equations (1, 2) for the isotropic case of last Section. They are:

Geometric postulate: *The rays $q(s)$ are continuous and piecewise differentiable. i.e.,*

$$\Delta q \doteq \dot{q}(s)\,\Delta s, \tag{9}$$

—which is in effect an identity— and, generalizing (6), *the momentum vector is*

$$p = n(q,\dot{q})\,\dot{q} + A(q,\dot{q}). \tag{10}$$

Dynamic postulate: *The momentum vector $p(s)$ responds to the space gradient of the refractive index $n(q,\dot{q})$. i.e.*, as before,

$$\Delta p \doteq \nabla_q n(q,\dot{q})\,\Delta s, \tag{11}$$

For $\Delta s \to 0$, the two Hamilton equations for anisotropic optical media thus generalize eq. (3) to

$$\begin{aligned} \dot{q}(s) \equiv \frac{dq}{ds} &= \frac{p-A}{|p-A|} = \frac{\partial \mathcal{H}}{\partial p}, \\ \frac{dp}{ds} &= \frac{\partial n(q,\dot{q})}{\partial q} = -\frac{\partial \mathcal{H}}{\partial q}, \end{aligned} \tag{12}$$

with the *anisotropic* Hamiltonian function

$$\mathcal{H} = |p-A| - n(q,\dot{q}). \tag{13}$$

The conservation of the Hamiltonian that lead in the isotropic case to the restriction of p to the Descartes sphere in eq. (5), now shows p is restricted to a *Descartes ovoid* (see Figures ahead), and

$$|p-A| = n(q,\dot{q}), \tag{14}$$

to the *ray surface* [5].

The *Ansatz* of equation (10) spells the difference between ray momentum p and ray tangent $\dot{q}$ in anisotropic media that is given by the anisotropy vector A. Attending covariance under rotations only, a cross-product vector term $\dot{q} \times A$ could be added to the right-hand side of eq. (10), but may be discarded on the ground that it would have parity different from the other two.

3. Three faces of Hamilton's equations

There are two other rather natural evolution parameters to compose Hamilton equations in optics besides arc lenght s: time t and distance along a chosen optical axis z. In each of these, the Lagrangian and the Hamiltonian take different forms [6]. The three faces of the Hamiltonian stem from the same extremal principle $\delta I = 0$, with

$$I \equiv \int_A^B dt = \int_A^B ds\, n(q,\dot{q}) = \int_A^B dz\, \sqrt{1 + |d\mathbf{q}/dz|^2}\, n(q,\dot{q}). \tag{15}$$

The first equality stems from the physical speed of light in a medium, $v = dq/dt$, $|v| = ds/dt = c/n$, while the second one derives from geometry when we choose to distinguish the z-axis writing $q = (q_x(s), q_y(s), q_z(s))^T = (\mathbf{q}(z), z)^T$ with $\mathbf{q}(z) = (q_x(z), q_y(z))^T$, and similarly for $\dot{q} = (\dot{\mathbf{q}}, \sqrt{1-|\dot{\mathbf{q}}|^2})$, changing independent variables for $dz/ds \neq 0$ through

$$ds = dz/\sqrt{1-|d\mathbf{q}/ds|^2} = dz\sqrt{1+|d\mathbf{q}/dz|^2} = dz \sec\theta,$$

with θ the angle between the ray and the z-axis ($|\theta| < \frac{\pi}{2}$).

Optical axis parameter: We followed [1] the route of the z-axis, with a Lagrangian function

$$L^{[z]} = \sqrt{1+|d\mathbf{q}/dz|^2}\, n(\mathbf{q}, z; d\mathbf{q}/dz). \tag{16}$$

The canonical momentum is then the two-vector $\mathbf{p} = (p_x, p_y)^T$, found as the derivative of $L^{[z]}$ with respect to $d\mathbf{q}/dz$, that may be written as

$$\mathbf{p} = n\dot{\mathbf{q}} + \mathbf{A}(\mathbf{q}, z, \dot{\mathbf{q}}), \quad \text{where} \quad \mathbf{A} = (1 - \dot{\mathbf{q}}\dot{\mathbf{q}}^T)\frac{\partial n(\mathbf{q}, z, \dot{\mathbf{q}})}{\partial \dot{\mathbf{q}}} \tag{17}$$

is the *anisotropy* two-vector [1]. Through the Legendre transformation, the Hamiltonian along z is written as

$$H^{[z]}(\mathbf{q}, z; \mathbf{p}) = -\sqrt{n^2 - |\mathbf{p}-\mathbf{A}|^2} + \frac{(\mathbf{p}-\mathbf{A})\cdot\mathbf{A}}{\sqrt{n^2-|\mathbf{p}-\mathbf{A}|^2}}. \tag{18}$$

Finally, the Hamilton equations can be written in *three*-vector form as

$$\begin{aligned} \frac{dq}{dz} &= \frac{\partial H^{[z]}}{\partial p} = \frac{p-A}{p_z - A_z}, \\ \frac{dp}{dz} &= -\frac{\partial H^{[z]}}{\partial q} = \frac{n}{p_z - A_z}\frac{\partial n}{\partial q}. \end{aligned} \tag{19}$$

The third component of the first equation is $1 = 1$, and that of the second equation identifies $p_z = -H^{[z]}$ and

$$A_z = -\frac{(\mathbf{p}-\mathbf{A})\cdot\mathbf{A}}{\sqrt{n^2-|\mathbf{p}-\mathbf{A}|^2}}. \tag{20}$$

Finally, passing from s- to z-derivatives leads from the Hamilton equations in the form (12) to those of the form (19).

Time parameter: Torres del Castillo [7], followed the t route of *time*, where the velocity three-vector is

$$v = \frac{dq}{dt} = \frac{c}{n}\frac{dq}{ds} = \frac{c}{n}\dot{q}, \tag{21}$$

of magnitude c/n. The Lagrangian is, as in the isotropic case,

$$L^{[t]} = \frac{n(q,\dot{q})^2}{2c}\, v\cdot v. \tag{22}$$

The extra direction dependence of the refractive index, however, yields the three-vector of momentum in the form

$$p = \frac{n^2}{c}v + A, \quad \text{where} \quad A = \frac{n}{c}(v\cdot v)\frac{\partial n}{\partial v} \tag{23}$$

is the *anisotropy* vector, which is orthogonal to v. Through the Legendre transformation, one obtains the Hamiltonian

$$H^{[t]} = \frac{n^2}{2c}\, v\cdot v = \frac{c}{2n^2}(p-A)\cdot(p-A). \tag{24}$$

Finally, the Hamilton equations in time are

$$\begin{aligned} v \equiv \frac{dq}{dt} &= \frac{c}{n^2}(p-A) = \frac{\partial H^{[t]}}{\partial p}, \\ \frac{dp}{dt} &= \frac{c}{n}\frac{\partial n(q,\dot{q})}{\partial q} = -\frac{\partial H^{[t]}}{\partial q}. \end{aligned} \tag{25}$$

It is also shown [7] that from eq. (25) follow equations (19). A change of independent variables yields the Hamilton equations (13) with respect to the arc lenght parameter s. Notice that if we try to use the extremal principle eq. (15) in s, the Lagrangian would be $n(q,\dot{q})$ and the conjugate momentum $p = \partial n(q,\dot{q})/\partial\dot{q}$. This is identically zero for any inhomogeneous but isotropic medium, rendering useless the Lagrangian formalism.

Thus, although in Section 2 we found the Hamiltonian for anisotropic media from the postulates by rather heuristic considerations, the results check with those obtained through the standard Lagrangian formalisms in distance along the optical axis and time. The weakness of the arc length parameter s for the Lagrangian description does not preclude its usefulness and naturality for the Hamiltonian formalism.

4. Multipolar media and Maxwell equations

Smooth functions over a sphere —and we assume the refractive index $n(q,\dot{q})$ of an anisotropic medium is such a function of ray direction— can be expanded in a spherical harmonic series, as a *multipole* expansion [8]:

$$n(q,\dot{q}) = \sum_{m\geq 0} n^{(m)}(q,\dot{q}), \tag{26}$$

where $n^{(m)}(q,\dot{q})$ is homogeneous of degree m in the components of $\dot{q}$,

$$n^{(m)}(q,\dot{q}) = \sum_{i_k} N^{(m)}_{i_1,i_2,\ldots i_m} \dot{q}_{i_1}\dot{q}_{i_2}\cdots\dot{q}_{i_m} = \dot{q}\cdot M^{(m)}. \tag{27}$$

where we define the three-vector

$$M^{(m)}_j(\dot{q}) = \sum_i N^{(m)}_{j,i_2,\ldots i_m}\dot{q}_{i_2}\cdots\dot{q}_{i_m}. \tag{28}$$

The last member of eq. (27) defines the three-vector $M^{(m)}$ which is $(m-1)$-linear in the components of $\dot{q}$. The rank-m tensor components $N^{(m)}_{i_1,i_2,\ldots i_m}$, $i_k = x,y,z$, are symmetric, $N^{(m)}_{i_1,\ldots,j,\ldots,k,\ldots i_m} = N^{(m)}_{i_1,\ldots,k,\ldots,j,\ldots i_m}$, and traceless, $\sum_j N^{(m)}_{i_1,\ldots,j,\ldots,j,\ldots,i_m} = 0$, containing $2m+1$ independent coefficients.

We next calculate the direction gradient of the refractive index

$$\nabla_{\dot{q}}\, n^{(m)}(q,\dot{q}) = m\, M^{(m)}(q,\dot{q}), \tag{29}$$

and find the anisotropy three-vector for each multipole m as

$$A^{(m)} = m(1-\dot{q}\dot{q}^T)M^{(m)} = m(M^{(m)} - n^{(m)}(q,\dot{q})\,\dot{q}). \tag{30}$$

There is no limitation in the postulates of Section 2 on the possible dependence of the refractive index on ray direction $\dot{q}$, except that it be smooth. The $m=0$ *monopole* term is independent of $\dot{q}$, $n^{(0)}(q,\dot{q}) = n^0(q)$ and corresponds to an inhomogeneous but isotropic medium. The $m=1$ *dipole* term can only have the form $n^{(1)}(q,\dot{q}) = D(q)\cdot\dot{q}$, with a 'dipole vector' $D(q)$ characterizing the possibly inhomogeneous anisotropy of the medium.

Quadrupolar anisotropy: it is described by the $m=2$ term in eq. (26). The refractive index of *quadrupolar media* has the form

$$n(q,\dot{q}) = n^0 + \dot{q}^T\hat{Q}\dot{q}, \tag{31}$$

containing a $\dot{q}$-quadratic form given by a 3×3 symmetric, traceless *optical quadrupole matrix* $\hat{Q} = ||Q_{j,k}||$ that has five independent quadrupolar coefficients (we omit the q-dependence of n^0 and $Q_{j,k}$ in the notation). When we refer the medium to its *principal axes*, three of these coefficients are set to zero so that $\hat{Q} = \mathrm{diag}\,(Q_x, Q_y, Q_z)$, $Q_x + Q_y + Q_z = 0$. The anisotropy three-vector (30) is

$$A^{(2)} = 2(1-\dot{q}\dot{q}^T)\hat{Q}\dot{q}, \qquad \text{(quadrupole media)} \tag{32}$$

with first- and third-degree terms in the components of the direction vector $\dot{q}$. Finally, the momentum three-vector in a quadrupole medium is

$$p = [n(q,\dot{q}) + 2(1-\dot{q}\dot{q}^T)\hat{Q}]\dot{q} = (n^0 + 2\hat{Q} - \dot{q}^T\hat{Q}\dot{q})\dot{q}. \tag{33}$$

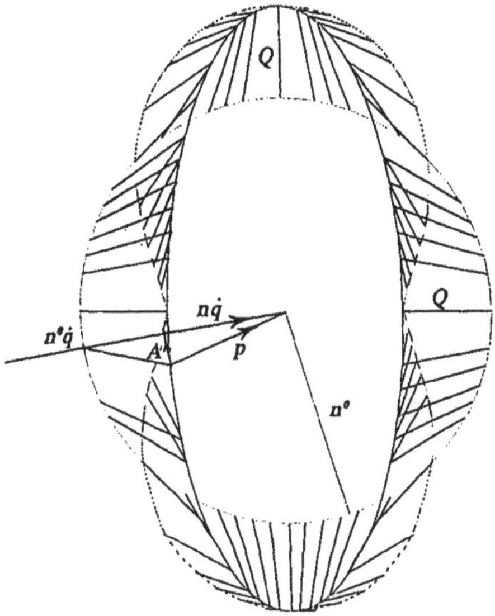

Figure 4: Quadrupole medium: As ray direction ranges over the sphere (circle in the two dimensions of the figure), the momentum vector draws the Descartes ovoid (in two dimensions it is an oval).

We show this in Figure 4 for two dimensions. In this case, anisotropy is described by a 2×2 symmetric traceless matrix having two independent coefficients characterizing the ellipticity and orientation of the ray surface. Refered to principal axes this is $\hat{Q} = \text{diag}\,(Q, -Q)$. In the Figure, the thin lines that join points on the circle with points on the oval map the ray direction on the corresponding momentum vector. This serves the purpose of a 'ruler, compass, and plotter' construction for the refracted ray in anisotropic media.

Higher anisotropic media are octupolar for $m = 3$, ekaidekapolar for $m = 4$, etc. Yet a selection rule applies when we require —on physical grounds— that the refractive index be independent on the *sense* of light transmission, *i.e.*, $n(q, \dot{q}) = n(q, -\dot{q})$. In this case, all odd-$m$ multipolar summands to $n(q, \dot{q})$ in (26) are forbidden, in particular the dipolar and octupolar terms. This leaves only monopole, quadrupole and higher even-m terms. Moreover, Maxwellian optics describes weak anisotropy in terms of the so-called first material equation that relates the displacement and electric vectors through the second-rank dielectric tensor [4]. Its trace is the average dielectric constant while the five remaining parameters, since they transform among themselves under rotations, because of Schur's lemma can only be linear combinations of those in our quadrupole tensor. Otherwise, Maxwell electromagnetism gives physical reality to ray direction $\dot{q}$ as the Poynting vector and direction of energy transmission, and to momentum p as the wavefront normal. In fact, the term $p - A$ must look familiar to the reader from covariant electromagnetism. Finally, when the anisotropy is not weak, the dielectric tensor can depend on the intensity of the electric field in a nonlinear way, perhaps giving rise to ekaidekapolar and higher-multipole contributions. This is still an open question.

5. Conclusions

We have constructed Hamilton equations for general inhomogeneous, anisotropic optical media. Anisotropy is a non-trivial extension of isotropy: if we knew only the isotropic Hamiltonian $H_i^{[z]}(\mathbf{q}, z; \mathbf{p}) = -\sqrt{n^2 - \mathbf{p}^2}$ we would be hard pressed to generalize it to the correct anisotropic form (18). This situation is common in high-energy physics, where *ad hoc* terms are often added to the Hamiltonian or Lagrangian to emulate new interactions that are restricted by symmetry and physical considerations. In the Hamiltonian system of geometric optics, the extension to anisotropy is well grounded and definite.

We studied [1] further the homogeneous uniaxial anisotropic media to find the optical properties of systems composed of such elements with refracting interfaces between them, and their third-order aberration coefficients. The Hamiltonian foundation of anisotropy was there obtained from the extremal principle through the Lagrangian formulation and its Legendre transformation, as sketched in eqs. (18)–(19). Here, the two rather self-evident postulates are shown to lead more succintly to the Hamiltonian setup (12)–(13). We deem the insights to be conceptually significant.

6. References

1. A.L. Rivera, S.M. Chumakov and K.B. Wolf, *JOSA A* **12** (1995).
2. K.B. Wolf and G. Krötzsch, *Eur. J. Phys.***15** (1994).
3. H. Goldstein, *Classical Mechanics*, 2ndEd. (Addison Wesley, Reading, Mass., 1980).
4. R.P. Feynman, R.B. Leighton and M. Sands, *The Feynman Lectures on Physics* (Addison-Wesley, Reading, Mass., 1963).
5. F.A. Jenkins and H.E. White, *Fundamentals of Optics*, 4th ed. (McGraw–Hill, New York, 1976).
6. G. Krötzsch and K.B. Wolf, *Rev. Mex. Física***36** (1990) 724–735.
7. G. Torres del Castillo, *Rev. Mex. Física* **40**accepted(1995).
8. R. Gilmore, *Lie Groups, Lie Algebras, and Some of Their Applications* (Wiley, New York, 1974).

SPECTRAL STABILITY OF RELATIVE EQUILIBRIA IN THE $N+1$ BODY PROBLEM

DIETER S. SCHMIDT
Department of Computer Science, University of Cincinnati
Cincinnati, Ohio 45221-0008, USA
E-mail: dieter.schmidt@uc.edu

ABSTRACT

The stability of regular N-gon configurations with a central mass is investigated. For $N \geq 7$ these configurations can be made stable if the central mass is large enough. When $3 \leq N \leq 6$ this can not be done, a fact which was already observed by R. Moeckel.

1. Introduction

Meyer and Schmidt [2] had shown that regular polygon configurations of N bodies of unit mass at the vertices of a regular polygon and a body of mass m at the center allow bifurcations to different types of relative equilibria, when the central mass is allowed to vary near some critical values. We have found that there are exactly $N/2$ such values, if we include those cases which occur for negative values of m.

Our method was based on the fact that relative equilibria are critical points of the potential function under the restriction that the moment of inertia remains constant. If the Hamiltonian function is

$$H(\mathbf{q}, \mathbf{p}) = \sum \frac{|p_i|^2}{2m_i} - U(\mathbf{q}) \tag{1}$$

with

$$U(\mathbf{q}) = \sum \frac{m_i m_j}{|q_i - q_j|} \tag{2}$$

then the problem is reduced to an algebraic one of finding extremas of U with

$$I(\mathbf{q}) = \frac{1}{2} \sum m_i |q_i|^2 = \text{const.}$$

In order to find the points of bifurcation we looked for degenerate critical values of the mass m. Degenerate means here in a reduced space where the center of mass is kept at the origin and the rotational symmetry of the problem is removed by placing one of the bodies on the positive x-axis. Since the problem stated in this way is a purely algebraic one, we used coordinate transformations which brought the quadratic terms into their Sylvester normal form, and we used Lie transformations to normalize the higher order terms, so that we could check that a certain term was nonzero and the bifurcation did indeed occur.

Since our problem was purely algebraic our transformation did not have to preserve the dynamics of the given problem and we could not answer questions concerning the stability of the relative equilibria. At the meeting on "Hamiltonian Dynamical Systems, History, Theory and Applications" R. Moeckel presented a complete account of the spectral stability of the regular polygon configuration for four cases: when the central mass is zero, small, equal to the other masses and finally very large. One observes that the changes in spectral stability is related to the number of bifurcations to new relative equilibria. Moeckel [3] used the existence of certain invariant subspaces to get his results whereas we [2] had determined the actual transformations in order to determine the invariant subspaces. In light of this we made our transformations in symplectic so that we could also obtain the results of Moeckel and correlate bifurcations to changes in stability. It turns out that the phenomena are related, but in a more complicated way than one would expect.

2. Spectral Stability

Consider the planar $N+1$ body problem

$$H = \frac{1}{2}\mathbf{p}^t\mathbf{M}^{-1}\mathbf{p} - U(\mathbf{q})$$

with $\mathbf{q}, \mathbf{p} \in \mathcal{R}^{2N+2}$ and

$$\mathbf{M} = \begin{pmatrix} m_0 & & & 0 & & \\ & \ddots & & & \ddots & \\ & & m_N & & & 0 \\ 0 & & & m_0 & & \\ & \ddots & & & \ddots & \\ & & 0 & & & m_N \end{pmatrix}.$$

Relative equilibria can only be found in a rotating coordinate system. Let $(\mathbf{Q}, \mathbf{P})$ be these uniformly rotating coordinates with the transformation to them generated by

$$W(\mathbf{p}, \mathbf{Q}) = -\mathbf{p}^t e^{-\nu t\mathbf{J}}\mathbf{Q}.$$

The matrix $\mathbf{J}$ is the usual symplectic matrix and the rate of rotation ν will be determined below. The transformation itself is then given by

$$\mathbf{q} \;=\; -\frac{\partial W}{\partial \mathbf{p}} = e^{-\nu t\mathbf{J}}\mathbf{Q}$$

$$\mathbf{P} = -\frac{\partial W}{\partial \mathbf{Q}} = e^{\nu t \mathbf{J}} \mathbf{p}.$$

With

$$\frac{\partial W}{\partial t} = \nu \mathbf{p}^t \mathbf{J} e^{-\nu t \mathbf{J}} \mathbf{Q} = \nu \mathbf{P}^t \mathbf{J} \mathbf{Q}$$

the Hamiltonian function in rotating coordinates is then found to be

$$H = \frac{1}{2} \mathbf{P}^t \mathbf{M}^{-1} \mathbf{P} + \nu \mathbf{P}^t \mathbf{J} \mathbf{Q} - U(\mathbf{Q}).$$

The corresponding differential equations are

$$\begin{aligned} \dot{\mathbf{Q}} &= \nu \mathbf{J} \mathbf{Q} + \mathbf{M}^{-1} \mathbf{P} \\ \dot{\mathbf{P}} &= DU(\mathbf{Q}) + \nu \mathbf{J} \mathbf{P}. \end{aligned}$$

Stationary solutions $(\mathbf{Q}_0, \mathbf{P}_0)$ have to satisfy

$$\begin{aligned} 0 &= \nu \mathbf{J} \mathbf{Q}_0 + \mathbf{M}^{-1} \mathbf{P}_0 \\ 0 &= DU(\mathbf{Q}_0) + \nu \mathbf{J} \mathbf{P}_0. \end{aligned}$$

or the set of $2n + 2$ equations

$$\nu^2 \mathbf{M} \mathbf{Q}_0 = -DU(\mathbf{Q}_0).$$

By taking the inner product with $\mathbf{Q}_0$ and using the fact that $U(\mathbf{Q})$ is a homogeneous function of degree -1 one finds that the rate of rotation has to be

$$\nu^2 = \frac{U(\mathbf{Q}_0)}{2I(\mathbf{Q}_0)} \tag{3}$$

where

$$I(\mathbf{Q}_0) = \frac{1}{2} \mathbf{Q}_0^t \mathbf{M} \mathbf{Q}_0 \tag{4}$$

is the moment of inertia.

In order to study the stability of a stationary solution one has to study the motion nearby. To do this set

$$\begin{aligned} \mathbf{Q} &= \mathbf{Q}_0 + \mathbf{X} \\ \mathbf{P} &= \mathbf{P}_0 + \mathbf{Y} \end{aligned}$$

so that the Hamiltonian function is

$$H = \frac{1}{2} (\mathbf{P}_0 + \mathbf{Y})^t \mathbf{M}^{-1} (\mathbf{P}_0 + \mathbf{Y}) + \nu (\mathbf{P}_0 + \mathbf{Y})^t \mathbf{J} (\mathbf{Q}_0 + \mathbf{X}) - U(\mathbf{Q}_0 + \mathbf{X}).$$

For a spectral stability analysis it suffices to consider the linearized system i.e. the second order terms

$$H_2 = \frac{1}{2}\mathbf{Y}^t\mathbf{M}^{-1}\mathbf{Y} + \nu\mathbf{Y}^t\mathbf{J}\mathbf{X} - \frac{1}{2}\mathbf{X}^t D^2U_0\mathbf{X} \tag{5}$$

where D^2U_0 stands for the Hessian of $U(\mathbf{Q})$ evaluated at the stationary point $\mathbf{Q}_0$. It gives rise to the following system of first order differential equations

$$\frac{d}{dt}\begin{pmatrix}\mathbf{X}\\ \mathbf{Y}\end{pmatrix} = \begin{pmatrix}\nu\mathbf{J} & \mathbf{M}^{-1}\\ D^2U_0 & \nu\mathbf{J}\end{pmatrix}\begin{pmatrix}\mathbf{X}\\ \mathbf{Y}\end{pmatrix} \tag{6}$$

which in turn corresponds to the following system of second order differential equations

$$\ddot{\mathbf{X}} - 2\nu\mathbf{J}\dot{\mathbf{X}} - \nu^2\mathbf{X} = \mathbf{M}^{-1}D^2U_0\mathbf{X}.$$

This system has nontrivial solutions of the form $\mathbf{X} = e^{\lambda t}\xi$ only when the matrix

$$(-\lambda^2 + \nu^2)\mathbf{I} + 2\nu\lambda\mathbf{J} + \mathbf{M}^{-1}D^2U_0 \tag{7}$$

is singular. With this we can state

Definition 1 *A relative equilibria solution* $\mathbf{Q}_0$ *is spectrally stable if all roots* λ *of*

$$|(\nu^2 - \lambda^2)\mathbf{I} + 2\nu\lambda\mathbf{J} + \mathbf{M}^{-1}D^2U_0| = 0 \tag{8}$$

are purely imaginary.

We will study the relative equilibria where N masses of unit size are placed at the vertices of a regular polygon and a mass of size m is at the origin. Let

$$\omega = e^{2\pi i/N}$$

so that the stationary solution is given by

$$\mathbf{Q}_0 = (\omega^0, \omega^1, \ldots, \omega^{N-1}, 0)^t.$$

With $\mathbf{Q} = (q_0, q_1, \ldots, q_{N-1}, q_N)^t$, $m_j = 1$ for $0 \le j \le N-1$, and $m_N = m$ the potential function can be written as

$$U = U_1 + mU_2 = \sum_{0\le i<j<N} \frac{1}{|q_i - q_j|} + m\sum_{i=0}^{N-1}\frac{1}{|q_i - q_N|}. \tag{9}$$

3. Stability of a Ring

First consider the case when $m = 0$. Without a central mass the relative equilibrium is just a ring. We can ignore the $(N+1)$-st coordinate and consider the problem in $\mathcal{R}^{2N}$. A transformation which brings $D^2U_1(\mathbf{Q}_0)$ into simpler form is $\mathbf{X} = \mathbf{W}\mathbf{z}$, that is

$$\mathbf{Q} = \mathbf{Q}_0 + \mathbf{W}\mathbf{z} \tag{10}$$

with

$$\mathbf{W} = \frac{1}{\sqrt{N}} \begin{pmatrix} 1 & 1 & \cdots & 1 \\ 1 & \omega & \cdots & \omega^{N-1} \\ \vdots & \vdots & & \vdots \\ 1 & \omega^{N-1} & \cdots & \omega^{(N-1)^2} \end{pmatrix}.$$

A similar transformation was already used by Palmore [4] and maybe others before him. Using the complex quantity ω makes it easier to write down and perform this transformation, but one should not forget that in the real space ω corresponds to a 2×2 submatrix of the form

$$\omega \leftrightarrow \begin{pmatrix} \cos 2\pi/N & \sin 2\pi/N \\ -\sin 2\pi/N & \cos 2\pi/N \end{pmatrix}.$$

Keeping this relationship in mind one sees that for $\mathbf{W}^T$ we have to use the conjugate transposed matrix, i.e. $\mathbf{W}^T = \overline{\mathbf{W}}^t$. Thus we have a unitary matrix with

$$\mathbf{W}^{-1} = \mathbf{W}^T.$$

The transformation Eq.(10) can be extended to a symplectic transformation for the entire phase space in $\mathcal{R}^{4N}$. Due to the structure of $\mathbf{W}$ we have in $\mathcal{R}^{2N}$

$$\begin{aligned} \mathbf{W}^T\mathbf{M}^{-1}\mathbf{W} &= \mathbf{M}^{-1} \\ \mathbf{W}^T\mathbf{J}\mathbf{W} &= \mathbf{J}. \end{aligned}$$

It means that, when the transformation is carried out in the Hamiltonian Eq.(5), the only task, albeit a tedious one, is to transform the potential function U. The task is simplified if one uses complex coordinates and notation for as long as possible. Here we give an outline of our approach as the details can be found in [5].

We use the potential function in the form

$$U_1(\mathbf{Q}_0 + \mathbf{W}\mathbf{z}) = \frac{1}{2}\sum_{j=1}^{N-1}\sum_{i=0}^{N-1} \frac{1}{|q_i - q_{i+j}|}$$

with

$$\begin{aligned} |q_i - q_{i+j}| &= |\omega^i - \omega^{i+j} + \frac{1}{\sqrt{N}}\sum_{k=0}^{N-1}(\omega^{ik} - \omega^{(i+j)k})z_k| \\ &= |d_j + \frac{1}{\sqrt{N}}\sum_{k=0}^{N-1} d_{jk}\omega^{i(k-1)}z_k|. \end{aligned}$$

We have set

$$d_j = 1 - \omega^j$$

so that

$$|d_j| = 2\sin \pi j/N \tag{11}$$

and compute

$$\begin{aligned} |q_i - q_{i+j}|^2 &= |d_j|^2 + \frac{1}{\sqrt{N}}\sum_{k=0}^{N-1}(\omega^{i(k-1)}d_{jk}d_{-j}z_k + \omega^{-i(k-1)}d_{-jk}d_j\bar{z}_k) \\ &+ \frac{1}{N}\sum_{k=0}^{N-1}\sum_{l=0}^{N-1} d_{jk}d_{-jl}\omega^{i(k-l)}z_k\bar{z}_l. \end{aligned}$$

The complex position vector $\mathbf{z}$ has components $z_k = x_k + \sqrt{-1}y_k$. We expand U_1 near $\mathbf{z} = 0$ into a Taylor series through terms of second order. To do this we compute all second order partial derivatives with respect to the real and imaginary components of $\mathbf{z}$ and then evaluate them for $\mathbf{z} = 0$. As an intermediate step one computes for example

$$\begin{aligned} \frac{\partial U_1}{\partial x_r} &= \sum_{i,j}\frac{-1}{4|q_i - q_{i+j}|^3}\{\frac{1}{\sqrt{N}}(\omega^{i(r-1)}d_{jr}d_{-j} + \omega^{-i(r-1)}d_{-jr}d_j) \\ &+ \frac{1}{N}\sum_k(\omega^{i(r-k)}d_{jr}d_{-jk}\bar{z}_k + \omega^{i(k-r)}d_{jk}d_{-jr}z_k)\} \end{aligned}$$

and then at $\mathbf{z} = 0$

$$\begin{aligned} \frac{\partial^2 U_1}{\partial x_r \partial x_s} &= \frac{1}{8N}\sum_{i,j}\{\frac{3}{|d_j|^5}(\omega^{i(r+s-2)}d_{jr}d_{js}d_{-j}^2 + \omega^{-i(r+s-2)}d_{-jr}d_{-js}d_j^2) \\ &+ \frac{1}{|d_j|^3}(\omega^{i(r-s)}d_{jr}d_{-js} + \omega^{-i(r-s)}d_{-jr}d_{js})\}. \end{aligned} \tag{12}$$

Since $\sum_{i=0}^{N-1}\omega^{ik} = 0$ except when $k = 0$ or a multiple of N, most of the second order derivatives of U_1 are zero. The only nonzero ones in Eq.(12) are

$$\frac{\partial^2 U_1}{\partial x_r \partial x_s} = \begin{cases} \displaystyle\frac{1}{8}\sum_{j=1}^{N-1}\frac{d_{jr}d_{-js} + d_{-jr}d_{js}}{|d_j|^3} & \text{for } r = s \\ \displaystyle\frac{3}{8}\sum_{j=1}^{N-1}\frac{d_{jr}d_{js}d_{-j}^2 + d_{-jr}d_{-js}d_j^2}{|d_j|^5} & \text{for } r + s = 2 \bmod N. \end{cases} \tag{13}$$

The nonzero terms in Eq.(13) will be denoted by b_r and c_r for $0 < r < N$. With Eq.(11) these terms are

$$
\begin{aligned}
b_r &= \frac{1}{8}\sum_{j=1}^{N-1}\frac{\sin^2 jr\pi/N}{\sin^3 j\pi/N}\\
c_r &= -\frac{3}{8}\sum_{j=1}^{N-1}\frac{\sin jr\pi/N\,\sin j(r-2)\pi/N}{\sin^3 j\pi/N}.
\end{aligned}
$$

The partial derivatives $\partial^2 U_1/\partial y_r\partial y_s$ are found to have the same values except for a change of sign in c_r. The mixed partial derivatives $\partial^2 U_1/\partial x_r\partial y_s$ are calculated by the same method and all turn out to be zero. Let $\mathbf{B}$ be the diagonal matrix with entries b_r and $\mathbf{C}$ a symmetric matrix with nonzero entries c_r just below the cross diagonal. Then we can write the Hessian in the new coordinates in the form

$$
D^2U_1(\mathbf{Q}_0) = \begin{pmatrix} \mathbf{B}_1+\mathbf{C}_1 & 0 \\ 0 & \mathbf{B}_1-\mathbf{C}_1 \end{pmatrix}
$$

with

$$
\mathbf{B}_1 \pm \mathbf{C}_1 = \begin{pmatrix}
0 & & 0 & & & & \\
 & b_1 \pm c_1 & & & & & \\
0 & & b_2 & & & & \\
 & & & b_3 & & & \pm c_3 \\
 & & & & b_4 & \pm c_4 & \\
 & & & & & \cdots & \\
 & & & & \pm c_4 & b_{N-2} & \\
 & & & \pm c_3 & & & b_{N-1}
\end{pmatrix}.
$$

Note that the two diagonals will cross in a single term of the matrix when N is even. We also remark that $c_0 = c_2 = b_0 = 0$ and that

$$
b_1 = \frac{1}{3}c_1 = \frac{1}{8}\sum_{j=1}^{N-1}\frac{1}{\sin j\pi/N}.
$$

Furthermore, we have the simple case $\mathbf{M} = \mathbf{I}$ so that

$$
\nu^2 = \frac{U_1(\mathbf{Q}_0)}{2I(\mathbf{Q}_0)} = \frac{1}{2N}\sum_{j=1}^{N-1}\sum_{i=0}^{N-1}\frac{1}{|\omega^i-\omega^{i+j}|} = \frac{1}{4}\sum_{j=1}^{N-1}\frac{1}{\sin j\pi/N} = 2b_1.
$$

With the form of the matrices given above it is seen that Eq.(8) decomposes nicely into 2×2 and 4×4 subdeterminants. We will discuss these cases now.

The case $r = 0$ gives rise to the determinant

$$\begin{vmatrix} \nu^2 - \lambda^2 & 2\nu\lambda \\ -2\nu\lambda & \nu^2 - \lambda^2 \end{vmatrix} = (\lambda^2 + \nu^2)^2 = 0$$

with the repeated pair of roots $\pm i\nu$.

The case $r = 1$ produces the determinant

$$\begin{vmatrix} 3\nu^2 - \lambda^2 & 2\nu\lambda \\ -2\nu\lambda & \nu^2 \end{vmatrix} = \lambda^2(\lambda^2 + \nu^2) = 0$$

with the double root $\lambda = 0$ and the pair $\pm i\nu$. The two cases account for the trivial characteristic exponents which can be anticipated from the general theory due to the existence of additional integrals. The triple pair $\pm i\nu$ accounts for the preservation of the angular momentum and the uniform motion of the center of mass. The zero exponents occur due to the special nature of the circular orbits in the two body problem, which are surrounded by elliptic orbits. For the trivial cases one finds that the matrix in Eq.(6) has non elementary divisors, so that the system can not be linearly stable.

The case $r = 2$ gives the following determinant

$$\begin{vmatrix} \nu^2 - \lambda^2 + b_2 & 2\nu\lambda \\ -2\nu\lambda & \nu^2 - \lambda^2 + b_2 \end{vmatrix} = (\nu^2 - \lambda^2 + b_2)^2 + 4\nu^2\lambda^2 = 0$$

with roots $\lambda^2 = -\nu^2 + b_2 \pm \nu\sqrt{-4b_2}$ or the four complex values $\lambda = \pm(\sqrt{b_2} \pm i\nu)$.

When N is even the case $r = (N+2)/2$ gives rise to another 2×2 determinant of the form

$$\begin{vmatrix} \nu^2 - \lambda^2 + b_r + c_r & 2\nu\lambda \\ -2\nu\lambda & \nu^2 - \lambda^2 + b_r - c_r \end{vmatrix} = (\nu^2 - \lambda^2 + b_r)^2 - c_r^2 + 4\nu^2\lambda^2 = 0$$

with roots $\lambda^2 = -\nu^2 + b_r \pm \sqrt{c_r^2 - 4\nu^2 b_r}$.

For the cases $3 \le r \le (N+1)/2$ with $s = N + 2 - r$ we encounter 4×4 determinants of the form

$$\begin{vmatrix} \nu^2 - \lambda^2 + b_r & c_r & 2\nu\lambda & 0 \\ c_r & \nu^2 - \lambda^2 + b_s & 0 & 2\nu\lambda \\ -2\nu\lambda & 0 & \nu^2 - \lambda^2 + b_r & -c_r \\ 0 & -2\nu\lambda & -c_r & \nu^2 - \lambda^2 + b_s \end{vmatrix} = 0. \tag{14}$$

This eighth order polynomial in λ can be written in a very special form

$$(\lambda^4 + (2\nu^2 - b_r - b_s)\lambda^2 + (\nu^2 + b_r)(\nu^2 + b_s) - c_r^2)^2 + 4(b_r - b_s)^2\nu^2\lambda^2 = 0. \tag{15}$$

and with the abbreviations

$$\begin{aligned} z &= \lambda^2 \\ \alpha &= 2\nu^2 - b_r - b_s \\ \beta &= (\nu^2 + b_r)(\nu^2 + b_s) - c_r^2 \\ \gamma &= 4(b_r - b_s)^2\nu^2 \end{aligned}$$

we are left to solve a quartic polynomial in z. As this polynomial was also used by Moeckel we can cite the following proposition from there [3]

Proposition 1 *Suppose $\gamma > 0$. Then the roots of the quartic*

$$Q(z) = (z^2 + \alpha z + \beta)^2 + \gamma z. \tag{16}$$

are all real and negative if and only if

$$\alpha > 0, \quad \beta \neq 0, \quad \alpha^2 - 4\beta \geq 0, \quad \Delta \geq 0$$

where the discriminant of $Q(z)$ is

$$\Delta = \frac{1}{256}\gamma^2(16\beta(\alpha^2 - 4\beta)^2 + 4\alpha(\alpha^2 - 36\beta)\gamma - 27\gamma^2).$$

The quartic has repeated roots if and only if $\Delta = 0$.

The discriminant Δ is the resultant of $Q(z)$ and $Q'(z)$. Additional details can be found in [1]. We will describe in section a numerical method to find the roots of Eq.(16) efficiently. We used this method to verify the results given by Moeckel [3]. We refer the reader to his paper for a listing of the numbers of imaginary, real, and complex characteristic exponents of ring.

4. Regular N-gon Configurations with a Central Mass

We include now a central mass of size m so that we have to consider the term U_2 from Eq.(9). The rate of rotation ν from Eq.(3) is

$$\nu^2 = 2b_1 + m.$$

The calculations of the previous section are extended without any additional work to $\mathcal{R}^{2N+2}$, as we have

$$\mathbf{Q} = \mathbf{Q}_0 + \mathbf{V}\mathbf{z} \tag{17}$$

with $\mathbf{Q}_0 = (1, \omega, \ldots, \omega^{N-1}, 0)^t$. In order for the transformation to be symplectic we choose

$$\mathbf{V} = \begin{pmatrix} \mathbf{W} & 0 \\ 0 & 1 \end{pmatrix}.$$

For

$$U_2(\mathbf{Q}) = \sum_{j=0}^{N-1} \frac{1}{|q_i - q_N|}$$

we find the Hessian at $\mathbf{z} = 0$. We use this time

$$\begin{aligned} |q_j - q_N|^2 &= |\omega^j + \frac{1}{\sqrt{N}} \sum_{k=0}^{N-1} \omega^{jk} z_k - z_N|^2 \\ &= 1 + \frac{1}{\sqrt{N}} \sum_k (\omega^{j(k-1)} z_k + \omega^{-j(k-1)} \overline{z}_k) + \frac{1}{N} \sum_{k,l} \omega^{j(k-l)} z_k \overline{z}_l \\ &\quad - \omega^j \overline{z}_N - \omega^{-j} z_N - \frac{1}{\sqrt{N}} \sum_k (\omega^{jk} z_k \overline{z}_N + \omega^{-jk} \overline{z}_k z_N) + z_N \overline{z}_N. \end{aligned}$$

and discover that the Hessian has a similar structure with the exception of the last row and column, that is

$$D^2 U_2(\mathbf{Q}_0) = \begin{pmatrix} \mathbf{B}_2 + m\mathbf{C}_2 & \mathbf{0} \\ \mathbf{0} & \mathbf{B}_2 - m\mathbf{C}_2 \end{pmatrix}$$

where

$$\mathbf{B}_2 = \begin{pmatrix} \frac{1}{2} & & & & & -\frac{\sqrt{N}}{2} \\ & \frac{1}{2} & & & & \\ & & \frac{1}{2} & & & \\ & & & \ddots & & \\ & & & & \frac{1}{2} & \\ -\frac{\sqrt{N}}{2} & & & & & \frac{N}{2} \end{pmatrix}$$

and

$$\mathbf{C}_2 = \begin{pmatrix} 0 & 0 & \frac{3}{2} & 0 & \cdots & 0 & 0 \\ 0 & \frac{3}{2} & 0 & 0 & \cdots & 0 & 0 \\ \frac{3}{2} & 0 & 0 & 0 & \cdots & 0 & -\frac{3\sqrt{N}}{2} \\ 0 & 0 & 0 & 0 & \cdots & \frac{3}{2} & 0 \\ \vdots & \vdots & \vdots & \vdots & & \vdots & \vdots \\ 0 & 0 & 0 & \frac{3}{2} & \cdots & 0 & 0 \\ 0 & 0 & -\frac{3\sqrt{N}}{2} & 0 & \cdots & 0 & 0 \end{pmatrix}.$$

The splitting of Eq.(8) into 2×2 and 4×4 subdeterminants still happens. One exception is caused by rows (and columns) with $r = 0$, 2, and N as they combine into a 6×6 subdeterminant. In order to see this structure we list the Hessian of $U(\mathbf{Q}_0)$ with respect to the x-components, i.e. with $\mathbf{B} = \mathbf{B}_1 + \mathbf{B}_2$

and $\mathbf{C} = \mathbf{C}_1 + \mathbf{C}_2$

$$\mathbf{B}+m\mathbf{C} = \begin{pmatrix} \frac{m}{2} & 0 & \frac{3m}{2} & 0 & \cdots & 0 & -\frac{m\sqrt{N}}{2} \\ 0 & 4b_1+2m & 0 & 0 & \cdots & 0 & 0 \\ \frac{3m}{2} & 0 & b_2+\frac{m}{2} & 0 & \cdots & 0 & -\frac{3m\sqrt{N}}{2} \\ 0 & 0 & 0 & b_3+\frac{m}{2} & \cdots & c_3+\frac{3m}{2} & 0 \\ \vdots & \vdots & \vdots & \vdots & & \vdots & \vdots \\ 0 & 0 & 0 & c_3+\frac{3m}{2} & \cdots & b_{N-1}+\frac{m}{2} & 0 \\ -\frac{m\sqrt{N}}{2} & 0 & -\frac{3m\sqrt{N}}{2} & 0 & \cdots & 0 & \frac{mN}{2} \end{pmatrix}. \tag{18}$$

5. Degenerate Relative Equilibria

The regular N–gon of the previous section was shown by Meyer and Schmidt to be degenerate for certain values of the central mass m. Because the notation in that paper [2] was different and the paper is not very detailed, we will show how to find the degenerate relative equilibria via the matrix Eq.(18).

Relative equilibria are extrema of the potential function Eq.(9) under the condition that the moment of inertia Eq.(4) is held constant. The relative equilibria is called degenerate if the extrema is degenerate in a reduced space, where the center of mass is kept at the origin and the rotational symmetry of the entire system has been removed.

The center of mass of the regular N–gon solution is at the origin if in the coordinates Eq.(17) we have

$$z_0 = -\frac{m}{\sqrt{N}} z_N. \tag{19}$$

If $\tilde{U}$ is the restriction of U given by Eq.(19) then the partial derivatives are

$$\begin{aligned} \frac{\partial^2 \tilde{U}}{\partial x_2 \partial x_N} &= -\frac{m}{\sqrt{N}} \frac{\partial^2 U}{\partial x_0 \partial x_2} + \frac{\partial^2 U}{\partial x_2 \partial x_N} &= -\frac{3m}{2\sqrt{N}}(m+N) \\ \frac{\partial^2 \tilde{U}}{\partial x_N^2} &= \frac{m^2}{N}\frac{\partial^2 U}{\partial x_0^2} - \frac{2m}{\sqrt{N}} \frac{\partial^2 U}{\partial x_0 \partial x_N} + \frac{\partial^2 U}{\partial x_N^2} &= \frac{m}{2N}(m+N)^2 \end{aligned}$$

and for all other cases

$$\frac{\partial^2 \tilde{U}}{\partial x_r \partial x_s} = \frac{\partial^2 U}{\partial x_r \partial x_s}.$$

The rotation of the configuration about a fixed angle can be eliminated by requiring that z_1 is real, i.e. by setting $y_1 = 0$. The moment of inertia of the unperturbed N-gon is $N/2$ and keeping it fixed leads to the equation

$$0 = 2\sqrt{N}x_1 + x_1^2 + \sum_{k=2}^{N-1}(x_k^2 + y_k^2) + \frac{m}{N}(m+N)(x_N^2 + y_N^2). \tag{20}$$

By the implicit function theorem this equation can be solved to give

$$x_1 = f(x_2, \ldots, x_N, y_2, \ldots, y_N)$$

near the origin. All of its first and second order partial derivatives are zero with the exceptions of

$$\frac{\partial^2 f}{\partial x_r^2} = \frac{\partial^2 f}{\partial y_r^2} = -\frac{1}{\sqrt{N}} \qquad \text{for} \quad 2 \leq r \leq N-1$$

and

$$\frac{\partial^2 f}{\partial x_N^2} = \frac{\partial^2 f}{\partial y_N^2} = -\frac{m}{\sqrt{N}}(1 + \frac{m}{N}).$$

If $\tilde{U}$ is the restriction of U given by Eq.(20) then only the partial derivatives along the main diagonal of the Hessian have to be modified, that is,

$$\frac{\partial^2 \tilde{U}}{\partial x_r^2} = \frac{\partial^2 U}{\partial x_r^2} + \frac{\partial U}{\partial x_1}\frac{\partial^2 f}{\partial x_r^2}, \qquad \frac{\partial^2 \tilde{U}}{\partial y_r^2} = \frac{\partial^2 U}{\partial y_r^2} + \frac{\partial U}{\partial x_1}\frac{\partial^2 f}{\partial y_r^2}.$$

The Hessian of $\tilde{U}$ under the restrictions of Eq.(19) and Eq.(20) with respect to $x_2, \ldots, x_N, y_2, \ldots, y_N$ is then found from Eq.(18) to have the form

$$D^2\tilde{U}(\mathbf{Q}_0) = \begin{pmatrix} \tilde{\mathbf{B}} + \tilde{\mathbf{C}} & 0 \\ 0 & \tilde{\mathbf{B}} - \tilde{\mathbf{C}} \end{pmatrix}.$$

With

$$\frac{\partial U}{\partial x_1} = -\sqrt{N}\left(\sum_{j=1}^{N-1} \frac{1}{4\sin \pi j/N} + m\right) = -\sqrt{N}(2b_1 + m)$$

and the abbreviation $b = 2b_1 + \frac{3m}{2}$ we compute the matrix $\quad \tilde{\mathbf{B}} + \tilde{\mathbf{C}} =$

$$\begin{pmatrix} b + b_2 & 0 & \cdots & 0 & -\frac{3m}{2\sqrt{N}}(m+N) \\ 0 & b + b_3 & \cdots & c_3 + \frac{3m}{2} & 0 \\ \vdots & \vdots & \ddots & \vdots & \vdots \\ 0 & c_3 + \frac{3m}{2} & \cdots & b + b_{N-1} & 0 \\ -\frac{3m}{2\sqrt{N}}(m+N) & 0 & \cdots & 0 & \frac{m}{N}(m+N)(b + \frac{N}{2}) \end{pmatrix}.$$

Table 1. Critical values of the mass for the regular N-gon

N	$m_{\nu=0}$	$m(2,N)$	$m(3,N)$	$m(4,N)$	$m(5,N)$	$m(6,N)$	$m(7,N)$
3	-0.577	0.770					
4	-0.957	2.379	-0.250				
5	-1.376	6.478	-0.244				
6	-1.828	20.906	-0.220	0.006			
7	-2.306	-643.284	-0.181	0.324			
8	-2.806	-37.929	-0.130	0.698	0.996		
9	-3.326	-25.443	-0.069	1.119	1.773		
10	-3.862	-21.723	0.001	1.581	2.641	3.012	
11	-4.416	-20.270	0.079	2.079	3.588	4.391	
12	-4.984	-19.738	0.165	2.610	4.605	5.893	6.337

For $\tilde{\mathbf{B}} - \tilde{\mathbf{C}}$ all terms on the cross diagonal change their signs. Thus $\tilde{\mathbf{B}} + \tilde{\mathbf{C}}$ and $\tilde{\mathbf{B}} - \tilde{\mathbf{C}}$ become singular for the same values of m. These critical values are found by evaluating the appropriate 2×2 subdeterminants. If $m(k, N)$ is this critical value then we have

$$m(r,N) = \begin{cases} \dfrac{(2b_1 + b_2)(N + 4b_1)}{3(N - 4b_1 - b_2)} & \text{for } r = 2, \\[2ex] \dfrac{2(c_r^2 - (2b_1 + b_r)(2b_1 + b_s))}{3(4b_1 + b_r + b_s - 2c_r)} & \text{for } 3 \le r \le (N+1)/2 \\ & \text{with } s = N + 2 - r \\[2ex] -\frac{1}{3}(2b_1 + b_r + c_r) & \text{for } r = (N+2)/2 \\ & \text{and } N \text{ even} \end{cases}$$

with the last case occurring when the two diagonals meet in the central element of the matrix. Tables with the numerical values for these critical masses have been given previously. We repeat them here for completeness sake. We have added a column, which gives the (repelling) mass when the rotation is zero. Any N-gon configuration with a mass smaller than the one listed under $m_{\nu=0}$ can not be realized, as the rotation would have to be imaginary. Of interest is the comparison with the values for $m(2, N)$, which shows that for $N \ge 7$ these critical values can not be realized even if repelling masses would exist.

6. Stability of a ring with a central mass

The spectral stability is determined by the roots of

$$\left|(\nu^2 - \lambda^2)\mathbf{I} + 2\nu\lambda\mathbf{J} + \mathbf{M}^{-1}D^2U(\mathbf{Q}_0)\right| = 0.$$

Due to the structure of Eq.(18) we have to consider only 2×2 or 4×4 subdeterminants. One exception is the case belonging to rows and columns $r = 0$, 2,

N, $N+1$, $N+3$, and $2N+2$. We consider first this case and list the terms in $\mathbf{M}^{-1}D^2U(\mathbf{Q}_0)$ which come into play

$$\begin{pmatrix} \frac{m}{2} & \frac{3m}{2} & -\frac{m\sqrt{N}}{2} & 0 & 0 & 0 \\ \frac{3m}{2} & \frac{m}{2}+b_2 & -\frac{3m\sqrt{N}}{2} & 0 & 0 & 0 \\ -\frac{\sqrt{N}}{2} & -\frac{3\sqrt{N}}{2} & \frac{N}{2} & 0 & 0 & 0 \\ 0 & 0 & 0 & \frac{m}{2} & -\frac{3m}{2} & -\frac{m\sqrt{N}}{2} \\ 0 & 0 & 0 & -\frac{3m}{2} & \frac{m}{2}+b_2 & \frac{3m\sqrt{N}}{2} \\ 0 & 0 & 0 & -\frac{\sqrt{N}}{2} & \frac{3\sqrt{N}}{2} & \frac{N}{2} \end{pmatrix}.$$

The corresponding determinant is a polynomial of degree 12 in λ. It has the factor $(\lambda^2+\nu^2)^2$ which accounts for some of the trivial exponents. The remaining polynomial can be written with $z=\lambda^2$ in the form Eq.(16), that is

$$(z^2+\alpha z+\beta)^2+\gamma z=0 \tag{21}$$

where

$$\begin{aligned} \alpha &= 4b_1-b_2-\frac{N}{2}+m \\ \beta &= (2b_1+\frac{N}{2})(2b_1+b_2)+3m(2b_1+\frac{b_2-N}{2}) \\ \gamma &= \frac{1}{4}(\frac{N}{2}-b_2)^2(2b_1+m). \end{aligned}$$

One sees at once that $\beta=0$ for the critical value of the mass $m(2,N)$. Before we discuss the behavior of the roots of Eq.(21) when m is near the critical values we will consider the other cases first. For $3\le r\le N/2+1$ the determinant is a modification of Eq.(14) i.e.

$$\begin{vmatrix} \nu^2+b_r+\frac{m}{2}-\lambda^2 & c_r+\frac{3m}{2} & 2\nu\lambda & 0 \\ c_r+\frac{3m}{2} & \nu^2+b_s+\frac{m}{2}-\lambda^2 & 0 & 2\nu\lambda \\ -2\nu\lambda & 0 & \nu^2+b_r+\frac{m}{2}-\lambda^2 & -c_r-\frac{3m}{2} \\ 0 & -2\nu\lambda & -c_r-\frac{3m}{2} & \nu^2+b_s+\frac{m}{2}-\lambda^2 \end{vmatrix}$$

We recall that $s=N+2-r$ and that $\nu^2=2b_1+m$ and find again that this eighth order polynomial in λ can be written in the form Eq.(21) with

$$\begin{aligned} \alpha &= 4b_1-b_r-b_s+m \\ \beta &= (2b_1+b_r)(2b_1+b_s)-c_r^2+3m(2b_1+\frac{b_r+b_s}{2}-c_r) \\ \gamma &= \frac{1}{4}(b_r-b_s)^2(2b_1+m). \end{aligned}$$

Again $\beta = 0$ at the critical value $m = m(r, N)$. When N is even and $r = (N+2)/2$ we have the 2×2 determinant

$$\begin{vmatrix} \nu^2 + b_r + c_r + 2m - \lambda^2 & 2\nu\lambda \\ -2\nu\lambda & \nu^2 + b_r - c_r - m - \lambda^2 \end{vmatrix} = 0$$

which is a second order polynomial in λ^2, that is,

$$z^2 + \alpha z + \beta = 0 \tag{22}$$

where

$$\begin{aligned} \alpha &= 4b_1 - 2b_r + m \\ \beta &= (2b_1 + b_r)^2 - c_r^2 + 3m(2b_1 + b_r - c_r). \end{aligned}$$

In this notation Eq.(22) has a simple root when $\beta = 0$ and this happens at the critical value $m = m(r, N)$.

Finally the case $r = 1$ has to be considered. It gives nothing new, just the missing trivial values which also were found in the case of a ring. The following summarizes our findings

Proposition 2 *For $2 \leq r \leq (N+1)/2$ the stability polynomial (8) will have a pair of zero roots when the mass m takes on the critical value $m(r, N)$. For values near the critical mass these two roots stay on the imaginary axis, that is, they do not become real. When $N > 2$ is even and $r = (N+2)/2$ the stability polynomial will have a pair of zero roots for $m = m(r, N)$. This time the pair will move from the imaginary axis onto the real axis as m goes through the critical value.*

Remark: The statement about the behavior of the characteristic exponents as m passes through its critical values follows at once from propositions 4 and 6 in the next section where we analyze the behavior of the roots of Eq.(21) and Eq.(22).

7. Discussion of the Spectral Stability

The previous section has shown that the roots of Eq.(21) determine the spectral stability of the regular N-gon. When N is even also Eq.(22) has to be considered, but we will postpone this special case until later. We want to investigate how the roots depend on m. Most of our work was done numerically, but we can state some general results. The coefficients of Eq.(21) depend linearly on m. Let us write

$$\alpha = \alpha_0 + m, \qquad \beta = \beta_0 + m\beta_1, \qquad \gamma = \gamma_0 + m\gamma_1.$$

The mass m should be positive, but we restrict ourselves to $\gamma \geq 0$. Since $\gamma_0 > 0$ and $\gamma_1 > 0$ we have $\gamma = 0$ when $m = -\gamma_0/\gamma_1 < 0$. For this value of m equation Eq.(21) has then repeated roots. When $r = 2$ the repeated roots lie in the complex plane and are given by

$$\frac{1}{4}\left(N + 2b_2 - 4b_1 \pm i\sqrt{72b_1(N - 2b_1) - (N - 2b_2)^2}\right).$$

When $r > 2$ and $s = N + 2 - r$ the repeated roots are given by

$$\frac{1}{2}\left(b_r + b_s - 2b_1 \pm \sqrt{(b_r - b_s)^2 + 4(3b_1 - c_r)^2}\right)$$

with one lying on the positive the other on the negative real axis. The last claim can be verified numerically.

Proposition 3 *With increasing m, that is with γ becoming positive, the roots of Eq.(21) on the positive real axis will move into the complex plane, whereas those on the negative axis will stay on that axis.*

Proof: The assumption is that $z^2 + \alpha z + \beta$ has real roots, i.e. $\alpha^2 - 4\beta > 0$. Let z_0 denote one of these roots. A Taylor expansion of Eq.(21) with γ as a small parameter gives

$$z = z_0 + \gamma\sqrt{-z_0/(\alpha^2 - 4\beta)} + \mathcal{O}(\gamma^2)$$

from which the statement follows.

Proposition 4 *For $\beta = 0$ equation Eq.(21) has a simple root $z = 0$ and for small values of β this root will stay on the negative real axis.*

Proof: The proposition requires $\gamma > 0$. A Taylor series expansion for small β gives $z = -\beta^2/\gamma + \mathcal{O}(\beta^3)$ from which the statement follows. It was shown earlier that $\beta = 0$ occurs when we have a degenerate relative equilibria. In the current notation this corresponds to $m = -\beta_0/\beta_1$.

Proposition 5 *As $m \to \infty$ all roots of Eq.(21) will be real and negative when $\beta_1 > 0$.*

Proof: A Taylor series expansion of the solutions of Eq.(21) in terms of $1/m$ gives

$$z = -\beta_1 \pm \sqrt{\frac{\beta_1\gamma_1}{m}} + \mathcal{O}(\frac{1}{m}) \qquad \text{and} \qquad z = -m + \beta_1 - \alpha_0 \pm \sqrt{\gamma_1} + \mathcal{O}(\frac{1}{m})$$

from which the result follows.

When N is even we have to consider the special case of the quadratic Eq.(22). If one looks at its coefficients α and β one notices that they follow from those of the quartic Eq.(21) when $r = s$. Under these conditions $\gamma = 0$. It is the case excluded by proposition 1. Fortunately it is an easier case as we only have to analyze the roots of the quadratic polynomial

$$z^2 + \alpha z + \beta = 0 \tag{23}$$

as m varies. As before we have set $\alpha = \alpha_0 + m$ and $\beta = \beta_0 + m\beta_1$. At the smallest acceptable value for m the roots are on the real axis. They are

$$2b_1 + b_r - c_r \qquad \text{and} \qquad -4b_1 - b_r + c_r$$

with the first one positive and the second one negative. The following is easily verified.

Proposition 6 *For small values of β the quadratic Eq.(23) has the root $z = -\beta/\alpha + \cdots$. As β passes through zero the root will change sign, under the generic assumption that $\alpha \neq 0$. The roots of Eq.(23) can move into the complex plane only for two values of m, that is, when*

$$m^2 + 2(\alpha_0 - 2\beta_1)m + \alpha_0^2 - 4\beta_0 = 0.$$

When $m \to \infty$ the roots of Eq.(23) are on the real axis and their Taylor series expansion is given by

$$z = -m - \alpha_0 + \beta_1 + \mathcal{O}\left(\frac{1}{m}\right) \qquad \text{and} \quad z = -\beta_1 + \mathcal{O}\left(\frac{1}{m}\right)$$

respectively.

The above propositions show that β_1 determines spectral stability or instability as m becomes very large. One finds numerically or otherwise that $\beta_1 < 0$ only when $N < 7$ and $r = 2$. In that case $\beta_0 > 0$ is easily verified. Thus we can make the following statement about Maxwell's ring, a fact which was already observed by Moeckel [3].

Proposition 7 *Maxwell's ring, a regular N-gon configuration with a very large central mass, is spectrally unstable for $N < 7$, i.e. when the critical value of the mass $m(2, N)$ is positive.*

8. Numerical Explorations

The roots of a quartic can be found with ease numerically if the following method [6] is used. The idea is to decompose the quartic into the product of two quadratic polynomials. Let

$$\begin{aligned} Q(z) &= z^4 + a_3 z^3 + a_2 z^2 + a_1 z + a_0 \\ &= (z^2 + b_1 z + c_1)(z^2 + b_2 z + c_2) \end{aligned}$$

so that by comparing coefficients we get the following set of nonlinear equations for b_1, b_2, c_1, and c_2

$$\begin{aligned} a_0 &= c_1c_2 & (24)\\ a_1 &= c_1b_2 + c_2b_1 & (25)\\ a_2 &= b_1b_2 + c_1 + c_2 & (26)\\ a_3 &= b_1 + b_2. & (27) \end{aligned}$$

Choose any initial guess for c_1, so that from Eq.(24), Eq.(25), and Eq.(27)

$$c_2 = \frac{a_0}{c_1}, \qquad b_1 = \frac{a_1 - a_3c_1}{c_2 - c_1}, \qquad b_2 = a_3 - b_1.$$

Finally from Eq.(26) find

$$F(c_1) = c_1 + c_2 + b_1b_2 - a_2.$$

Typically it will not be zero, but via a Newton iteration c_1 can be modified so that $F(c_1) = 0$ and with it all four equations Eq.(24–27) will be satisfied. Since $a_0 = c_1c_2$ and since it does not matter which one of the two terms is called c_1 we can restrict the initial value of c_1 to the interval $-\sqrt{|a_0|} \le c_1 \le \sqrt{|a_0|}$. When $a_0 > 0$ this interval can be restricted even further. We have either the case $c_1 > 0$ and $c_2 > 0$ or $c_1 < 0$ and $c_2 < 0$, but in the later case the roots of $Q(z) = 0$ are all real with $z_1 > 0$, $z_2 < 0$ and $z_3 > 0$, $z_4 < 0$. By pairing these roots differently we can achieve $c_1 > 0$ for $F(c_1) = 0$. Furthermore, $F(c_1) \to \infty$ as $c_1 \to 0+$, $F(c_1) \to -\infty$ as $c_1 \to \sqrt{|a_0|}$ and $F(c_1)$ is monotonically decreasing in $(0, \sqrt{|a_0|})$ so that $F(c_1) = 0$ has a unique solution in this interval. Therefore finding the desired value for c_1 numerically poses no difficulties and with it the roots of $Q(z)$ can be determined.

With the above method we have plotted the roots of the stability polynomial in the complex plane. The behavior of these roots as m varies can be understood in conjunction with the results of the previous section. The roots of Eq.(23) are the easiest one to describe and therefore we start with them and figure 1.

For the smallest allowable value of m one root is on the negative axis the other on the positive real axis. With increasing m both roots move towards the origin. Shortly after one root has gone through the origin the two roots coincide. They then move into the complex plane with their real parts going into the direction of the negative real axis. The two roots will coalesce again on the negative real axis. They will then stay on that axis with one root tending to $-\infty$, whereas the other one will tend to the value given in proposition 6.

Since equation Eq.(21) is to some degree an extension of the 2×2 case we will have four separate traces in the complex plane as m increases from its lowest

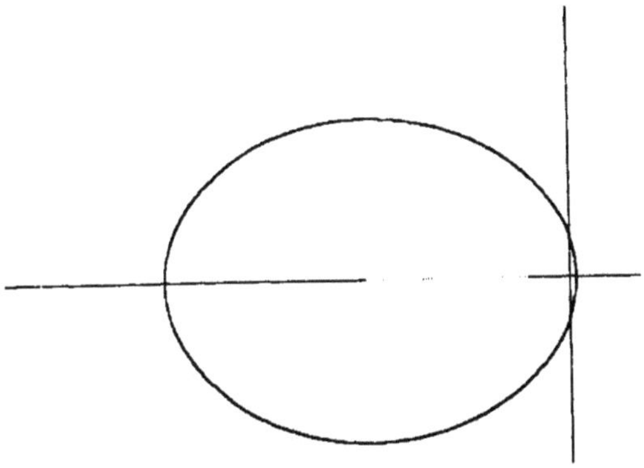

Fig. 1. Roots of equation (22) as m varies

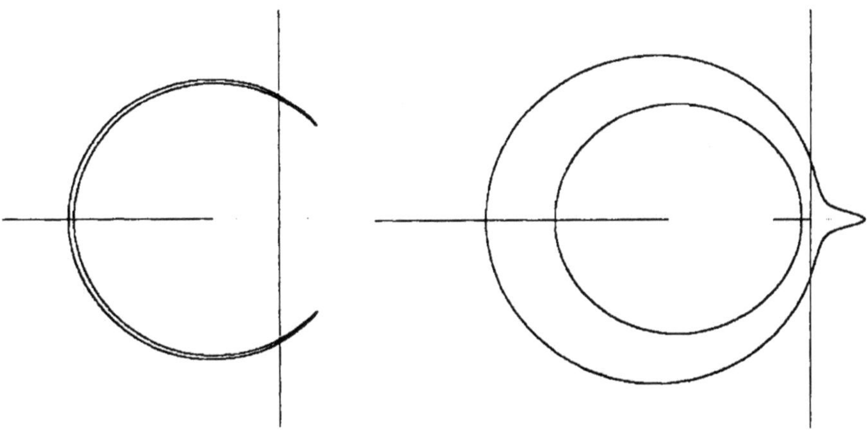

Fig. 2. Roots of equation (21) for $N > 7$ a) $r = 2$ b) $r > 2$

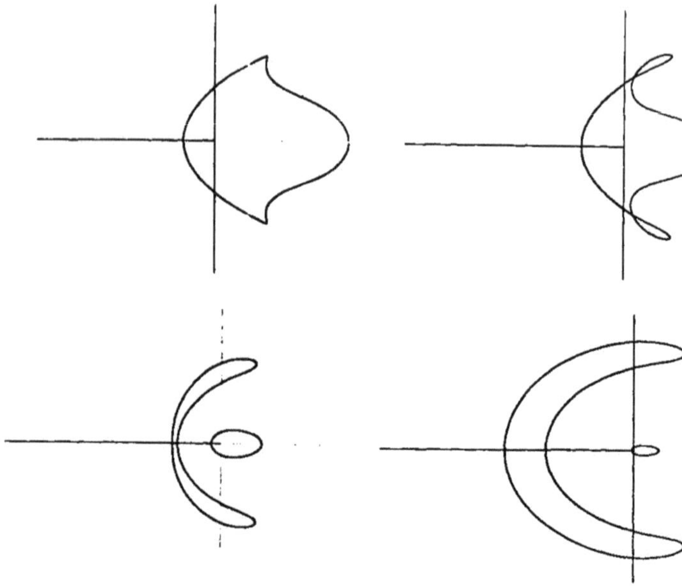

Fig. 3. Roots of equation (21) for $r = 2$ top row $N = 3$ and $N = 4$ bottom row $N = 5$ and $N = 6$

admissible value. We will start with the generic case when $N \geq 7$ and $r \neq 2$, see figure 2b). The pair of roots on the negative real axis will move towards the origin with different speeds. After one root has reached the origin it turns around. Shortly afterwards it meets the other root at which point the two roots move into the complex plane. The two roots loop back with their real parts becoming increasingly more negative. Eventually the two roots meet again on the negative real axis, where one root then goes to $-\infty$ the other to $-\beta_1$. The other pair of roots started on the real axis and moved at once into the complex plane. Their real values are decreasing until they meet on the negative real axis. At this point they follow the other two roots which have reached the real axis at a slightly smaller value of m.

When $N \geq 7$ and $r = 2$ the two pairs of roots start in the complex plane, see figure 2a). The two sets of roots move towards the negative real axis, where each meets its conjugate complex root. Every time a double root has happened one root will move towards $-\infty$ the other towards $-\beta_1$.

When $N < 7$ and $r = 2$ the behavior is slightly different for each value of N. We have plotted all these cases in figures 3. Since $\beta_1 < 0$ in all these cases one pair of roots tends from the complex plane to a value on the real axis. The other pair of roots goes to $-\infty$. How this happens is different when $N \leq 4$ or $N = 5$ or 6. For $N \leq 4$ one pair of conjugate complex roots always stays in the complex plane and heads directly for the real axis. The other conjugate complex pair tends to the negative real axis. There one root will first visit the origin before it will follow the other one towards $-\infty$.

For $N = 5$ or $N = 6$ both pairs of complex conjugate roots will reach the negative real axis. One root of each pair goes to $-\infty$. The other root will move towards the origin on the negative real axis. After one has turned around at the origin it will collide with the other one. They will then branch into the complex plane with their final destination the positive real axis.

The plots for $N < 7$ and $r \neq 2$ show nothing substantially different from those for $N \geq 7$ and for this reason have been omitted.

9. Acknowledgement

We would like to thank Professor Hildeberto Cabral for a very careful reading of the original manuscript and finding our misprints.

10. References

1. W. S. Burnside. *The Theory of Equations.* Volume 1, Dublin University Press, 1928.
2. K. R. Meyer and D. S. Schmidt. Bifurcations of relative equilibria in the N-body and Kirchhoff problems. *SIAM Journal on Math. Anal.*, 19:1295–1313, 1988.
3. R. Moeckel. Linear stability analysis of some symmetrical classes of relative equilibria. In H. S. Dumas, K. R. Meyer, and D. S. Schmidt, editors, *Hamiltonian Dynamical Systems, History, Theory, and Applications*, pages 291–317, Springer, 1994.
4. J. I. Palmore. Measure of degenerate relative equilibria I. *Ann. of Math.*, 140:421–429, 1976.
5. D. S. Schmidt. Central configurations and relative equilibria for the N–body problem. *Lecture Notes, Departmento de Matemática, Universidade Federal de Pernambuco, Recife, Brasil*, 1:, 1993.
6. R. Zurmühl. *Praktische Mathematik für Ingenieure und Physiker.* Springer Verlag, 3rd edition, 1961.

ANALYTIC CONTINUATION OF CIRCULAR KEPLER MOTION TO THE GENERAL THREE-DIMENSIONAL THREE-BODY PROBLEM

JAUME SOLER
Departament d'Informàtica i Matemàtica Aplicada, Universitat de Girona
Avda. Lluís Santaló, s/n, 17071 Girona, Spain.
E-mail: jsmat@xauxa.udg.es

ABSTRACT

Families of symmetric periodic solutions of the general three-body problem in three-dimensional space are shown to exist in a rotating frame by analytic continuation of circular Keplerian orbits. The parameter along each family is the relative inclination of the orbital planes.

1. Introduction

The main purpose of this work is to show that there exist families of solutions of the general three-body problem in which two bodies of small mass move in almost circular orbits around a central body of much larger mass, the angle between the orbital planes being arbitrary (see Theorem 2). The method used is the classical analytic continuation applied to the equations which stem from the symmetries of the problem.

This work is a straightforward generalization of a similar one by Jefferys[1] on the three-dimensional restricted three-body problem, in which one of the small bodies is actually of zero mass. As in the restricted case, a rotating frame of reference is used although its rotation is not uniform. The reason is that the z-axis is taken parallel to the angular momentum vector of the system in order to simplify the symmetry conditions (see Theorem 1) and this rules out the possibility of uniform rotation when the elimination of one of the angular variables is performed.

Mention must be made of a result of Poincaré[3] in which a continuity argument (based on a theorem of Kronecker) is used to show the existence of symmetric periodic solutions of the problem. Poincaré defines symmetric conjunction at time t_0 as a position in which the velocities are perpendicular to the plane defined by the three bodies. If at time $t_0 + T$ the system is again in symmetric conjuction, then at time $t_0 + 2T$ the three bodies will be in the same relative position of time T_0, with the same velocities, although the plane can have a different orientation in space. Then the motion is $2T$-periodic in a suitable rotating frame.

2. Lagrangian formulation in a rotating frame

We consider three bodies P_0, P_1, P_2 with masses m_0, m_1, m_2, respectively, moving in the space under their mutual gravitational attraction. We consider an inertial reference frame $OXYZ$ with origin at the center of mass O of the system and Z-axis parallel to the constant angular momentum M of the system.

We take a reference frame $QX'Y'Z'$ whose origin is at the center of mass of P_0 and P_1 and whose axes are always parallel to the $OXYZ$ axes; we consider also a new, rotating frame $Qxyz$ with origin Q and whose z-axis coincides with the Z'-axis. The rotating motion is such that the xz-plane always contains P_0 and P_1.

If R_i is the position vector of P_i in $OXYZ$, then the Lagrangian is given by

$$\mathcal{L} = \sum_{i=0}^{2} \frac{1}{2} m_i \|\dot{R}_i\|^2 + \sum_{0\le i<j\le 2} \frac{K^2 m_i m_j}{\|R_i - R_j\|}$$

where K^2 is the constant of gravitation. As is well-known, by using the linear momentum integral we can write the Lagrangian in the $QX'Y'Z'$ system as

$$\mathcal{L} = \frac{1}{2}\nu_1 \|\dot{r}_1\|^2 + \frac{1}{2}\nu_2 \|\dot{r}_2\|^2 + \sum_{0\le i<j\le 2} \frac{K^2 m_i m_j}{\|r_i - r_j\|} \tag{1}$$

where r_i, for $i = 1, 2$, is the position vector of P_i, ν_1, ν_2 are ficticious masses given by

$$\nu_1 = \frac{m_1(m_0 + m_1)}{m_0}, \qquad \nu_2 = \frac{m_2(m_0 + m_1)}{m_0 + m_1 + m_2},$$

and $r_0 = -(m_1/m_0)r_1$.

We select normalizing units in such a way that $m_0 + m_1 + m_2 = 1$ and $K^2 = 1$. In the rotating $Qxyz$ frame we take x_1, z_1, x_2, y_2, z_2, θ as generalized coordinates, where (x_1, z_1) is the position of P_1 in the xz-plane, (x_2, y_2, z_2) is the position of P_2 and θ is the angle between the x-axis and the X'-axis. Then the Lagrangian is given by

$$\begin{aligned} \mathcal{L} &= \frac{1}{2}\nu_1(\dot{x}_1^2 + x_1^2\dot{\theta}^2 + \dot{z}_1^2) \\ &\quad + \frac{1}{2}\nu_2(\dot{x}_2^2 + \dot{y}_2^2 + \dot{z}_2^2 + (x_2^2 + y_2^2)\dot{\theta}^2 + 2(x_2\dot{y}_2 - y_2\dot{x}_2)\dot{\theta}) \\ &\quad + \sum_{0\le i<j\le 2} \frac{m_i m_j}{r_{ij}}, \end{aligned} \tag{2}$$

where

$$r_{ij} = \sqrt{(x_i - x_j)^2 + (y_i - y_j)^2 + (z_i - z_j)^2},$$

with

$$y_1 = 0, \quad (x_0, y_0, z_0) = -(m_1/m_0)(x_1, y_1, z_1).$$

From this Lagrangian, expressions for the momenta $P_\sigma = \partial\mathcal{L}/\partial\dot{\sigma}$ can be obtained, where σ is any of the generalized coordinates. As θ is an ignorable coordinate, it follows that P_θ is a first integral. Actually $P_\theta = M_Z$, where M_Z stands for the Z-component of the angular momentum. For later use, we include the expressions for all the components of the angular momentum $M = (M_X, M_Y, M_Z)$

$$\begin{aligned}
M_X &= \nu_1[(x_1\dot{z}_1 - \dot{x}_1 z_1)\sin\theta - x_1 z_1\dot{\theta}\cos\theta] \\
&\quad +\nu_2[(x_2\dot{z}_2 - \dot{x}_2 z_2)\sin\theta + (y_2\dot{z}_2 - \dot{y}_2 z_2)\cos\theta \\
&\quad +z_2\dot{\theta}(y_2\sin\theta - x_2\cos\theta)], \\
M_Y &= \nu_1[(-x_1\dot{z}_1 + \dot{x}_1 z_1)\cos\theta - x_1 z_1\dot{\theta}\sin\theta] \\
&\quad +\nu_2[(-x_2\dot{z}_2 + \dot{x}_2 z_2)\cos\theta + (y_2\dot{z}_2 - \dot{y}_2 z_2)\sin\theta \\
&\quad -z_2\dot{\theta}(y_2\cos\theta + x_2\sin\theta)], \\
M_Z &= P_\theta = \nu_1 x_1^2\dot{\theta} + \nu_2[(x_2^2 + y_2^2)\dot{\theta} + x_2\dot{y}_2 - y_2\dot{x}_2].
\end{aligned}$$

It can be easily seen that the equations of motion are invariant under the symmetries

$$t \to -t,\ \theta \to -\theta,\ x_1 \to x_1,\ z_1 \to z_1,\ x_2 \to x_2,\ y_2 \to -y_2,\ z_2 \to z_2, \tag{3}$$

and

$$t \to -t,\ \theta \to -\theta,\ x_1 \to x_1,\ z_1 \to -z_1,\ x_2 \to x_2,\ y_2 \to -y_2,\ z_2 \to -z_2. \tag{4}$$

The symmetry (3) is a mirror reflection with respect to the xz-plane and (4) is a symmetry with respect to the x-axis, and the two symmetries can be combined to obtain three-dimensional periodic solutions which are symmetric with respect to the xz-plane and to the xy-plane. The following theorem holds (see Michalodimitrakis[2]).

Theorem 1 *Let $\Gamma(t)$ be a trajectory of the Lagrangian given by Eq. (2) such that $M_X = 0$, $M_Y = 0$. If $\Gamma(t)$ has initial conditions of the form*

$$\dot{x}_1(0) = 0, \dot{x}_2(0) = 0, y_2(0) = 0, z_2(0) = 0,$$

and after some time $t = T/4$ satisfies

$$\dot{x}_1(T/4) = 0, \dot{x}_2(T/4) = 0, y_2(T/4) = 0, \dot{z}_2(T/4) = 0,$$

then $\Gamma(t)$ *is periodic of period* T.

From the hypothesis $M_X = 0$, $M_Y = 0$ we have $z_1(0) = 0$, so that at time $t = 0$ the body P_1 is on the x-axis with zero velocity in the x direction, while P_2 is also on the x-axis with velocity perpendicular to this axis. By the same reason, at $t = T/4$ we have $\dot{z}_1 = 0$, so P_1 has zero velocity in its motion in the xz-plane while P_2 hits perpendicularly the same plane. The two symmetries applied to this piece of trajectory complete to a periodic solution. For more details about the proof see Michalodimitrakis[2].

3. Hamiltonian equations of motion

After the reduction of the linear momentum, the problem described by Lagrangian (1) is equivalent to the motion of two interacting particles P_1 and P_2 of mass ν_1 and ν_2 in a central force field. Let (q, ϕ, θ) and (r, β, γ) be, respectively, the spherical coordinates of P_1 and P_2, and let $\nu_1 = \mu$, $\nu_2 = \alpha\mu$, where μ is a small parameter and α is of the order of unity. Then the Hamiltonian of the problem, in the reference frame $OX'Y'Z'$ of Section 2, is

$$\mathcal{H} = \mathcal{H}_1 + \mathcal{H}_2 + \mu\mathcal{V}, \tag{5}$$

where $\mathcal{V}$ is the perturbing potential due to the gravitational interaction between P_1 and P_2 and $\mathcal{H}_1$, $\mathcal{H}_2$ are the Hamiltonians of two uncoupled Kepler problems, given by

$$\begin{aligned} \mathcal{H}_1 &= \frac{1}{2\nu_1}\left(p_q^2 + \frac{p_\phi^2}{q^2} + \frac{p_\theta^2}{q^2\cos^2\phi}\right) - \frac{\nu_1}{q}, \\ \mathcal{H}_2 &= \frac{1}{2\nu_2}\left(p_r^2 + \frac{p_\beta^2}{r^2} + \frac{p_\gamma^2}{r^2\cos^2\beta}\right) - \frac{\nu_2}{r}. \end{aligned}$$

When $\mu = 0$ we denote by $M_{(i)} = (M_{(i)X}, M_{(i)Y}, M_{(i)Z})$ the angular momentum of P_i and by M_i its modulus. The following relations are readily found between the components of the angular momentum and the momenta conjugate to the spherical coordinates

$$M_{(1)X} = p_\phi \sin\theta - p_\theta \tan\phi \cos\theta, \tag{6}$$
$$M_{(1)Y} = -p_\phi \cos\theta - p_\theta \tan\phi \sin\theta, \tag{7}$$
$$M_{(1)Z} = p_\theta; \tag{8}$$

similar expressions can be found for $M_{(2)}$.

Given a Hamiltonian $H(q_i, p_i)$, a canonical transformation of multiplier μ is a change of variables that is the identity on the coordinates q_i and scales the momenta p_i by a factor $1/\mu$, i.e.

$$\bar{q}_i = q_i, \quad \bar{p}_i = p_i/\mu;$$

then the new Hamiltonian $\bar{H}$ is given by

$$\bar{H}(\bar{q}_i, \bar{p}_i) = \frac{1}{\mu} H(\bar{q}_i, \mu \bar{p}_i).$$

If we apply this transformation to the Hamiltonian (5) and drop the bars, we get the same expression as given in Eq. (5) with $\mathcal{H}_1$ and $\mathcal{H}_2$ given by

$$\begin{aligned}
\mathcal{H}_1 &= \frac{1}{2}\left(p_q^2 + \frac{p_\phi^2}{q^2} + \frac{p_\theta^2}{q^2 \cos^2 \phi}\right) - \frac{1}{q}, \\
\mathcal{H}_2 &= \frac{1}{2\alpha}\left(p_r^2 + \frac{p_\beta^2}{r^2} + \frac{p_\gamma^2}{r^2 \cos^2 \beta}\right) - \frac{\alpha}{r}.
\end{aligned}$$

In order to write the Hamiltonian in the rotating coordinates of Section 2, we must perform the change of variables

$$\begin{aligned}
\bar{\gamma} &= \gamma - \theta, \\
\bar{\theta} &= \theta,
\end{aligned}$$

which is just the classical elimination of the node. The transformation is the identity on all the coordinates except that the angle $\gamma - \theta$ substitutes the former γ. In order to keep the Hamiltonian character of the equations, the conjugate momenta must transform according to (see Szebehely[3])

$$\begin{aligned}
p_\gamma &= p_{\bar{\gamma}}, \\
p_\theta &= p_{\bar{\theta}} - p_{\bar{\gamma}}.
\end{aligned}$$

The bars are dropped again for commodity of notation, but it must be kept in mind that, although the new variable θ is the same as before, its conjugate momentum p_θ is now the Z-component of the total angular momentum of the system.

We take now a combination of Poincaré and Delaunay variables to describe the motion of P_2. Let ℓ, g, v, be the usual Delaunay elements (mean anomaly, longitude of the pericenter and longitude of the node), and L, G, V its conjugate momenta. We recall that $G = M_2$ and $V = M_{(2)Z}$ and that Delaunay elements are related to the orbital elements through

$$\begin{aligned}
L &= \alpha a^{1/2} \\
G &= \alpha a^{1/2}(1 - e^2)^{1/2} \\
V &= \alpha a^{1/2}(1 - e^2)^{1/2} \cos J
\end{aligned}$$

where a, e and J are, respectively, the major semiaxis, the eccentricity and the inclination of the orbit; in these variables, the Hamiltonian for P_2 is given by

$$\mathcal{H}_2 = -\frac{\alpha^3}{2L^2}.$$

Let us introduce new variables Y_1, Y_2, Y_3, and conjugate momenta X_1, X_2, X_3 defined by

$$\begin{aligned}
X_1 &= L - G + V \\
X_2 &= \sqrt{2(L-G)}\cos(g+v) \\
X_3 &= G - V \\
Y_1 &= l + g + v \\
Y_2 &= -\sqrt{2(L-G)}\sin(g+v) \\
Y_3 &= l + g,
\end{aligned}$$

(see Szebehely[3] and Jefferys[1]); in terms of the new variables, the classical Delaunay elements are given by

$$L = X_1 + X_3, \tag{9}$$
$$G = X_1 + X_3 - \tfrac{1}{2}(X_2^2 + Y_2^2), \tag{10}$$
$$V = X_1 - \tfrac{1}{2}(X_2^2 + Y_2^2), \tag{11}$$

and we finally obtain the following expression for the Hamiltonian of the problem

$$\mathcal{H} = \frac{1}{2}\left[p_q^2 + \frac{p_\phi^2}{q^2} + \frac{\left(p_\theta - X_1 + \frac{1}{2}(X_2^2 + Y_2^2)\right)^2}{q^2\cos^2\phi}\right] - \frac{1}{q} - \frac{\alpha^3}{2L^2} + \mu V. \tag{12}$$

In these variables, the conditions of symmetry stated in Theorem 1 are

$$Y_1(0) = i\pi,\ Y_2(0) = 0,\ Y_3(0) = j\pi, \tag{13}$$
$$\dot{q}(0) = 0, \tag{14}$$
$$Y_1(T/4) = (i+k)\pi,\ Y_2(T/4) = 0,\ Y_3(T/4) = (j + m + \tfrac{1}{2})\pi, \tag{15}$$
$$\dot{q}(T/4) = 0, \tag{16}$$

where i, j, k, m are positive integers. The conditions of symmetry stated in Eq. (13) and Eq. (15) are the same found in Jefferys[1] for the restricted case, which in the general case must be supplemented with Eq. (14) and Eq. (16). The equations of motion corresponding to Hamiltonian (12) are

$$\dot{q} = p_q + O(\mu), \tag{17}$$

$$\dot{p}_q = -\frac{1}{q^2} + \frac{p_\phi^2}{q^3} + \frac{\left[p_\theta - X_1 + \frac{1}{2}(X_2^2 + Y_2^2)\right]^2}{q^3 \cos^2\phi} + O(\mu), \tag{18}$$

$$\dot{\phi} = \frac{p_\phi}{q^2} + O(\mu), \tag{19}$$

$$\dot{p}_\phi = -\frac{\left[p_\theta - X_1 + \frac{1}{2}(X_2^2 + Y_2^2)\right]^2}{q^3 \cos^3\phi} \sin\phi + O(\mu), \tag{20}$$

$$\dot{X}_1 = 0 + O(\mu), \tag{21}$$

$$\dot{X}_2 = \frac{p_\theta - X_1 + \frac{1}{2}(X_2^2 + Y_2^2)}{q^3 \cos^2\phi} Y_2 + O(\mu), \tag{22}$$

$$\dot{X}_3 = 0 + O(\mu), \tag{23}$$

$$\dot{Y}_1 = \alpha^3 (X_1 + X_3)^{-3} - \frac{p_\theta - X_1 + \frac{1}{2}(X_2^2 + Y_2^2)}{q^3 \cos^2\phi} + O(\mu), \tag{24}$$

$$\dot{Y}_2 = -\frac{p_\theta - X_1 + \frac{1}{2}(X_2^2 + Y_2^2)}{q^3 \cos^2\phi} X_2 + O(\mu), \tag{25}$$

$$\dot{Y}_3 = \alpha^3 (X_1 + X_3)^{-3} + O(\mu). \tag{26}$$

4. Flow near the circular Keplerian orbits

For $\mu = 0$ we consider a family of solutions in which P_1 moves in a circular orbit of radius one, P_2 moves in a resonant circular orbit of radius L^{*2}/α^2 in such a way that the conditions in Eq. (13) and Eq. (15) hold. The orbital planes of P_1 and P_2 in the $OX'Y'Z'$ system intersect along the X'-axis, so that $M_X = 0$, and the angle between these planes is the parameter along the family. We can still choose the inclination of the orbit of one of the bodies, say P_2, and we do so in such a way that $M_Y = 0$, hence the angular momentum vector of the system is parallel to the Z-direction.

In order to show the existence of a family of periodic solutions when $\mu \neq 0$, which for $\mu \to 0$ tends to the former Keplerian family, we make use of the usual analytic continuation argument applied to the symmetry conditions in Eq. (13)–(16). We expand the solution of the equations of motion in powers of small increments of the initial conditions near the above mentioned family and show that a non-vanishing Jacobian can be found; as all the functions involved are analytic, it is sufficient if we show this for $\mu = 0$. We stress the fact that for Theorem 1 to hold it is necessary that $M_X = 0$ and $M_Y = 0$; the expression for the angular momentum, however, does not depend on μ, so we need only make sure that we take the appropriate initial conditions.

Let L^* be a value of L such that, for $\mu = 0$, P_2 is in a given resonance with P_1. If J is the inclination of the orbital plane of P_2 then

$$X_3(0) = G - V = G(1 - \cos J),$$

and $X_3(0) = X_{30}$ is taken as a freely varying parameter. Notice that $X_{30} \simeq 0$ when the inclination of this plane is small. The initial value of X_1 is taken to be

$$X_1(0) \quad = \quad L^* - X_{30} + \delta X_{10},$$

δX_{10} being small. Then $L = X_1 + X_2 = L^* + \delta X_{10}$ and the orbital radius of P_2 is allowed to vary slightly in order to obtain a periodic solution. Finally, we take $X_2(0) = \delta X_{20}$, a small quantity. Then, the modulus M_2 of the angular momentum of P_2 is given by

$$M_2 = G = X_1 + X_3 - \frac{1}{2}(X_2^2 + Y_2^2) = L^* + \delta X_{10} - \frac{1}{2}\delta X_{20}^2, \tag{27}$$

and these initial values give an almost circular orbit because $G \simeq L^*$.

The initial values of the Y_j variables must be

$$Y_1(0) = i\pi, Y_2(0) = 0, Y_3(0) = j\pi, \tag{28}$$

for the hypothesis of Theorem 1 to hold; with these initial conditions, we have $M_{(2)X} = 0$. As for the initial values for the body P_1, we must take

$$\begin{aligned} p_q(0) &= 0, \\ \phi(0) &= 0, \end{aligned}$$

because of the symmetry conditions ($\phi(0) = 0$ meaning that P_1 is on the x-axis). We take also $\theta(0) = 0$, so that $M_{(1)X} = 0$, and take $q(0) = 1 + \delta q_0$, δq_0 small.

We now choose $p_\phi(0)$ in such a way that M_Y vanishes. We have from Eq. (6)–(8)

$$\begin{aligned} M_Y &= M_{(1)Y} + M_{(2)Y} \\ &= -p_\phi \cos\theta - (p_\theta - p_\gamma)\frac{\sin\phi}{\cos\phi}\sin\theta + \sqrt{M_2^2 - M_{(2)Z}^2} \end{aligned}$$

and, as $\theta(0) = 0$, M_2 is given by Eq. (27) and $M_{(2)Z}$ by Eq. (11), we take

$$p_\phi(0) \quad = \quad \sqrt{X_3(2L^* + 2\delta X_{10} - \delta X_{20}^2)}.$$

We finally choose M_Z in such a way that $M_1 = 1$. We have

$$M_1^2 = M_{(1)Y}^2 + M_{(1)Z}^2,$$

so we must see that

$$\begin{aligned} M_{(1)Z} &= \sqrt{1 - M_{(1)Y}^2} \\ &= \sqrt{1 - X_3(2L^\star + 2\delta X_{10} - \delta X_{20}^2)}. \end{aligned} \tag{29}$$

Actually $M_{(1)Z}$ is not a variable of Hamiltonian (12) but $p_\theta = M_{(1)Z} + M_{(2)Z}$, so we choose p_θ in such a way that Eq. (29) holds.

Let ψ be the angle of position of P_1 in its orbital plane, measured from the X'-axis, and p_ψ the associated momentum (i.e. the angular momentum if we consider its motion restricted to a plane); then Eq. (17) and Eq. (18) can be written

$$\dot{q} = p_q, \tag{30}$$

$$\dot{p}_q = -\frac{1}{q^2} + \frac{1}{q^3}\left(p_\phi^2 + \frac{M_{(1)Z}^2}{\cos^2\phi}\right) = -\frac{1}{q^2} + \frac{p_\psi^2}{q^3}, \tag{31}$$

Because of the chosen initial values we have $p_\psi \equiv 1$ and, expanding the solutions of Eq. (30) and Eq. (31) with $p_q(0) = 0$ we get

$$\delta q = \delta q_0 \cos t + O_2(\delta q_0), \tag{32}$$

$$p_q = -\delta q_0 \sin t + O_2(\delta q_0), \tag{33}$$

where $\delta q = q - 1$ is a new variable, $\delta q_0 = \delta q(0)$ and O_2 stands for second order terms. On the other side, since $p_\psi = q^2\dot{\psi}$, we have

$$\begin{aligned} \dot{\psi} &= \frac{1}{(1+\delta q)^2} \\ &= -1 - 2\delta q + O_2(\delta q), \end{aligned}$$

and

$$\psi = t - 2\delta q_0 \cos t + O_2(\delta q_0),$$

because $\psi(0) = 0$.

Let ζ be the inclination of the orbital plane of P_1 in the $OX'Y'Z'$ system. It is clear that

$$\sin\phi = \sin\psi \sin\zeta,$$

and, as ζ is also the angle between the Z'-axis and the angular momentum vector of P_1, we have

$$\begin{aligned}\sin\zeta &= \sqrt{1 - M_{(1)Z}^2} \\ &= \sqrt{X_3(2L^* + 2\delta X_1 - \delta X_2^2)}\end{aligned}$$

and

$$\begin{aligned}\cos\phi &= \sqrt{1 - \sin^2\zeta \sin^2\psi} \\ &= \sqrt{1 - k^2 \sin^2(t - 2\delta q_0 + O_2(\delta q_0))},\end{aligned}$$

where $k = \sin\zeta$.

The equations of motion (21)–(26) for P_2 may be written as

$$\dot{X}_1 = 0, \tag{34}$$

$$\dot{X}_2 = \frac{M_{(1)Z}}{q^2\cos^2\phi} Y_2, \tag{35}$$

$$\dot{X}_3 = 0, \tag{36}$$

$$\dot{Y}_1 = \alpha^3 (X_1 + X_3)^{-3} \frac{M_{(1)Z}}{q^2\cos^2\phi}, \tag{37}$$

$$\dot{Y}_2 = -\frac{M_{(1)Z}}{q^2\cos^2\phi} X_2, \tag{38}$$

$$\dot{Y}_3 = \alpha^3 (X_1 + X_3)^{-3}, \tag{39}$$

where $M_{(1)Z}/(q^2\cos^2\phi)$ is a given function of time

$$\frac{M_{(1)Z}}{q^2\cos^2\phi} = \frac{\sqrt{1-k^2}}{1 - k^2\sin^2 t} + O(\delta q_0),$$

and k^2 is a function of the initial values

$$k^2 = X_{30}(2L^* + 2\delta X_{10} - \delta X_{20}^2).$$

The solution of equations (34)–(39), recalling that $Y_3(0) = 0$ and $X_1(0) + X_3(0) = L^* + \delta X_{10}$, is then

$$X_1 = X_{10}, \tag{40}$$

$$X_2 = \delta X_{20} \cos\left(\int_0^t \frac{\sqrt{1-k^2}\,d\tau}{1 - k^2\sin^2\tau + O(\delta q_0)}\right), \tag{41}$$

$$X_3 = X_{30}, \tag{42}$$

$$\begin{aligned}
Y_1 &= \alpha^3(L^* + \delta X_{10})^{-3}t - \int_0^t \frac{\sqrt{1-k^2}d\tau}{1 - k^2\sin^2\tau + O(\delta q_0)} + i\pi, && (43)\\
Y_2 &= -\delta X_{20}\sin\left(\int_0^t \frac{\sqrt{1-k^2}d\tau}{1 - k^2\sin^2\tau + O(\delta q_0)}\right), && (44)\\
Y_3 &= \alpha^3(L^* + \delta X_{10})^{-3}t + j\pi. && (45)
\end{aligned}$$

The motion of P_2 in the rotating system is the same as in Jefferys[1] except for the fact that the rotation of the node is not uniform but the projection of an inclined uniform rotation.

5. Existence of periodic solutions

The following theorem shows the existence of the periodic orbits mentioned in Section 4.

Theorem 2 *Let us consider a family of circular Keplerian solutions in which P_1 moves in a circular orbit of radius one and P_2 in a circular orbit of radius L^{*2}/α^2, the inclination of P_1 being arbitrary and the inclination of P_2 such that $M_X = 0$, $M_Y = 0$. Let the periodicity conditions Eq. (13)–(16) be satisfied with $\frac{1}{4}(n-1)T = k\pi$ and $\frac{1}{4}nT = (m+\frac{1}{2})\pi$, where $n = \alpha^3 L^{\star-3}$. Then, for $\mu \neq 0$ sufficiently small, this family can be analytically continued to a solution of the general 3-body problem by slightly varying T, δX_{10}, δX_{20}, and δq_0.*

We need only show that the Jacobian determinant

$$\frac{\partial(p_q, Y_1, Y_2, Y_3)}{\partial(\delta q_0, t, \delta X_{10}, \delta X_{20})}, \qquad (46)$$

where p_q is given by Eq. (33) and X_1, X_2 and X_3 are given by Eq. (43)–(45), evaluated at the point

$$t^* = T/4,\ \delta q_0 = 0,\ \delta X_{20} = 0,\ \delta X_{10} = 0,$$

where T is the Keplerian period, does not vanish.

The entries in the Jacobian (46) are shown in the table below

	$\dfrac{\partial}{\partial \delta q_0}$	$\dfrac{\partial}{\partial t}$	$\dfrac{\partial}{\partial \delta X_{10}}$	$\dfrac{\partial}{\partial \delta X_{20}}$
p_q	$-\sin t^*$	$-\delta q_0 \cos t^* = 0$	0	0
Y_1	$\star$	$\alpha^3 L^{\star-3} - \dfrac{\sqrt{1-k^2}}{1-k^2\sin^2 t^\star}$	$-3\alpha^3 L^{\star-4} t^\star$ $-\dfrac{\omega_1(t^*)}{1-k^2}\dfrac{\partial k^2}{\partial \delta X_{10}}$ $-\omega_2(t^*)\dfrac{\partial k^2}{\partial \delta X_{10}}$	0
Y_2	$\star$	$\star$	$\star$	$-\sin \omega_1(t^*)$
Y_3	$\star$	$\alpha^3 L^{\star-3}$	$-3\alpha^3 L^{\star-4} t^\star$	0

where

$$\begin{aligned}
\frac{\partial k^2}{\partial \delta X_{10}} &= 2X_{30}, \\
\omega_1(t^*) &= \int_0^{t^*} \frac{\sqrt{1-k^2}\, d\tau}{1-k^2\sin^2\tau}, \\
\omega_2(t^*) &= \int_0^{t^*} \frac{\sqrt{1-k^2}\, \sin^2\tau\, d\tau}{(1-k^2\sin^2\tau)^2}.
\end{aligned}$$

As $\sin t^* = 1$, this determinant reduces (except for a sign) to

$$\frac{\partial(Y_1, Y_2, Y_3)}{\partial(t, \delta X_{10}, \delta X_{20})}; \tag{47}$$

if we now take X_{30} moderately small, then

$$\sqrt{1-k^2} \simeq 1, \quad \frac{\partial k^2}{\partial \delta X_{10}} \simeq 0, \quad k^2 \simeq 0, \quad \omega_1(t) \simeq t,$$

and Eq. (47) is readily seen to be different from zero, because it is close to the determinant used by Jefferys[1] in the restricted case. As X_{30} increases, it is possible that Eq. (47) vanishes for some particular values of the parameters, giving rise to bifurcation phenomena.

6. References

1. W. H. Jefferys, *Astron. J.*, **70** (1965) 393-394.
2. M. Michalodimitrakis, *Celest. Mech.*, **19** (1979) 263-277.
3. H. Poincaré, *Bulletin Astronomique*, **1** (1884) 65-74.
4. V. Szebehely, *Theory of Orbits*, (Academic Press, New York, 1967).

ON PERTURBATIONS OF HAMILTONIAN SYSTEMS GENERATED BY CONTRACTIONS OF LIE ALGEBRAS*

YURII VOROBJEV
Department of Applied Mathematics
Moscow Institute of Electronics and Mathematics
Moscow 109028, Russia
E-mail: vorob@aplmat.miemstu.msk.su

and

RUBEN FLORES ESPINOZA
Department of Mathematics, University of Sonora
C.P.83000 Hermosillo, Sonora, Mexico
E-mail: rflorese@fisica.uson.mx

ABSTRACT

We consider the class of perturbations of Euler equations generated by contractions of Lie algebras. In the case of the contraction of the Lie algebra $\mathbf{so}(4)$ into $\mathbf{e}(3)$ we investigate the connection between flows of corresponding Euler equations on $\mathbf{so}(4)^*$ and $\mathbf{e}(3)^*$ (with one and the same Hamiltonian function) when the contraction parameter is sufficiently small. For this purpose we use a homotopy of the Lie–Poisson structures on $\mathbf{e}(3)^*$ and $\mathbf{so}(4)^*$.

1. Introduction

Consider a dynamical system in $\mathbb{R}^6$,

$$\begin{aligned} \frac{dy}{dt} &= y \times \frac{\partial H}{\partial y} + z \times \frac{\partial H}{\partial z}, \\ \frac{dz}{dt} &= z \times \frac{\partial H}{\partial y} + \varepsilon y \times \frac{\partial H}{\partial z}, \end{aligned} \tag{1.1}$$

where $y = (y_1, y_2, y_3), z = (z_1, z_2, z_3) \in \mathbb{R}^3$, the sign $\times$ denotes the vector product in $\mathbb{R}^3$, $H = H(y, z) \in C^\infty(\mathbb{R}^6)$, $\varepsilon \geq 0$ is a parameter. As is known, system (1.1) is a *Hamiltonian* system with respect to the *Lie–Poisson bracket* on $\mathbb{R}^6$

$$\{f, g\}_\varepsilon = \left\langle \frac{\partial f}{\partial y}, \frac{\partial g}{\partial y} \times y + \frac{\partial g}{\partial z} \times z \right\rangle + \left\langle \frac{\partial f}{\partial z}, \frac{\partial g}{\partial y} \times z + \varepsilon \frac{\partial g}{\partial z} \times y \right\rangle \tag{1.2}$$

*Research partially supported by the Consejo Nacional de Ciencia y Technologia of Mexico under grants 489 100-2 and 489 100-5-4106 E.

and with respect to the Hamiltonian function H.

If $\varepsilon > 0$, system (1.1) is known as the *Euler equations* asociated with the Lie algebra $\mathbf{so}(4)$. For $\varepsilon = 0$, system (1.1) is reduced to the Euler equations

$$\begin{aligned} \frac{dy}{dt} &= y \times \frac{\partial H}{\partial y} + z \times \frac{\partial H}{\partial z}, \\ \frac{dz}{dt} &= z \times \frac{\partial H}{\partial y}, \end{aligned} \tag{1.3}$$

associated with the Lie algebra $\mathbf{e}(3)$ of the *Euclidean* motion group in $\mathbb{R}^3$. System (1.3) is a Hamiltonian system with respect to the limiting Lie–Poisson bracket $\{,\}_0 = \lim_{\varepsilon\to 0}\{,\}_\varepsilon$ and to the same Hamiltonian function H. Note that the ε-regular family of Lie–Poisson brackets (1.2) is generated by the well-known procedure of the *contraction* of the Lie algebra $\mathbf{so}(4)$ into $\mathbf{e}(3)$ (see[1,2] and Section 3).

The Euler equations (1.1), (1.3) arise in various problems of classical mechanics, in particular, in the description of rigid body motion and spin dynamics. The integrability of (1.1), (1.3) is well understood for quadratic functions H (about integrable tops, for example, see[2–5,14–16]).

We shall suppose that system (1.3) is integrable, and some domain of $\mathbb{R}^6$ is fibrated by 2-dimensional invariant tori. Under these assumptions we are interested in the qualitative behavior of the flow of system (1.1) as $\varepsilon \to 0$. For this purpose, system (1.1) can be considered as a perturbation of (1.3) with *perturbation vector field*

$$P_H = \varepsilon\left\langle y \times \frac{\partial H}{\partial z}, \frac{\partial}{\partial z}\right\rangle. \tag{1.4}$$

In general, P_H is a *non-Hamiltonian field* with respect to the Lie–Poisson bracket $\{,\}_0$. This statement is certainly true when the function H satisfies, for example, the following condition

$$\left\langle \frac{\partial H}{\partial z}, y \times z \right\rangle \neq 0. \tag{1.5}$$

So, we have the non-Hamiltonian perturbation P_H of the Hamiltonian system (1.3). Hence, we cannot directly apply the *KAM theorem*[5,6] to this situation.

Note that the behavior of invariant quasiperiodic tori under general perturbations (in particular, non-Hamiltonian) was investigated in[7]. An application of the KAM theorem to a class of non-Hamiltonian perturbations of a system with central forces was considered in[8].

We shall follow the idea proposed in[9,10] to separate the application of the *averaging method* for systems of type (1.1) into two steps. The first step concerns the *deformation* of the Lie–Poisson bracket $\{,\}_0 \to \{,\}_\varepsilon$. The second step is the application of the averaging method to a system like (1.3), Hamiltonian with respect to the bracket $\{,\}_0$, but generated by a "new" Hamiltonian function H_ε instead of H.

More precisely, below in Section 4 we construct a *homotopy* γ_ε of the Lie–Poisson structures $\{,\}_0$ and $\{,\}_\varepsilon$ using the general approach[9,10]. A homotopy γ_ε allows us to reduce the original system (1.1) to the Euler equations (1.3) corresponding to the *perturbed Hamiltonian function* $H_\varepsilon = H \circ \gamma_\varepsilon$, $H_0 \equiv H$. So, the perturbation field P_H (1.4) is transformed into the Hamiltonian field with respect to the Lie–Poisson bracket $\{,\}_0$ corresponding to the function $H_\varepsilon - H = O(\varepsilon)$.

In the second step, we apply the KAM theorem to the Euler equations (1.3) corresponding to the perturbed Hamiltonian function H_ε and obtain information about the existence of invariant tori with respect to the flow of (1.1) (see Section 5).

At first, we introduce the necessary notions in Section 2 and then investigate the Euler equations by means of contractions of Lie algebras in Section 3.

2. Main Notions

Here we recall the necessary facts from the Poisson geometry (for more details, see[2–4,11–13]).

Let $\mathcal{N}$ be a manifold. An antisymmetric bilinear operation $\{,\} : C^\infty(\mathcal{N}) \times C^\infty(\mathcal{N}) \to C^\infty(\mathcal{N})$ is called a *Poisson bracket* (*structure*) if it satisfies the *Jacobi identity* and the *Leibnitz rule*. In this case $\mathcal{N}$ is called a *Poisson manifold.* A vector field v_H on $\mathcal{N}$ is called a *Hamiltonian* field of a function $H \in C^\infty(\mathcal{N})$ if $v_H(f) \equiv \langle df, v_H \rangle = \{H, f\}$ for all $f \in C^\infty(\mathcal{N})$. A *Casimir function* on $\mathcal{N}$ is a function k such that $\{k, f\} = 0$ for all $f \in C^\infty(\mathcal{N})$, equivalent condition is that k is constant along the trajectories of all Hamiltonian fields.

In local coordinates $\xi = (\xi^1, \ldots, \xi^n)$ on $\mathcal{N}$ the Poisson bracket has the form

$$\{f, g\} = \sum_{ij} \Psi^{ij}(\xi) \frac{\partial f}{\partial \xi^i} \frac{\partial g}{\partial \xi^j}, \tag{2.1}$$

where the functions $\Psi^{ij}(\xi)$ are the components of an antisymmetric contravariant 2-tensor field on $\mathcal{N}$

$$\Psi : T^*\mathcal{N} \to T\mathcal{N}, \qquad \Psi(\xi) = ((\Psi^{ij}(\xi)))$$

satisfying the Jacobi identity. In these terms every Hamiltonian field v_H has the representation

$$v_H = -\sum_{ij} \Psi^{ij}(\xi) \frac{\partial H}{\partial \xi^j} \frac{\partial}{\partial \xi^i}. \tag{2.2}$$

A Poisson structure for which rank $\Psi(\xi)$ everywhere equal to the dimension of $\mathcal{N}$ is called *nondegenerate* and $\mathcal{N}$ is called *symplectic*. Otherwise, a Poisson structure is called *degenerate*.

By the Jacobi identity it follows that the plane distribution Ker $\Psi(\xi)^{\perp} \subset T_\xi\mathcal{N}$, $\xi \in \mathcal{N}$ is *integrable* in the sense of Frobenius. Hence, an integral manifold $\Omega(\xi)$ of this distribution passes through each point $\xi \in \mathcal{N}$. This manifold is called a *symplectic leaf* on $\mathcal{N}$. The dimension of $\Omega = \Omega(\xi)$ equals rank $\Psi(\xi)$ and the *symplectic structure* on Ω is given by the *Kirillov form* ω_Ω which is compatible with the Poisson bracket in the following way

$$\omega_\Omega(v_f, v_g) = \{f, g\}.$$

Note that Casimir funcitons are constant along each symplectic leaf. Every Hamiltonian field on $\mathcal{N}$ is *tangent* to symplectic leaves everywhere. So, the restriction of a Hamiltonian field v_H to a symplectic leaf Ω gives the Hamiltonian system with $(\frac{1}{2}\dim\Omega)$-degrees of freedom.

A point $\xi \in \mathcal{N}$ is *regular* if rank $\Psi(\xi)$ is *maximal*. A symplectic leaf passing through a regular point is also called regular and has the maximal dimension.

If we have two manifolds $\mathcal{N}_0$ and $\mathcal{N}$ with Poisson brackets $\{,\}_0$ and $\{,\}$, then a smooth mapping $\gamma : \mathcal{N}_0 \to \mathcal{N}$ is a *Poisson mapping* if $\{f \circ \gamma, g \circ \gamma\} = \{f, g\} \circ \gamma$ for all $f, g \in C^\infty(\mathcal{N})$. Obviously, if γ is a diffeomorphism, then γ^{-1} is also a Poisson mapping.

A vector field w on $\mathcal{N}$ is called a *Poisson field* (or an *infinitesimal Poisson automorphism*) if its (local) flow is a Poisson mapping. Note that every Hamiltonian field on $\mathcal{N}$ is a Poisson field. But there may exist Poisson fields *transversal* to symplectic leaves (for example, see[10,11]).

A *Poisson submanifold* is a *submanifold* $\mathcal{M}$ in a Poisson manifold $\mathcal{N}$ with a Poisson bracket for which the inclusion is a Poisson mapping. Such a bracket, if it exists, is unique. Let $\mathcal{M}$ be a submanifold of $\mathcal{N}$ such that for every symplectic leaf Ω of $\mathcal{N}$ the intersection $\mathcal{M} \cap \Omega$ is a symplectic submanifold of Ω. Then $\mathcal{M}$ is a Poisson submanifold of $\mathcal{N}$. In particular, if $\dim\mathcal{M} = \dim\mathcal{N}$, then this condition holds and $\mathcal{M}$ is a Poisson submanifold. In this case, for any $H \in C^\infty(\mathcal{N})$, each trajectory on $\mathcal{M}$ of the Hamiltonian filed v_H coincides with a trajectory of the Hamiltonian field on $\mathcal{M}$ with Hamiltonian function $H|_{\mathcal{M}}$.

A family of Poisson brackets $\{,\}_\varepsilon$ on $\mathcal{N}$ smoothly depending on a parameter $\varepsilon \geq 0$ is a *deformation* of a given Poisson bracket $\{,\}$ on $\mathcal{N}$ if $\{,\}_0 \equiv \{,\}$. Let $\mathcal{M}$ be a submanifold of $\mathcal{N}$ and $\dim \mathcal{M} = \dim \mathcal{N}$. Then a family of mappings $\gamma_\varepsilon : \mathcal{M} \to \mathcal{N}$ smoothly depending on ε is called a *homotopy* of Poisson brackets $\{,\}_0$ and $\{,\}_\varepsilon$ if there exists $\delta > 0$ such that the following conditions hold

(i) for every $\varepsilon \in [0, \delta]$ the mapping γ_ε is a diffeomorphism onto its image $\mathcal{M}_\varepsilon = \gamma_\varepsilon(\mathcal{M})$ and $\gamma_0 = id_{\mathcal{M}}$ is the identity mapping of $\mathcal{M}$;

(ii) for every $\varepsilon \in [0, \delta]$ the mapping γ_ε is Poisson,

$$\{f \circ \gamma_\varepsilon, g \circ \gamma_\varepsilon\}_0 = \{f, g\}_\varepsilon \circ \gamma_\varepsilon, \qquad \forall f, g \in C^\infty(\mathcal{N}). \tag{2.3}$$

In terms of components $(\gamma_\varepsilon^i(\xi))$ and $(\Psi_\varepsilon^{ij}(\xi))$ of the mapping γ_ε and the 2-tensor Ψ_ε of $\{,\}_\varepsilon$ corresponding to some local coordinates on $\mathcal{N}$, condition (2.3) can be written as

$$\left\{\gamma_\varepsilon^i(\xi), \gamma_\varepsilon^j(\xi)\right\}_0 = \Psi_\varepsilon^{ij}(\gamma_\varepsilon(\xi)), \qquad \xi \in \mathcal{M}, \quad i, j = 1, \dots, \dim \mathcal{N}. \tag{2.3$'$}$$

Denote by v_H^ε the Hamiltonian field of $H \in C^\infty(\mathcal{N})$ with respect to the Poisson bracket $\{,\}_\varepsilon$, $\varepsilon \geq 0$.

Proposition 2.1. *Let γ_ε be a homotopy of the Poisson brackets $\{,\}_0$ and $\{,\}_\varepsilon$. Then for every $H \in C^\infty(\mathcal{N})$ the trajectories on $\mathcal{M}_\varepsilon$ of the Hamiltonian field v_H^ε are the projections under γ_ε of the trajectories on $\mathcal{M}$ of $v_{H_\varepsilon}^0$. Here*

$$H_\varepsilon \stackrel{\text{def}}{=} H \circ \gamma_\varepsilon = H + \varepsilon\langle dH, \Gamma\rangle + O(\varepsilon^2) \qquad (\text{as } \varepsilon \to 0) \tag{2.4}$$

and the vector field Γ on $\mathcal{M}$ is an infinitesimal generator of γ_ε,

$$\Gamma(m) = \frac{d}{d\varepsilon}\bigg|_{\varepsilon=0} \gamma_\varepsilon(m), \qquad m \in \mathcal{M}. \tag{2.5}$$

Moreover, such correspondence between trajectories is one-to-one.

Remark 2.1. If a homotopy γ_ε is globally defined, i.e., $\mathcal{M} = \mathcal{M}_\varepsilon = \mathcal{N}$, then symplectic leaves corresponding to the Poisson structures $\{,\}_0$ and $\{,\}_\varepsilon$ must be symplectomorphic. But in some interesting cases, it is not true. For example, the symplectic leaves of Lie–Poisson structures (1.2) for $\varepsilon > 0$ and for $\varepsilon = 0$ have the different topological structure (see Section 4).

Note that the problem of finding a homotopy γ_ε can be reduced to solving the linear equations with Schouten bracket[9,10]. We shall discuss these equations in the case of Lie–Poisson brackets in Section 3.

Let $\mathcal{N} = \mathbb{R}^n$ be a vector space with some coordinates $\xi = (\xi^1, \ldots, \xi^n)$. Then formula (2.1) gives the global representation of a given Poisson bracket. If $\Psi^{ij}(\xi) = \sum_s \lambda^{ijs} \xi^s$ are linear functions, then we obtain an important class of (*degenerate*) Poisson brackets on $\mathbb{R}^n$ which are called the *Lie–Poisson brackets* associated to the Lie algebra g with structure constants λ^{ijs}.

In this case symplectic leaves coincide with *coajoint orbits* (orbits of the *coadjoint action* of the Lie group corresponding to g). Dynamical systems in $\mathbb{R}^n$ generated by a Hamiltonian field v_H (2.2) relative to a Lie–Poisson structure are just *Euler equations*.

3. Euler Equations and Contractions of Lie Algebras

Any contraction of a Lie algebra generates a family of Lie–Poisson brackets. This family can be considered as a deformation of the limiting bracket. The natural problem is to reduce the dynamics of the original Euler equations to the dynamics of the limiting equations in some natural sense. For this purpose we can choose various contractions of a given Lie algebra.

Let g be a real Lie algebra. Suppose that g splits into the direct sum of two subspaces $\mathcal{Y}$ and $\mathcal{Z}$,

$$g = \mathcal{Y} \oplus \mathcal{Z}. \tag{3.1}$$

For every $\varepsilon > 0$ consider the linear transformation $\varphi_\varepsilon : g \to g$ defined by

$$\varphi_\varepsilon(Y) = Y, \qquad \varphi_\varepsilon(Z) = \varepsilon^{1/2} Z \qquad \forall Y \in \mathcal{Y},\ Z \in \mathcal{Z}.$$

Then the original Lie bracket $[,]$ on g and the transformation φ_ε generate a"new" Lie bracket on g by the formula

$$[X, X']_\varepsilon = \varphi_\varepsilon^{-1}\left([\varphi_\varepsilon(X), \varphi_\varepsilon(X')]\right) \qquad (X, X' \in g). \tag{3.2}$$

Denote by g_ε the Lie algebra corresponding to $[,]_\varepsilon$. It is clear that g_ε is isomorphic to g for every $\varepsilon > 0$. Suppose that the limit

$$[X, X']_0 = \lim_{\varepsilon \to 0} [X, X']_\varepsilon \quad \text{exists for all } X, X' \in g. \tag{3.3}$$

Then we obtain the limiting Lie algebra g_0 with the same vector space structure as on g and with a new Lie bracket $[,]_0$ which is not isomorphic to $[,]$ in general. Such procedure is called the *contraction* of the Lie algebra g into g_0 (see[1,2]).

Fix a basis $X^1, \ldots, X^n$ of g satisfying

$$[X^i, X^j]_\varepsilon = \sum_k \lambda_\varepsilon^{ijk} X^k, \tag{3.4}$$

where the *structure constants* $\lambda_\varepsilon^{ijk}$ of g_ε have the following dependence on the parameter ε

$$\lambda_\varepsilon^{ijk} = \lambda_0^{ijk} + \varepsilon^{1/2} \lambda_1^{ijk} + \varepsilon \lambda_2^{ijk}. \tag{3.5}$$

Let g^* be a *dual space* (a *Lie coalgebra*) of g. Consider the Euler equations on g^* associated with the Lie bracket $[,]_\varepsilon$,

$$\frac{d\xi}{dt} = -\Psi_\varepsilon(\xi) \frac{\partial H}{\partial \xi}, \qquad \xi \in g^*, \quad H \in C^\infty(g^*). \tag{3.6}$$

Here the antisymmetric matrix $\Psi_\varepsilon(\xi) = ((\Psi_\varepsilon^{ij}(\xi)))$ has the elements

$$\Psi_\varepsilon^{ij}(\xi) = \sum_k \lambda_\varepsilon^{ijk} \xi^k, \tag{3.7}$$

and $\xi = (\xi^1, \ldots, \xi^n)$ are the coordinates on g^* corresponding to the basis $\Xi_1, \ldots, \Xi_n$ on g^* dual to $(X^i) : \langle \Xi_i, X^j \rangle = \delta_i^j$. So, the Euler equations (3.6) is the Hamiltonian system relative to the Lie–Poisson bracket $\{,\}_\varepsilon$ defined by (2.1) and the matrix $\Psi_\varepsilon(\xi)$ (3.7), i.e., (3.6) can be written as $\dot{\xi}_i = \{H, \xi^i\}_\varepsilon$ $(i = 1, \ldots, n)$.

The decomposition (3.5) induces the following

$$\Psi_\varepsilon(\xi) = \Psi_0(\xi) + \varepsilon^{1/2} \Phi_1(\xi) + \varepsilon \Phi_2(\xi). \tag{3.8}$$

For the limiting value $\varepsilon = 0$, system (3.6) is transformed into the Euler equations

$$\frac{d\xi}{dt} = -\Psi_0(\xi) \frac{\partial H}{\partial \xi}, \tag{3.9}$$

which is the Hamiltonian system relative to the limiting Lie–Poisson structure on g^*

$$\{f, g\}_0 = \sum_{i,j} \Psi_0^{ij}(\xi) \frac{\partial f}{\partial \xi^i} \frac{\partial g}{\partial \xi^j} \tag{3.10}$$

associated with the Lie algebra g_0.

Thus, for small ε, we can interpret the original Euler equations (3.6) as a perturbation of (3.9) with perturbation vector field

$$P_H^\varepsilon = -\varepsilon^{1/2}\left\langle \Phi_1 \frac{\partial H}{\partial \xi}, \frac{\partial}{\partial \xi} \right\rangle - \varepsilon\left\langle \Phi_2 \frac{\partial H}{\partial \xi}, \frac{\partial}{\partial \xi} \right\rangle. \tag{3.11}$$

Consider the following important case[2–4]. Suppose that subspaces $\mathcal{Y}$ and $\mathcal{Z}$ from (3.1) satisfy the relations

$$\begin{aligned} [\mathcal{Y}, \mathcal{Y}] &\subset \mathcal{Y}, \\ [\mathcal{Y}, \mathcal{Z}] &\subset \mathcal{Z}, \\ [\mathcal{Z}, \mathcal{Z}] &\subset \mathcal{Y}. \end{aligned} \tag{3.12}$$

Then the Lie algebra g is called *$\mathbb{Z}_2$-graded*[3]. In this case the limit (3.3) exists always and hence there exists the contraction of g into the Lie algebra g_0 which is a semidirect sum of the subalgebra $\mathcal{Y}$ and the commutative ideal $\mathcal{Z}$ (see[13]). It is easy to see the matrix Φ_1 incoming in formula (3.11) for the perturbation field vanishes, $\lambda_1^{ijk} = 0 \Rightarrow \Phi_1(\xi) \equiv 0$.

In this case we can look for a homotopy γ_ε of the Lie–Poisson structures $\{,\}_0$ and $\{,\}_\varepsilon$ in two steps.

First step. Following[9,10] let us consider the following equations with respect to a vector field $u = \sum_{i=1}^n u^i(\xi)\frac{\partial}{\partial \xi^i}$ on $g^* \approx \mathbb{R}^n$:

$$\sum_k \left(\Psi_\varepsilon^{ik}(\xi)\frac{\partial u^j}{\partial \xi^k} - \Psi_\varepsilon^{jk}(\xi)\frac{\partial u^i}{\partial \xi^k} - \lambda_\varepsilon^{ijk} u^k \right) = \Phi_2^{ij}(\xi), \tag{3.13}$$

where $i, j = 1, \ldots, n$. The left-hand side of (3.12) is just the *Schouten bracket* between the 2-tensor field Ψ_ε (3.7) and the vector field u. The general approach to solving the equations of type (3.13) (for nonlinear Poisson brackets) was suggested in[9,10]. In our (linear) case we can reduce the solving of (3.13) to finding the Poisson vector fields relative to the Lie–Poisson bracket $\{,\}_\varepsilon$. Indeed, let us take u in the form

$$u = \frac{1}{2\varepsilon}\left\langle \Pr(\xi), \frac{\partial}{\partial \xi} \right\rangle + w, \tag{3.14}$$

where $\Pr : g^* \to \mathcal{Z}^*$ is the projection along $\mathcal{Y}^*$ induced by the decomposition $g^* = \mathcal{Y}^* \oplus \mathcal{Z}^*$ corresponding to the original (3.1). It is not hard to see that the first summand in (3.14) is a quotient solution of (3.13) and hence a vector field w must satisfy equation (3.13) but with zero right-hand side. This means that w is a Poisson vector field relative to $\{,\}_\varepsilon$. Since we need to find a vector field u from (3.13) smoothly depending on ε, the main point is to investigate all Poisson vector fields of $\{,\}_\varepsilon$ which have the same type of singularity (at $\varepsilon = 0$) as the first summand of (3.14).

Second step. Suppose that we have a ε-regular vector field $u(\xi,\varepsilon)$ satisfying (3.13). Consider the ordinary differential equation corresponding to u,

$$\frac{d\Xi}{d\varepsilon} = u(\Xi;\varepsilon), \qquad \Xi\big|_{\varepsilon=0} = \xi. \tag{3.15}$$

Let $\Xi(\varepsilon;\xi)$ be a smooth solution of (3.15) which is well-defined for $0 \le \varepsilon \le \delta$ and all initial points ξ from a certain domain $\mathcal{M} \subset g^*$. If $\mathcal{M}$ is a submanifold of g^* and $\dim\mathcal{M} = \dim\mathcal{N}$, then[10,11] the formula

$$\gamma_\varepsilon(\xi) \stackrel{\text{def}}{=} \Xi(\varepsilon;\xi)$$

defines the mapping γ_ε satisfying (2.3) and hence γ_ε is the desired homotopy. It is easy to verify this fact by differentiating both sides of (2.3′) with respect to ε.

Note that if $g = \mathbf{so}(4)$ then the decomposition (3.1) satisfying (3.12) is induced by the decomposition of $\mathbf{so}(4)$ into the direct sum of Lie algebras,

$$\mathbf{so}(4) = \mathbf{so}(3) \oplus \mathbf{so}(3). \tag{3.17}$$

Indeed, choosing the standard basis of $\mathbf{so}(3)$ and using (3.17), we obtain the basis M_1, M_2, M_3; N_1, N_2, N_3 of $\mathbf{so}(4)$ satisfying

$$\begin{aligned}
[M_i, M_j] &= \epsilon_{ijk} M_k,\\
[M_i, N_j] &= 0,\\
[N_i, N_j] &= \epsilon_{ijk} N_k,
\end{aligned}$$

where ϵ_{ijk} is the completely antisymmetric 3-tensor, $\epsilon_{123} = 1$.

Then the subspaces $\mathcal{Y}$, $\mathcal{Z}$ from (3.1) satisfying (3.12) can be defined as

$$\mathcal{Y} = \mathrm{Span}\,\{M_i + N_i\}, \qquad \mathcal{Z} = \mathrm{Span}\,\{M_i - N_i\}, \quad (i = 1,2,3).$$

In this case the "extension" φ_ε from (3.2) gives the following family of Lie brackets

$$\begin{aligned}
[\widetilde{M}_i, \widetilde{M}_j]_\varepsilon &= \epsilon_{ijk} \widetilde{M}_k,\\
[\widetilde{M}_i, \widetilde{N}_j]_\varepsilon &= \epsilon_{ijk} \widetilde{N}_k,\\
[\widetilde{N}_i, \widetilde{N}_j]_\varepsilon &= \varepsilon\epsilon_{ijk} \widetilde{N}_k,
\end{aligned} \tag{3.18}$$

where $\widetilde{M}_i = M_i + N_i$, $\widetilde{N}_i = \varepsilon^{1/2}(M_i - N_i)$. Thus for $\varepsilon = 0$ (3.18) is transformed into the Lie bracket corresponding to the Lie algebra $\mathbf{e}(3)$ which is the semidirect sum[2,3,13] of $\mathbf{so}(3)$ and $\mathbb{R}^3$,

$$\mathbf{e}(3) = \mathbf{so}(3) \oplus \mathbb{R}^3. \tag{3.19}$$

4. A Homotopy of Lie–Poisson Brackets on $\mathbf{e}(3)^*$ and $\mathbf{so}(4)^*$

Consider the standard coordinates $y = (y_1, y_2, y_3)$, $z = (z_1, z_2, z_3)$ on the dual space $\mathbf{e}(3)^*$ of $\mathbf{e}(3)$ corresponding to decomposition (3.19). Then the Lie–Poisson bracket in $\mathbf{e}(3)^*$ relative to these coordinates is given by

$$\begin{aligned} \{y_i, y_j\} &= \epsilon_{ijk} y_k, \\ \{y_i, z_j\} &= \epsilon_{ijk} z_k, \\ \{z_i, z_j\} &= 0. \end{aligned} \tag{4.1}$$

The bracket (4.1) is just the limiting bracket $\{,\}_0$ from (1.2). Below we identify $\mathbf{e}(3)^*$ with $\mathbb{R}^6$ by means of the coordinates y, z.

It is well-known that there are two Casimir functions of the Lie–Poisson bracket (4.1),

$$\begin{aligned} k_1 &= y \cdot z \equiv \sum_{i=1}^{3} y_i z_i, \\ k_2 &= \frac{1}{2}|z|^2 \equiv \frac{1}{2}\sum_{i=1}^{3} z_i^2, \end{aligned} \tag{4.2}$$

which are functionally independent on the set of all *regular points* of (4.1):

$$\mathcal{N}_0 = \{(y, z) \in \mathbb{R}^6 : z \neq 0\}.$$

The restriction of (4.1) to $\mathcal{N}_0$ gives the degenerate Poisson bracket of constant rank.

Indeed, the antisymmetric matrix-valued function Ψ_0 (3.7) associated to (4.1) has the form

$$\Psi_0 = -\begin{bmatrix} \Lambda \circ y & \Lambda \circ z \\ \Lambda \circ z & 0 \end{bmatrix}, \qquad \Lambda \circ y \overset{\text{def}}{=} \begin{bmatrix} 0 & -y_3 & y_2 \\ y_3 & 0 & -y_1 \\ -y_2 & y_1 & 0 \end{bmatrix}. \tag{4.3}$$

Then

$$\text{rank } \Psi_0 = 4 \quad \text{everywhere on } \mathcal{N}_0$$

and the gradients of the Casimir functions k_1, k_2 generate the kernel of Ψ_0. Moreover, the level sets of k_1, k_2

$$\Omega_{\lambda_1,\lambda_2} = \{(y,z) \in \mathcal{N}_0 : k_1(y,z) = \lambda_1, k_2(z) = \lambda_2\} \tag{4.4}$$

coincide with the *regular coadjoint orbits* (the *regular symplectic leaves*) of $\mathbf{e}(3)^*$, for any constants $\lambda_1 \in \mathbb{R}$, $\lambda_2 > 0$. Note that any Hamiltonian field v on $\mathcal{N}_0$ is *tangent* to every orbit $\Omega_{\lambda_1,\lambda_2}$, i.e., $v(k_1) = v(k_2) = 0$. This implies that if condition (1.5) holds, then the vector field P_H (1.4) is transverse to $\Omega_{\lambda_1,\lambda_2}$ and hence P_H is non-Hamiltonian with respect to (4.1).

Each orbit $\Omega_{\lambda_1,\lambda_2}$ is *diffeomorphic* to the cotangent bundle over the 2-sphere[13],

$$\Omega_{\lambda_1,\lambda_2} \approx T^*\mathbf{S}^2, \qquad \dim \Omega_{\lambda_1,\lambda_2} = 4. \tag{4.5}$$

Indeed, if we consider the embedding of $T^*\mathbf{S}^2$ into $\mathbb{R}^6 = \mathbb{R}^{3*} \times \mathbb{R}^3$ given by

$$T^*\mathbf{S}^2 = \{(y,z) \in \mathbb{R}^6 : y \cdot z = 0,\ |z| = 1\},$$

then the restriction of the mapping

$$\ell : \mathcal{N}_0 \to \mathcal{N}_0, \qquad \ell(y,z) = \left(y - \frac{(y \cdot z)}{|z|^2} z, \frac{z}{|z|}\right) \tag{4.6}$$

onto $\Omega_{\lambda_1,\lambda_2}$ gives the diffeomorphism (4.5). So, the set $\mathcal{N}_0$ of all regular points of the Lie bracket (4.1) can be described as the fibration

$$T^*\mathbf{S}^2 \to \mathcal{N}_0 \to \mathbb{R} \times (0,\infty).$$

Note that for every $\varepsilon > 0$ the regular coajoint orbits of the Lie–Poisson bracket (1.2) are *compact* and diffeomorphic to $\mathbf{S}^2 \times \mathbf{S}^2$. So, the coadjoint orbits corresponding to the Lie–Poisson brackets (1.2) and (4.1) have different topological structures. By this reason we construct a homotopy γ_ε of (4.1) and (1.2) which is well-defined only in a certain domain of $\mathcal{N}_0$.

Fix $\delta > 0$ and define a domain $\mathcal{M}^\delta \subset \mathcal{N}_0$ by

$$\mathcal{M}^\delta = \{(y,z) \in \mathcal{N}_0 : |y| < \delta^{-1/2}|z|\}. \tag{4.7}$$

It is not hard to see that $\dim \mathcal{M}^\delta = 6$ and the intersection of $\mathcal{M}^\delta$ and every orbit $\Omega_{\lambda_1,\lambda_2}$ is a 4-dimensional submanifold. Hence, $\mathcal{M}^\delta$ is the Poisson submanifold of $\mathcal{N}_0$ (see Section 2).

Define the family of mappings

$$\gamma_\varepsilon(y,z) = \left(y, \left(1 - \varepsilon\frac{|y|^2}{|z|^2}\right)^{1/2} z + \frac{y \cdot z}{|y|^2}\left(1 - \left(1 - \varepsilon\frac{|y|^2}{|z|^2}\right)^{1/2}\right) y\right). \tag{4.8}$$

It is clear that $\gamma_\varepsilon(y,z)$ is well-defined for any $(y,z) \in \mathcal{M}^\delta$, $0 \le \varepsilon \le \delta$.

Theorem 4.1. *For every $\delta > 0$ the family of mappings $\gamma_\varepsilon : \mathcal{M}^\delta \to \mathbb{R}^6$ $(0 \le \varepsilon \le \delta)$ defined by (4.8) is the homotopy of the Lie–Poisson brackets (4.1) and (1.2) (in the sense of the definition of Section 2).*

The proof is given in the Appendix. It is based on solving equations (3.13) that correspond to the matrix-valued function $\Psi_\varepsilon = \Psi_0 + \varepsilon\Phi_2$, where Ψ_0 is defined by (4.3) and

$$\Phi_2 = -\begin{bmatrix} 0 & 0 \\ 0 & \Lambda \circ y \end{bmatrix}.$$

Note that the domain $\mathcal{M}^\delta$ (4.7) possesses the following property. For any *compact* submanifold $\mathcal{K} \subset \mathcal{N}_0$ there exists δ such that

$$\mathcal{K} \subset \mathcal{M}^\delta. \tag{4.9}$$

Indeed, we have the inclusion

$$\mathcal{K} \subset \{(y,z) \in \mathbb{R}^6 : |y| < C_1, |z| > C_2\} \tag{4.10}$$

for some positive constants $C_1, C_2 > 0$. Choosing the parameter δ from the condition

$$\delta < \frac{C_2^2}{C_1^2},$$

we obtain the inclusion (4.9). So, the homotopy γ_ε is well-defined in an arbitrary large compact submanifold $\mathcal{K}$ of $\mathcal{N}_0$ for a small enough ε.

Consider the intersection

$$\mathcal{O}^\delta_{\lambda_1,\lambda_2} = \Omega_{\lambda_1,\lambda_2} \cap \mathcal{M}^\delta. \tag{4.11}$$

It is easy to see that the image of $\mathcal{O}^\delta_{\lambda_1,\lambda_2}$ under the diffeomorphism ℓ (4.6):

$$\widetilde{\mathcal{O}}^\delta_{\lambda_1,\lambda_2} = \ell(\mathcal{O}^\delta) \subset T^*\mathbf{S}^2$$

is the fibration over $\mathbf{S}^2$

$$D^2 \to \widetilde{\mathcal{O}}^\delta_{\lambda_1,\lambda_2} \to \mathbf{S}^2,$$

whose fibers are isomorphic to the open disk D^2 with radius r defined by

$$r = \frac{2\lambda_2^2}{\delta} - \frac{\lambda_1^2}{2\lambda_2}, \qquad r \to \infty \quad \text{as} \quad \delta \to 0.$$

Thus, in the limit as $\delta \to 0$ the domain $\mathcal{O}^\delta_{\lambda_1,\lambda_2}$ coincides with the whole orbit $\Omega_{\lambda_1,\lambda_2}$. In other words, for each bounded domain $\mathcal{U} \subset \Omega_{\lambda_1,\lambda_2}$, there exists a sufficiently small δ such that $\mathcal{U} \subset \mathcal{O}^{\delta}_{\lambda_1,\lambda_2}$. We will use this key property of γ_ε in Section 5.

Now using Theorem 4.1 let us discuss the connection between the dynamics of (1.1) and (1.3). By Proposition 2.1 we have that the original Euler equations (1.1) is equivalent (in some domain) to the Euler equations of type (1.3)

$$\begin{aligned} \frac{dy}{dt} &= y \times \frac{\partial H_\varepsilon}{\partial y} + z \times \frac{\partial H_\varepsilon}{\partial z}, \\ \frac{dz}{dt} &= z \times \frac{\partial H_\varepsilon}{\partial y}, \end{aligned} \tag{4.12}$$

where the "new" Hamiltonian function H_ε is generated by the homotopy γ_ε (4.8),

$$H_\varepsilon(y,z) = H(\gamma_\varepsilon(y,z)) \equiv H(y,z) + \varepsilon H^{(1)}(y,z) + O(\varepsilon^2).$$

In this case the infinitesimal generator Γ (2.5) of γ_ε and the first correction $H^{(1)}$ to the original Hamiltonian function H have the form

$$\Gamma = \frac{1}{2|z|^2}\left\langle ((y\cdot z)y - |y|^2 z), \frac{\partial}{\partial z}\right\rangle, \tag{4.13}$$

$$H^{(1)}(y,z) = \frac{k_1}{4k_2}\left\langle y, \frac{\partial H}{\partial z}\right\rangle - \frac{|y|^2}{4k_2}\left\langle z, \frac{\partial H}{\partial z}\right\rangle. \tag{4.14}$$

It is of interest to note that the infinitesimal generator (4.13) (which gives a homotopy of brackets (1.2) and (4.1) up to $O(\varepsilon^2)$) is well-defined in the whole set $\mathcal{N}_0$ (for more details, see[9–11]).

So, for sufficiently small ε, the homotopy γ_ε (4.8) transforms any trajectory of (4.12), lying in a compact set, into a trajectory of (1.1). On the other hand, as $\varepsilon \to 0$, system (4.12) is a perturbation of (1.3) with perturbed Hamiltonian function $H_\varepsilon(y,z)$. Hence, under the assumption about the integrability of (1.3) we can apply the KAM theorem to (4.12).

5. An Application of the KAM–theorem

Consider the Euler equations (1.3). The Casimir functions k_1, k_2 (4.2) give the first integrals of motion of (1.3) in $\mathcal{N}_0$, i.e., the orbits $\Omega_{\lambda_1,\lambda_2}$ are invariant with respect to the flow of (1.3). The restriction of the Lie–Poisson bracket (4.1) to every orbit $\Omega_{\lambda_1,\lambda_2}$ is the nondegenerate Poisson bracket associated to the symplectic (Kirillov) form[13]:

$$\omega_{\lambda_1,\lambda_2} = \ell^*\left(d\beta + \frac{\lambda_1}{(2\lambda_2)^{1/2}}\sigma\right).$$

Here ℓ is the diffeomorphism (4.6), $\beta = p_\theta\, d\theta + p_\varphi\, d\varphi$ is the canonical 1-form on $T^*\mathbf{S}^2$, (θ, φ) are the standard spherical coordinates on $\mathbf{S}^2$, $\sigma = \sin\theta\, d\theta \wedge d\varphi$ is the area 2-form. Thus, the restriction of the Euler equations (1.3) to the orbit $\Omega_{\lambda_1,\lambda_2}$ is the Hamiltonian system with two degrees of freedom, corresponding to the Hamiltonian function $H_{\lambda_1,\lambda_2} = H\big|_{\Omega_{\lambda_1,\lambda_2}}$.

Suppose that for fixed $\lambda_1 \in \mathbb{R}$, $\lambda_2 > 0$ the function H_{λ_1,λ_2} is analytic in an open 4-dimensional *bounded* domain $\mathcal{U}$ of $\Omega_{\lambda_1,\lambda_2}$ and the following conditions hold.

(i). $\mathcal{U}$ is diffeomorphic to $\mathbf{T}^2 \times \mathcal{B}$, where $\mathcal{B}$ is an open subset of $\mathbb{R}^2$. Moreover, there are coordinates $I = (I_1, I_2) \in \mathcal{B}$, $\alpha = (\alpha_1, \alpha_2) \pmod{2\pi} \in \mathbf{T}^2$ in $\mathcal{U}$ such that

$$\omega_{\lambda_1,\lambda_2}\big|_{\mathcal{U}} = dI_1 \wedge d\alpha_1 + dI_2 \wedge d\alpha_2, \qquad H_{\lambda_1,\lambda_2}\big|_{\mathcal{U}} = H^0(I).$$

(ii). $\det\left(\left(\dfrac{\partial^2 H^0}{\partial I_i \partial I_j}\right)\right) \neq 0.$

The condition (i) means that the domain $\mathcal{U}$ is fibrated by the two dimensional tori

$$T_b = \{I_1 = b_1, I_2 = b_2\} \approx \mathbf{T}^2 \times \{b\}, \qquad b = (b_1, b_2) \in \mathcal{B} \tag{5.1}$$

invariant with respect to the flow of (1.3). Moreover, the flow of (1.3) on T_b is *quasiperiodic*[5,6] with *frequency vector* $\omega(b) = (\omega_1(b), \omega(b_2)) = \partial H^0/\partial I\big|_{I=b}$. So, system (1.3) is *integrable* on $\mathcal{U}$ and (I, α) are the corresponding *action–angle* variables. Consider the set

$$\mathcal{B}_{c,\nu} = \left\{b \in \mathcal{B} :\ |\omega(b) \cdot m| \geq c|m|^{-\nu}\ \forall m \in \mathbb{Z}^2 \setminus \{0\}\right\}$$

where $\nu > 1$, c are some positive constants.

Under these assumptions we have the following statement about the existence of invariant tori of the original system (1.1) lying on the coadjoint orbit

$$\Omega^\varepsilon_{\lambda_1,\lambda_2} = \left\{(y, z) \in \mathbb{R}^6 :\ y \cdot z = \lambda_1,\ \frac{|z|^2}{2} + \varepsilon\frac{|y|^2}{2} = \lambda_2 + \varepsilon\frac{\lambda_1^2}{4\lambda_2}\right\}$$

of the Lie–Poisson bracket (1.2).

Theorem 5.1. *If ε is small enough, then for every $b \in \mathcal{B}_{c,\nu}$ there exists a torus $T^\varepsilon \subset \Omega^\varepsilon_{\lambda_1,\lambda_2}$ invariant with respect to the flow of* (1.1)*, with quasiperiodic motion corresponding to the frequency vector $\omega(b)$.*

PROOF: By assumption the domain $\mathcal{U} \subset \Omega_{\lambda_1,\lambda_2}$ is bounded. Hence, we can fix $\delta > 0$ such that

$$\mathcal{O}^\delta_{\lambda_1,\lambda_2} \supset \mathcal{U},$$

where $\mathcal{O}^\delta_{\lambda_1,\lambda_2}$ is defined by (4.11). Thus, we have also the inclusion $\mathcal{O}^{\delta/2}_{\lambda_1,\lambda_2} \supset \mathcal{U}$ and the homotopy γ_ε (4.8) is well-defined in $\mathcal{O}^{\delta/2}_{\lambda_1,\lambda_2}$ for $\varepsilon \in [0, \delta/2]$. Under assumptions (i), (ii) we can consider system (4.12) on $\mathcal{U}$ as a perturbation of the integrable system (1.3) with the Hamiltonian function $H^0(I)$. Applying the KAM theorem[5,6] to (4.12), we obtain that for sufficiently small $\varepsilon < \delta/2$ and for every $b \in \mathcal{B}_{c,\mu}$, there exists a *Kolmogorov torus* $\widetilde{T}^\varepsilon \subset \mathcal{O}^{\delta/2}_{\lambda_1,\lambda_2}$ of (4.12) with frequency vector $\omega(b)$ of quasiperiodic motion. By Proposition 2.1 the image of $\widetilde{T}^\varepsilon$ under the homotopy γ_ε is the required torus $T^\varepsilon = \gamma_\varepsilon(\widetilde{T}^\varepsilon)$.

So, the above observation can be used in the case when system (1.3) is integrable, but there is no information about the integrability of (1.1). For example, for quadratic functions $H = H(y, z)$ we have this situation if (1.3) corresponds to the integrable Clebsch case [3,4,6], but the coefficients of quadratic form H do not satisfy the conditions corresponding to the integrable Manakov case[14–16].

6. Appendix

Here we give the proof of Theorem 4.1. The first step is to solve equations (3.13) associated with the family (1.2). Let us look for a vector field u from (3.13) in the form

$$u = \sum_{i=1}^{3} a_i(y, z) \frac{\partial}{\partial z_i}.$$

Then equations (3.13) are equivalent to the following equations with respect to the vector-function $\mathbf{a} = (a_1, a_2, a_3)$,

$$\{y_i, a_j\} = \epsilon_{ijk} a_k, \tag{6.1}$$

$$\{z_i, a_j\} - \{z_j, a_i\} + \varepsilon \sum_{s,k} y_k \left(\epsilon_{isk} \frac{\partial a_j}{\partial z_s} - \epsilon_{jsk} \frac{\partial a_i}{\partial z_s} \right) = \epsilon_{ijk} y_k. \tag{6.2}$$

Here $i, j = 1, 2, 3$ and $\{,\}$ is the Lie–Poisson bracket (4.1).

Note that the functions

$$\tau_1 = \frac{1}{2}|y|^2, \qquad \tau_2 = \frac{1}{2}|z|^2, \qquad \tau_3 = y \cdot z$$

satisfy the following relations relative to the bracket (4.1)

$$\{y_i, \tau_j\} = 0 \qquad (i, j = 1, 2, 3).$$

Using this fact, we obtain that for any function $f = f(\tau_1, \tau_2, \tau_3)$ the vector-function

$$\mathbf{a} = f\, y \times (z \times y) \tag{6.3}$$

satisfies equations (6.1) automatically. Substituting (6.3) into (6.2), we obtain the following equation for f:

$$\left(\frac{\partial}{\partial \tau_1} - \varepsilon \frac{\partial}{\partial \tau_2}\right)\big((\tau_3^2 - 4\tau_1\tau_2)f - \tau_1\big) = 0.$$

It is easy to see that this equation has the solution

$$f = \frac{c - \tau_1}{4\tau_1\tau_2 - \tau_3^2}$$

for any constant $c \in \mathbb{R}$. So, the vector-functions

$$\mathbf{a} = \frac{(2c - |y|^2)}{2|y \times z|^2} y \times (z \times y) \tag{6.4}$$

are the solutions of (6.1), (6.2) parametrized by $c \in \mathbb{R}$.

The second step is to solve the Cauchy problem (3.15). In our case system (3.15) corresponding to (6.4) has the form

$$\frac{dY}{d\varepsilon} = 0, \qquad\qquad Y\big|_{\varepsilon=0} = y,$$
$$\frac{dZ}{d\varepsilon} = \frac{(2c - |Y|^2)}{2|Y \times Z|^2} Y \times (Z \times Y), \qquad Z\big|_{\varepsilon=0} = z.$$

Integrating this system, we obtain the solutions

$$Y(\varepsilon; y, z, c) = y,$$
$$Z(\varepsilon; y, z, c) = \kappa z + \frac{y \cdot z}{|y|^2}(1 - \kappa) y, \qquad \kappa \stackrel{\text{def}}{=} \left(1 + \frac{2c\varepsilon|y|^2 - \varepsilon|y|^4}{|y|^2|z|^2 - (y \cdot z)^2}\right)^{1/2},$$

which give the family of homotopies of the Lie–Poisson brackets (4.1) and (1.2):

$$\gamma_\varepsilon(y,z,c) \stackrel{\text{def}}{=} \big(Y(\varepsilon;y,z,c), Z(\varepsilon;y,z,c)\big). \tag{6.5}$$

Thus, $\gamma_\varepsilon(y,z,c)$ satisfies conditions (2.3′) for any constant c. Note that if in (6.5) we replace c by any function of τ_2, τ_3, then we again obtain a homotopy of the Poisson brackets (4.1) and (1.2), since τ_2, τ_3 are the Casimir functions of (4.1), and the conditions (2.3′) are again satisfied. Finally, substituting $c = \tau_3^2/4\tau_2$ into (6.5), we obtain (4.8).

Note that if we start with representation (3.14) then the first summand in (3.14), corresponding to a quotient solution of (6.1), (6.2), has the form $\frac{1}{2\varepsilon}\langle z, \partial/\partial z\rangle$, and there exists a Poisson vector field w of (1.1) satisfying the property required in the context of Section 3.

7. Acknowledgements

We would like to thank Professor M. V. Karasev for the fruitful discussions, Professor E. A. Lacomba for the interest in our work and the referees for useful comments and inspiration.

8. References

1. R. Hermann, *Lie Groups for Physicists* (INC, New York, Amsterdam, 1966).
2. V. Guillemin and S. Sternberg, *Symplectic Techniques in Physics* (Cambridge Univ. Press, 1984).
3. A. T. Fomenko and V. V. Trofimov, *Integrable Systems on Lie Algebras and Symplectic Spaces* (Gordon and Breach Sci. Publ., Adv. Studies in Contemporary Math., Vol. 2, New York, 1988).
4. A. M. Perelomov, *Integrable Systems of Classical Mechanics and Lie Algebras*. Vol. I (Birkhäuser Verlag, 1990).
5. V. I. Arnold and A. Avez, *Ergodic Problems of Classical Mechanics*, (W. A. Benjamin, INC, New York, Amsterdam, 1968).
6. V. I. Arnold, V. V. Kozlov, and A. I. Neishtadt, *Mathematical Aspects of Classical and Celestial Mechanics. Encyclopedia of Math. Sci.*, vol.3 (Dynamical Systems III) (Springer Verlag, Berlin and New York, 1988).
7. J. Moser, *Ann. Math.* **169** (1967) 136–176.

8. E. A. Lacomba, J. Llibre and A. Nunes, in *The Geometry of Hamiltonian Systems* (Proc. Workshop, June 1989), ed. T. Ratiu (Springer Verlag, 1991) 373–385.
9. Yu. M. Vorobjev and M. V. Karasev, *Dokl. Akad. Nauk SSSR*, **297** (1987) 1294–1298 (Russian); English transl. in *Soviet Math. Dokl.* **36** (1988).
10. Yu. M. Vorobjev and M. V. Karasev, *Functional Anal. Appl.* **22** (1988) 1–9.
11. M. V. Karasev and V. P. Maslov, *Nonlinear Poisson Brackets. Geometry an Quantization* (A.M.S., Providence, R.I., 1993)
12. A. Weinstein, *J. Diff. Geom.* **18** (1983) 3523–3557.
13. J. E. Marsden, T. S. Ratiu, and A. Weinstein, *Cont. Math., AMS* **28** (1984) 55–100.
14. S.V.Manakov, *Functional Anal. Appl.* **10** (1976) 328–329.
15. A. P. Veselov, Dokl. Akad. Nauk SSSR **270** (1983) 1298–1300 (Russian).
16. M. Adler and P. van Moerbeke, *Invent. Math.* **67** (1982) 297–331.

THE HORSESHOE MAP IN THE COLLINEAR THREE-BODY PROBLEM

QIUDONG WANG
Department of Mathematics, Vanderbilt University
Nashiville, TN 37240, U. S. A.
E-mail: wang@athenea.cas.vanderbilt.edu

ABSTRACT

In this paper, we prove the existence of a globally embedded horseshoe structure inside of the phase space of the collinear three-body problem. The method developed here can also be applied to other gravitational systems with two degrees of freedom.

1. Introduction

The study of the general three-body problem is on the motion of three Newtonian gravitational particles in a 3-Euclian space. It is a very hard dynamical system partly because the dimension of the phase space is relatively high. To make the problem easier, many relatively lower dimensional gravitational models have been proposed in the history of celestial mechanics.

One of the natural reduction is to assume lower dimensional physical space. One obtains the planar three-body problem if the physical space is two dimensional, and the collinear three-body problem if that dimension is one.

There have been many important studies on the collinear three-body problem. It was the first problem on which McGehee applied his famous "blow up" transformation [3]. The first existence proof of the non-collision singularity [2] and that of the super-hyperbolic motion [7] in the Newtonian gravitational system were based on the study of this specific case of the three-body problem.

This paper follows the lines of [4,5,6] where we developed a systematic method to study the phase structure of some low dimensional gravitational models, and applied it to the collinear three-body problem. Our discussion in [6] was divided into three-steps. Initially we did not regularize the binary collision. A set of transformation was introduced to create fictitious boundaries to make the phase space compact. Then we regularized one of the binary collision and analyze the phase structure of this "half regularized" system. Finally, we regularized the second binary collision, and studied how the phase structure is transfered. In that study, we were mostly concerned with the global geometry of the phase flow of the collinear three-body problem. We did not give much attention on any particular solutions with special properties.

Based on the study of the phase structure in [6], we show the existence of a globally embeded horseshoe structure inside of the phase space of the collinear three-body problem in this paper. We first give a brief description of the phase structure obtained in [6] for the self-containedness, then devote the rest of the paper to create the desired horseshoe structure.

2. The global phase structure

2.1. The equations of motion for the unregularized system

Three particles, m_1, m_2 and m_3, are on a line with coordinates x_1, x_2 and x_3 respectively. Without loss of generality, let $x_1 < x_2 < x_3$. The system has potential function

$$U = \frac{m_1 m_2}{r_{12}} + \frac{m_2 m_3}{r_{23}} + \frac{m_1 m_3}{r_{13}}, \tag{1}$$

where $r_{12} = x_2 - x_1, r_{13} = x_3 - x_1, r_{23} = x_3 - x_2$.

The equations of motion are

$$\frac{dq}{dt} = M^{-1}p \qquad \frac{dp}{dt} = \nabla U(q), \tag{2}$$

where $q = (x_1, x_2, x_3)^T, M = diag(m_1, m_2, m_3)$. We can fix the center of mass at the origin and set the total linear momentum to zero, i.e.

$$m_1 x_1 + m_2 x_2 + m_3 x_3 = 0, \tag{3}$$

$$p_1 + p_2 + p_3 = 0. \tag{4}$$

The system also admits the integral of energy

$$\frac{1}{2} p^T M^{-1} p - U(q) = h. \tag{5}$$

We will only consider the case of negative energy in this paper.

A set of coordinates and time transformation has been introduced to reduce the dimension of the problem and compactify the phase space in [6]. The relations between the new variables (θ, Z_1, Z_2, Z_3) and the original variables (q, p) are

$$\theta = tan^{-1}(\frac{r_{12}}{r_{23}}), \qquad Z_3 = \frac{1}{\sqrt{2U(q)}} \tag{6}$$

$$Z_1 = Z_3(\frac{p_2}{m_2} - \frac{p_1}{m_1}), \qquad Z_2 = Z_3(\frac{p_3}{m_3} - \frac{p_2}{m_2}) \tag{7}$$

and the new time variable is defined by

$$\frac{dt}{d\tau} = (2U(q))^{-\frac{3}{2}} \tag{8}$$

The equations of motion for the new variables (θ, Z_1, Z_2, Z_3) with respect to τ are

$$\frac{d\theta}{d\tau} = \sin\theta\cos\theta(\sin\theta + \cos\theta)(Z_1\cos\theta - Z_2\sin\theta)(2\Psi(\theta))^{-1} \tag{9}$$

$$\frac{dZ_3}{d\tau} = -PZ_3 \tag{10}$$

$$\begin{aligned}\frac{dZ_1}{d\tau} = \{&-(m_1+m_2)\cos^2\theta(\sin\theta+\cos\theta)^2 \\ &+m_3\sin^2\theta(\sin\theta+\cos\theta)^2 \\ &-m_3\cos^2\theta\sin^2\theta\}(2\Psi(\theta))^{-2} - PZ_1\end{aligned} \tag{11}$$

$$\begin{aligned}\frac{dZ_2}{d\tau} = \{&m_1\cos^2\theta(\sin\theta+\cos\theta)^2 \\ &-(m_3+m_2)\sin^2\theta(\sin\theta+\cos\theta)^2 \\ &-m_1\cos^2\theta\sin^2\theta\}(2\Psi(\theta))^{-2} - PZ_2\end{aligned} \tag{12}$$

and the integral of energy becomes

$$T(Z_1, Z_2) + 2 \mid h \mid Z_3^2 = 1 \tag{13}$$

where

$$\begin{aligned}P = \{&Z_1 m_1 m_2\cos^2\theta(\sin\theta+\cos\theta)^2 \\ &+Z_2 m_2 m_3\sin^2\theta(\sin\theta+\cos\theta)^2 + \\ &(Z_1+Z_2)m_1m_3\sin^2\theta\cos^2\theta\}(2\Psi(\theta))^{-2}\end{aligned} \tag{14}$$

$$\begin{aligned}\Psi(\theta) = &(m_1m_2\cos\theta + m_2m_3\sin\theta)(\sin\theta+\cos\theta) \\ &+ m_1m_3\cos\theta\sin\theta\end{aligned} \tag{15}$$

and

$$\begin{aligned}T(Z_1, Z_2) = \frac{1}{2(m_1+m_2+m_3)^2}\{&m_1(m_3Z_2 + (m_2+m_3)Z_1)^2 \\ &+m_2(m_1Z_1 - m_3Z_2)^2 \\ &+m_3(m_1Z_1 + (m_1+m_2)Z_2)^2\}\end{aligned} \tag{16}$$

For the details of the derivation of these equations, refer to [6].

2.2. The phase space

The function $T(Z_1, Z_2)$ is a positive definite quadratic form and the variable Z_3 is non-negative. So the equation (13) defines a two-dimensional disk D^2 in (Z_1, Z_2, Z_3) space with the circle $Z_3 = 0$ as its boundary.

According to (6), binary collisions occur when $\theta = 0$ or $\frac{\pi}{2}$. Since $2Z_3^2$ is the inverse of the potential function, $Z_3 = 0$ also corresponds to singularities of the system. The original equations of motion (2) are not well defined on the above values. However, they are not singular anymore in the new system (9-13). The phase flow has been smoothly extended onto these singularities so a compact phase space is obtained in new coordinates and time.

Refer to Figure 1 for the new phase space and the fictitious flow on its boundary. On the boundaries of $\theta = 0$ and $\theta = \frac{\pi}{2}$, the fictitious flows are integrable. On the boundary of $Z_3 = 0$, it is equivalent to the flow on the McGehee's total collision manifold. See [6] for more details.

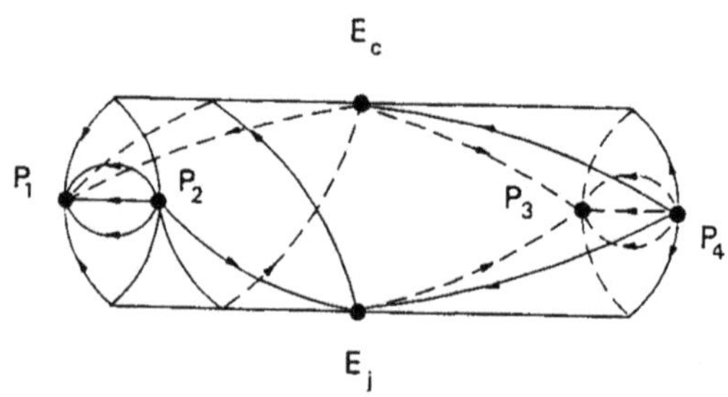

Figure 1: The phase space and the flows on its boundary.

In Figure 1, $P_1, P_2, P_3, P_4, E_j, E_c$ are rest points. P_1, P_3 are sinks and P_2, P_4 are sources. All the solutions of binary collision of m_1 and m_2 tend to P_1 and all the solutions of binary ejection of these two start from P_2. Similarly P_3 stands for the collision of m_2 and m_3, and P_4, the corresponding ejection.

E_c and E_j are hyperbolic fixed points corresponding to the Euler collinear central configuration. E_c has a two-dimensional stable manifold inside of the phase space and one-dimensional unstable manifold on the collision boundary. Similarly, E_j has two dimensional unstable manifold inside and one dimensional stable manifold on the boundary. Physically, all the real solutions of triple collision form the stable manifold of E_c and that of triple ejection form the unstable manifold of E_j.

Since all the solutions of collinear three-body problem with negative energy end up with collision (triple or binary), we have

Proposition 1 *Any given solution of equations (9-13) takes one of the rest points on the boundary as its α (ω) limit set.*

2.3. The Poincaré section

Refer to Figure 2. Take a sufficiently small sphere (two-dimensional surface) centered at P_2 and denote the intersection of this sphere with the boundary $\theta = 0$ by l and that of the sphere with the collision manifold by l'. Call the portion of the sphere inside of the phase space as S. We see that S is an quarter sphere which takes l and l' as its boundary. Denote the interior of S by S* $(= S\backslash(l \cup l'))$.

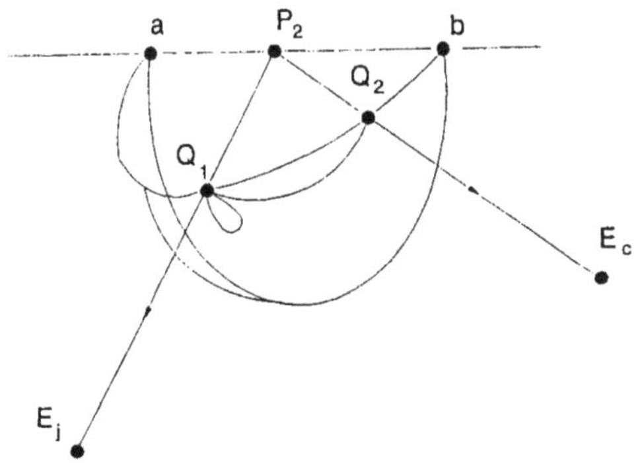

Figure 2: The Poincaré section.

Since P_2 is a source, we can choose the sphere such that it transverses to the vector field so all the solutions go outward on S. Let $W^s(P)$ and $W^u(P)$ be the stable and unstable manifold of a hyperbolic fixed point P and

$$A = S \cap W^s(P_1) \qquad W = S \cap W^s(P_3) \qquad C = S \cap W^s(E_c).$$

We see that $(A \cup W \cup C) \cap S^* = S^*$. A solution in A will end at the binary collision of m_1 and m_2 and a solution in W will end at the binary collision of m_2 and m_3. Likewise, a solution in C will end at a triple collision. We now need some new geometric terminologies for a detailed description of the sets A,W and C.

For an one-dimensional curve **c** in S*, take a parametrization of **c** as $c(t) : (-\infty, +\infty) \mapsto S^*$. Define the α limit and ω limit set of **c** by

$$\alpha = \{p \in S\text{: there exist } t_n \to -\infty\text{, such that } c(t_n) \to p\ \}$$

and

$$\omega = \{p \in S\text{: there exist } t_n \to +\infty\text{, such that } c(t_n) \to p\ \}$$

We also say that **c** connects α and ω.

A simple curve on S* is *Normal* if both of its α (and ω) limit sets are a single point on l'. For a normal curve, if its α and ω limit sets are the same, we call it a *Loop*. If they are different, we call it a *Regular* curve.

On the curve l', denote $Q_1 = W^s(E_j) \cap l', Q_2 = W^s(E_c) \cap l'$. We have

Proposition 2 *a). The set C is a collection of one dimensional curves on S^*.*

b). each curve in C takes either Q_1 or Q_2 as its α (and ω) limit set.

c). C consists of only one regular curve connecting Q_1 and Q_2, and possibly other loops at Q_1.

See [6] for the proof of this proposition. Also see figure 2 for the picture of the set C. The branches of C divide the whole surface S* into two-dimensional regions. These regions are components of the sets A or C.

2.4. The half regularized system

For the unregularized system, the geometric structure of A, W, C on the surface S* is relatively simple. This structure becomes richer after the regularization of binary collisions. However, to keep the complexity under control, we regularize only one of the binary collisions at this stage: the collision between m_2 and m_3.

Refer to Figure 3 for the phase space and the flow on the boundary after the regularization. $\hat{\gamma}^-$ and $\hat{\gamma}^+$ are two branches of $W^u(E_c)$. Notice that there is no binary collision of m_2 and m_3 on $\hat{\gamma}^-$ and only one such collision on $\hat{\gamma}^+$. This property of $W^u(E_c)$ is quite remarkable. It makes the application of symbolic dynamics possible.

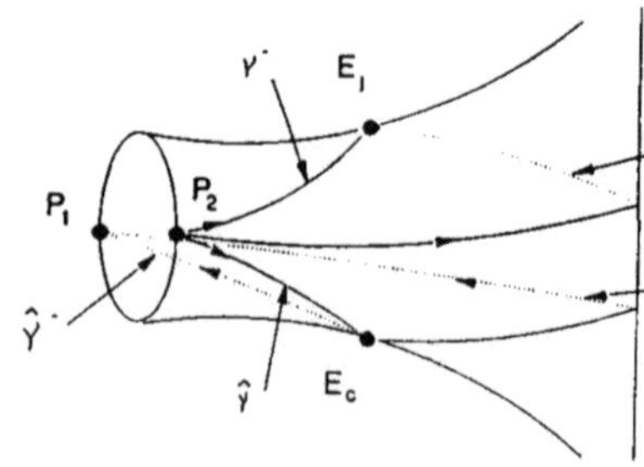

Figure 3: The half regularized phase space.

The final evolution of the solutions of this half regularized system is more complicated. Now there are four possible destinations for a given solution:

a). End at binary collision of m_1 and m_2.

b). End at triple collision.

c). The particle m_1 approaches infinity with non-zero velocity. (the hyperbolic solution).

d). The particle m_1 approaches infinity with zero velocity. (the parabolic solution).

We call a point p *hyperbolic* (or *parabolic*) if p is on a hyperbolic (or parabolic) solution. Otherwise we call the solution an *elliptic* solution and the corresponding points *elliptic* points.

For any point p inside of the phase space, let

$O(\tau,p) = \{$ the solution at time τ which starts from p at $\tau = 0$ $\}$.

$$O(p) = \{\ O(\tau,p): \qquad 0 < \tau < \infty\ \}$$

Definition 1 *For a given point p inside of the phase space, we call the number of binary collisions of m_2 and m_3 on $O(p)$ as the* **characteristic number** *of p, and denote it as $N(p)$*

Also let

H ={ p: p is a hyperbolic point }

P = { p: p is a parabolic point }

E = { p: p is a elliptic point }.

We see that if $p \in H$ (or P), $N(p) = \infty$, and if $p \in E$, $N(p) < \infty$.

2.5. The geometric structure

In this subsection, we give a list of propositions about the geometric structure on S* without detailed proof. Related concepts are introduced to make all the statements understandable. See [6] for the proofs of these propositions.

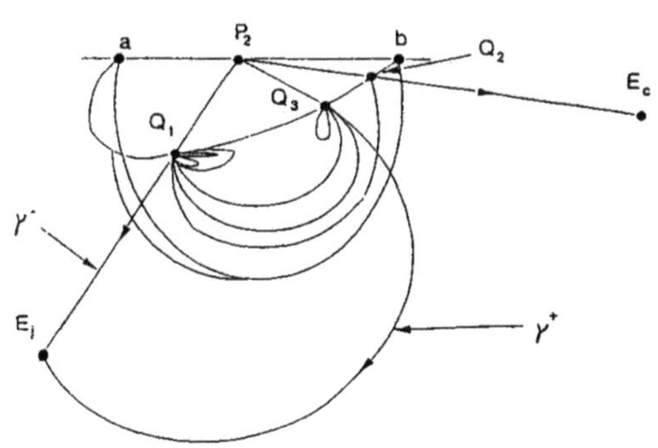

Figure 4: The geometric structure

Refer to Figure 4. Take again the quarter sphere S around P_2. Denote its interior as S* and the two boundaries as l and l'. On l', the points a, b, Q_1, Q_2 are as before, and Q_3 is the intersection of γ^+ with l'. Let

$$S^* \cap P = \Sigma_P \qquad S^* \cap W^s(E_c) = \Sigma$$

We have

Proposition 3 *(a). Σ_P is a collection of regular curves connecting Q_1 and Q_3, or loops at Q_1, Q_3.*

(b). Σ are collection of regular curves connecting Q_1 and Q_2, Q_1 and Q_3, or loops at Q_1, Q_3. Furthermore, there is only one such curve ending at Q_2.

All these curves divide the surface S* into simply connected two dimensional regions. We call such a region a *Unit*.

Definition 2 *(a). A unit is called a strip if the boundary of this unit includes regular curves. A strip must have two and only two regular curves on its boundary. If these two regular curves have different character number, we call this strip a regular strip.*

(b). A unit is called a ring if the boundary consists only loops.

We have

Proposition 4 *(a). all the points on the same connected branch of Σ have the same character number.*

(b). All the points in the same unit have the same character number.

(c). In a neighborhood of a point of Σ, the character number changes at most by one.

According to this proposition, we can represent a unit by a triplet (k_1, k_2, k_3). where k_2 is the character number of the unit, and $k_1 \leq k_3$ are the other possible character numbers on the boundary of this unit.

We further have

Proposition 5 *(a). Units on S^* satisfy the restriction:*

$$k_1 \leq k_3 \leq k_2 \leq k_1 + 1.$$

That is, any unit on S^* has one of the following representations

$$(k, k+1, k+1); \qquad (k, k, k); \qquad (k, k+1, k),$$

where k is either a non-negative integer or $+\infty$

(b). For any positive integer k, there is a regular strip with representation $(k, k+1, k+1)$.

This proposition, especially the part (b), provides us the ingredients to create a horseshoe structure. Refer to [6] for further discussion on the geometric structure on S*, and the relationship between this structure and the intersection of $W^u(E_j)$ and $W^s(E_c)$.

2.6. The flow action

Mimicking the construction of S at P_2, we take a quarter sphere $\hat{S}$ at P_1. Refer to Figure 5. Denote the interior of $\hat{S}$ as $\hat{S}^*$, and the intersection of $\hat{S}$

with the boundary of collision manifold and the boundary of $\theta = 0$ as $\hat{l}'$, $\hat{l}$ respectively. We see that $\hat{\gamma}^+$ and $\hat{\gamma}^-$ are two branches of $W^u(E_c)$ and $\hat{\gamma}$ is one of the branches of $W^u(E_j)$ on the collision boundary. Define

$$\hat{Q}_1 = \hat{\gamma}^+ \cap \hat{l}'; \qquad \hat{Q}_3 = \hat{\gamma}^- \cap \hat{l}'; \qquad \hat{Q}_2 = \hat{\gamma} \cap \hat{l}'.$$

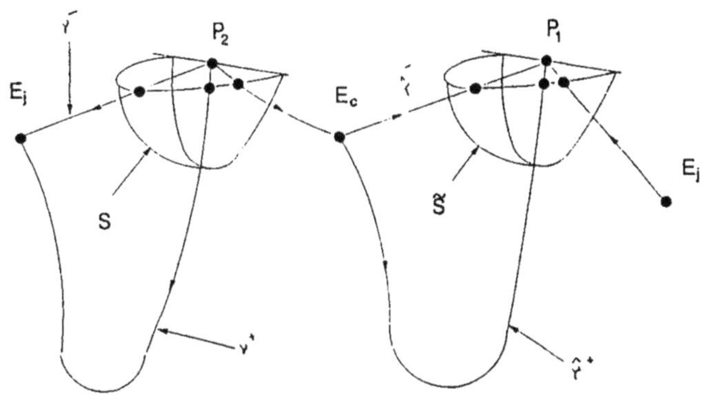

Figure 5: The surface $\hat{S}$ at P_1

Following the flow backward, we also have four possible final evolutions: the binary ejection of m_1 and m_2; the triple ejection ; m_1 coming in from infinity with positive velocity (hyperbolic catch); m_1 coming in from infinity with zero velocity (parabolic catch). We can again define the sets $\hat{H}, \hat{P}, \hat{E}$ analogous to H, P, E on S, and let

$$\hat{S}^* \cap \hat{P} = \hat{\Sigma}_P \qquad \hat{S}^* \cap W^u(E_j) = \hat{\Sigma}.$$

Also, we have the corresponding definition of the character number N(p) for a given point $p \in \hat{S}^*$. However, we now count the number of collision of m_2, m_3 for $\tau < 0$.

All the discussion on the geometric structure of S* can be carried over. This gives a similar structure on $\hat{S}^*$. We can use the same terminologies to describe this new structure on $\hat{S}^*$. In fact, since the collinear three-body system is reversible, we have

Proposition 6 *The geometric structure on $\hat{S}^*$ is exactly the same as that on S^*.*

If the character number for a unit U of S* is finite, the flow carries U over onto $\hat{S}^*$. Denote this flow defined map from S* to $\hat{S}^*$ as $F : S^* \mapsto \hat{S}^*$. We have

Proposition 7 *(a). For any unit U on S^* with finite character number, its image $F(U)$ is a unit on $\hat{S}^*$.*

(b). If U is a regular strip with a representation $(k, k+1, k+1)$*, F(U) is a regular strip on* $\hat{S}^*$ *with exactly the same representation.*

We have more about the regular strips and the flow defined map F.

Proposition 8 *For any regular strip U with representation* $(k, k+1, k+1)$*, there is a regular curve in U, which divides its boundary into two consecutive parts, such that the character number of one part is k and that of the other part is k+1.*

We define

Definition 3 *For a regular strip U on* S^* *with representation (k, k+1, k+1), take a simple curve* **c** *inside of U.*

(a). **c** *is a horizontal-path if* **c** *is a regular curve which divides the boundary of U into two consecutive parts as described in the proceeding Proposition.*

(b). **c** *is a vertical-path if* **c** *connects two regular branches of U.*

(c). **c** *is a vertical-curve if* **c** *connects two points* p_1, p_2 *on the boundary of U, where* p_1, p_2 *have different character numbers.*

See Figure 6 for the geometric meaning of this definition . We will use $l_v(U)$ for a vertical curve of U, and $l_h(U)$ for a horizontal path.

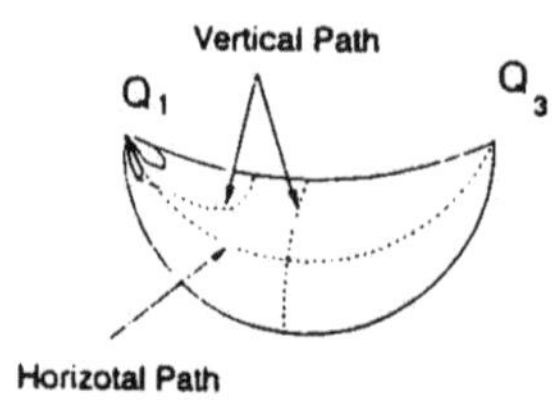

Figure 6: The horizontal and vertical path.

Next proposition reveals the "hyperbolic behavior" of the flow defined map F.

Proposition 9 *Take a regular strip U on* S^* *with representation (k, k+1, k+1).*

(a). The image of a vertical-path of U is a horizontal-path in F(U), and the image of a horizontal-path of U is a vertical-path of F(U).

(b) The image of a vertical curve is a regular curve in F(U) connecting $\hat{Q}_1, \hat{Q}_3$.

Though no metric is involved in this statement, one can still recognize the "topological hyperbolicity" of the map F on U. This property is crucial for the existence of a horseshoe map. For the detailed proofs of all the claims here and further discussion of the geometric structure, see [6].

3. The horseshoe map

Now we are ready to regularize the second binary collision, the collision of

m_1 and m_2. The regularization gives a map R from $\hat{S}^*$ to S^*. Combining F and R, we obtain the flow- induced map $T = R \circ F : S \mapsto S$.

T is not a continous map on S. The discontinuity occurs on Σ. Unfortunately, one has to iterate the map T more than once in general to create the desired horseshoe structure. So, it requires extra cautious to trace the image of each regular strip under the map T.

Since the map T depends on the fictitious flow on the McGehee's collision manifold, we first discuss the related properties of this well-known system.

3.1. McGehee's flow and the map T on the boundary

The triple collision manifold for the collinear three-body problem, after the regularization of both binary collisions, is as in Figure 7 according to McGehee.
3

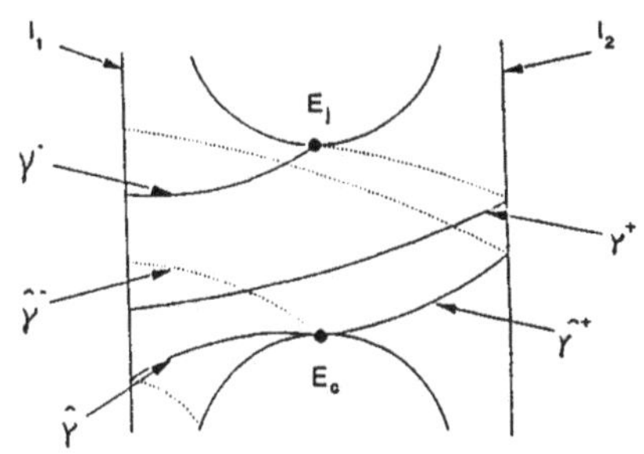

Figure 7: The flow on the McGehee's collision manifold

The lines l_1, l_2 correspond to the binary collision of m_1, m_2 and m_2, m_3 respectively. E_c and E_j are as before. On this two dimensional surface, two branches of $W_u(E_c)$ are denoted as $\hat{\gamma}^+$ and $\hat{\gamma}^-$. In the picture, $\hat{\gamma}^+$ goes to l_2 and $\hat{\gamma}^-$ goes to l_1, then they go back and forth between l_1 and l_2 and finally end up on one of the arms. For the rest of the discussion, we assume that $\hat{\gamma}^+$ and $\hat{\gamma}^-$ go up to different arms on the collision manifold. This is our only assumption to ensure the existence of a horseshoe structure. Also denote the number of times $\hat{\gamma}^+$ hit l_2 by μ. $\mu \geq 1$. μ is an integer valued function of the masses m_1, m_2 and m_3.

Compare Figure 2 with Figure 7. Since the open arc $l' = (a, b)$ on S is on the collision boundary $Z_3 = 0$, it certainly has a corresponding image on the McGehee's manifold. This image is close to the line l_1. It intersects with all the solutions newly coming out of the binary collision of m_1 and m_2, and the intersection is transversal. So this curve is in the front of the collision manifold, all the way from the downward infinity to the upward infinity, and close to the line l_1. Similarly, the corresponding image of the open arc $\hat{l}' = (\hat{a}, \hat{b})$ is the symmetric image of that of l' with respect to the line l_1 on the back of the collision manifold. Refer to Figure 8.

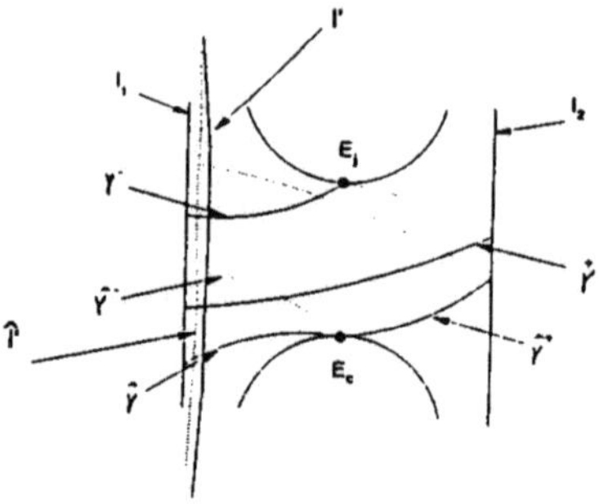

Figure 8: The image of l' and $\hat{l}'$ on collision manifold.

Denote these images on the collision manifold again by l' and $\hat{l}'$ respectively. The flow defined map F maps l' onto $\hat{l}'$. The regularization of the binary collision of m_1 and m_2 induces a map R, which maps $\hat{l}'$ back onto l'. The composition of these two maps: $T = R \circ F$, restricted on the collision boundary, is a map from l' to l'. Take a point p on l', one can follow the flow on McGehee's manifold to find F(p) on $\hat{l}'$, and the image of $T(p) = R \circ F(p)$ on l'.

3.2. The horseshoe structure: $\mu = 1$

Take a regular strip $U = (k, k+1, k+1)$ on S*. The boundary of U consists of two regular curves connecting Q_1 Q_3, and a set of loops. According to Proposition 8, a horizontal path in S divides the boundary into two parts. We denote the part with character number k+1 as b_{k+1}, and the other part as b_k. Refer to Figure 9(a).

The strip U can be associated with a standard strip V without loops. As in Figure 9(b), the standard strip V takes two regular branches c_{k+1}, c_k as its boundary. The part c_{k+1} of the boundary of V is obtained by first listing all the boundary branches in b_{k+1} in order, then connecting them by adding extra points. The other part c_k of the boundary V is obtained from b_k in the same way.

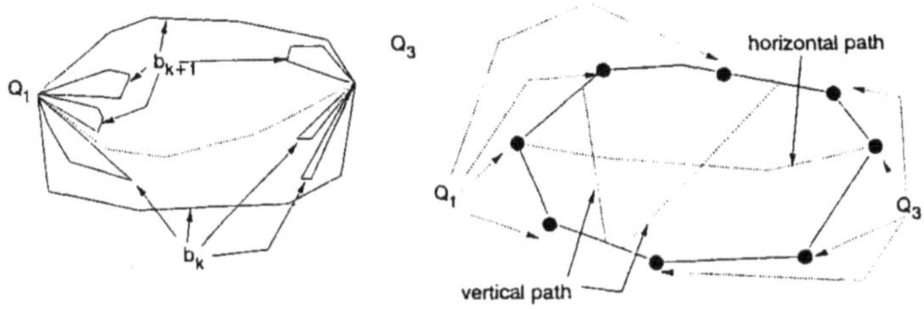

Figure 9: The standard strip associated with U

The interior of U and V are topologically equivalent. By such a topological equivalence, a vertical-curve in U corresponds to a vertical curve in V. Also, a horizontal path in U corresponds to a horizontal-path in V.

The correspondence between the boundaries of U and that of its image F(U) on $\hat{S}$ is now easier to understand in terms of the standard strip V. The regular branches of the boundary of F(U) are the images of the vertices of V, and each loop on the boundary of F(U) is the image of one of the connecting points we added on the boundary of V. Other points on the c_{k+1} are mapped to $\hat{Q}_3$ and those on c_k are mapped to $\hat{Q}_1$. Refer to Figure 10.

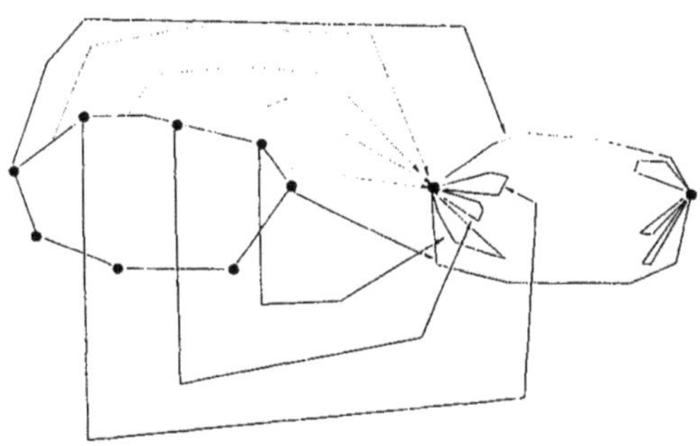

Figure 10: The image of a regular strip under F

Though F is not even an one to one map on the closure of U, it is analytic on the interior of U.

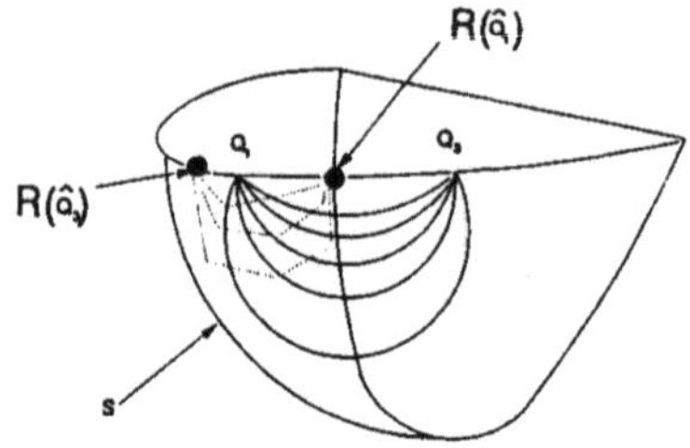

Figure 11: The horseshoe structure for $\mu = 1$

The map R induced by the regularization of the binary collision of m_1 and m_2 is also analytic. So $R(F(U))$ is a strip connecting $R(\hat{Q}_1)$ and $R(\hat{Q}_3)$. Furthermore, $R(l_h(\hat{U}))$ is a regular curve on S connecting $R(\hat{Q}_1)$ and $R(\hat{Q}_3)$, and $R(l_v(\hat{U}))$ is a simple curve connecting $R(b_{k+1}(\hat{U}))$ and $R(b_k(\hat{U}))$.

In the case of $\mu = 1$, the images of $\hat{Q}_1$, $\hat{Q}_3$ and $\hat{Q}_2$ under R are as in Figure 11. Notice that Q_1 separates R($\hat{Q}_3$) and R($\hat{Q}_1$).

According to Proposition 3, for any non-negative integer k, there exists at least one regular strip U_{k+1} on S with expression $(k, k+1, k+1)$. Let Ω be the collection of all the regular strips on S

$$\Omega = \{ \text{ U} \mid \text{U is a regular strip on S } \}$$

For any given regular strip $U \in \Omega$, $T(U) = R \circ F(U)$ crossly overlaps *all* the other regular strips in Ω, as shown in Figure 9. This overlap creates a typical horseshoe structure. Therefore, we have

Theorem 1 *Let $\hat{\Gamma}$ be the set of all finite, one-side infinite and two-side infinite sequences of positive integers. That is*

$$\hat{\Gamma} = \{s = (\ldots s_{-1}, s_0; s_1, s_2, \ldots) \mid s_j \in Z^+\}.$$

We introduce two additional symbols $\{c, e\}$ to augment the set $\hat{\Gamma}$. If a given sequence $s \in \hat{\Gamma}$ is terminated at the left side (or right side), we add symbol c or e in front of s (or end of s) to form new sequences. Denote the resulted set as Γ.

For any given $s \in \Gamma$, there exists a solution of the collinear three-body problem corresponding to s in the case of $\mu = 1$. Terminated symbols correspond to solutions of triple collision(terminated on the right with c), triple ejection(terminated on the left with c), catching of m_1 from infinity (terminated at left with e) and escaping of m_1 to infinity (terminated at right with e) respectively. For this given s, the integers in this sequence represent consecutive *segments of the corresponding solution of s. Each integer represents a segment started from binary ejection of m_1 and m_2, ended at binary collision of m_1 and m_2. Furthermore, the integer s_j stands for the number of binary collisions of m_2 and m_3 on the specific segment of the solution corresponding to s_j.*

Remarks:

(a). This theorem is by no means a completed classification of all the solutions of the collinear three-body problem. This is because the symbols of the sequence s in Γ can not be zero. It gives no answer on the existence of the solutions corresponding to a given sequence if zero is included. In fact, such a solution might not exist for some of these sequences. For instance, the sequence with all symbols zero has no corresponding solution.

(b). Here we only considered the regular strips and their images under T. The actual structure on S might also include strips with expression (k, k+1, k), (k, k, k) as well as rings. All these units are transfered similarly by the map T (see [6]). One sees that the geometry induced by T might be far more complicated than that of a standard horseshoe map.

3.3. The case of $\mu > 1$: horseshoe structure

It is easy to create a horseshoe structure in the case of $\mu = 1$ because Q_1 separates $R(\hat{Q}_1)$ and $R(\hat{Q}_3)$ on l'. The image of every regular strip under T is forced to crossly overlap all the other regular strips on S. In the case of $\mu > 1$, one does not have such a separation. $R(\hat{Q}_1)$ and $R(\hat{Q}_3)$ are both on the arc Q_3Q_2.

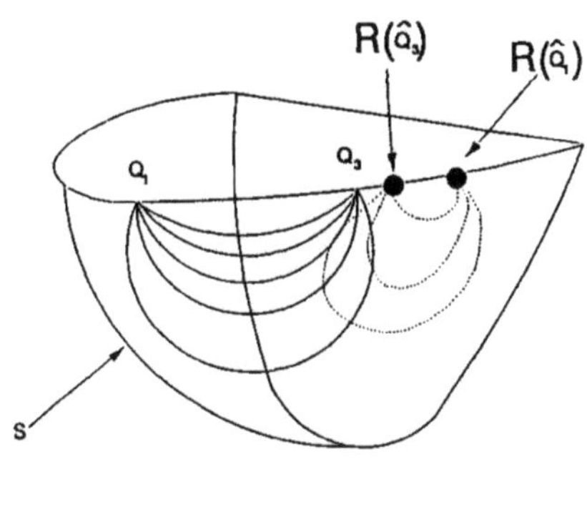

Figure 12: The image of strips under map $T = R \circ F$.

We can still trace the images of $R(\hat{Q}_1)$ and $R(\hat{Q}_3)$ under iteration of T by the help of the flow on the McGehee's collision manifold. Without loss of generality, we assume that $\hat{\gamma}^+$ goes up on the right arm and $\hat{\gamma}^-$ goes up on the left arm on the collision manifold. Under this assumption, $R(\hat{Q}_1)$ and $R(\hat{Q}_3)$ move toward the point Q_3 as T is iterated. They stay inside of the arc Q_3Q_2 under the first $\mu - 2$ iterations. Then $T^{\mu-1}(R(\hat{Q}_3))$ jumps into the arc Q_1Q_3, while $T^{\mu-1}(R(\hat{Q}_1))$ stays on the arc Q_3Q_2. Refer to Figure 13.

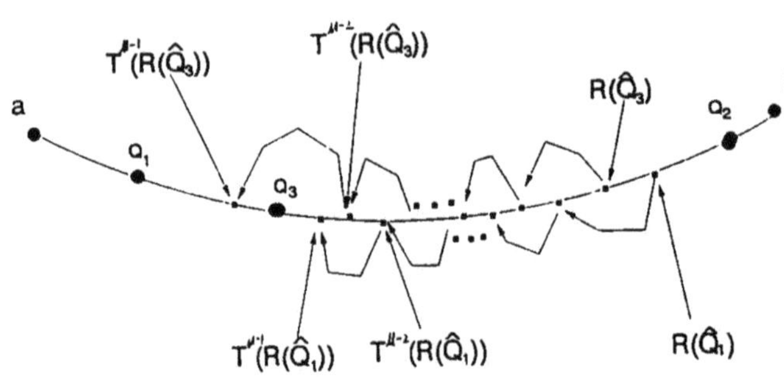

Figure 13: The image of $\hat{Q}_1$ and $\hat{Q}_3$ under the iteration of T

$T^{\mu-1}(R(\hat{Q}_3))$ and $T^{\mu-1}(R(\hat{Q}_3))$ now are separated by Q_3 on l'. However, as we pointed out at the very beginning of this section, horseshoe structure does not come out immediately as in the case of $\mu = 1$. Since T is not continous, one has no control on the image of the whole strip when T is iterated more than once. So the overlap illustrated in Figure 9 does not occur for T^{μ}.

The major idea here is that, instead of using the image of the whole strip, we only consider the part of a given strip whose image is close to $T^i(R(\hat{Q}_3))$ for $1 \leq i \leq \mu - 2$. The map T is continuous on those points so we have control on their images under T. The overlap of these sub-strips with their images under T^{μ} will create a horseshoe structure.

In the case of $\mu > 1$, the strip on S which takes the arc Q_3Q_2 of l' as part of its boundary plays an important role since part of the images of U under the first $\mu - 1$ iterations of T will be in this strip. We call this strip (as well as its counter part on $\hat{S}$), as the *Basic strip* and denote it as B(S) ($B(\hat{S})$).

Now take a regular strip U on S. $T^{\mu-1}(U) \cap S(B)$ is not empty. There are two cases to be considered.

(a). The whole image is kept inside of S(B) for all the iterations. In this case, we have complete control on the image of U. Apply T again on $T^{\mu-1}(U)$ will give the required overlap.

(b). In the process of the iterations of T, part of the image goes out of B(S). Since T is not continous on the boundary of B(S), one loses control on the part of the image which goes out of B(S) in the next iteration of T. However, after the $\mu - 1$ iterations of T, there are still some parts of the strip U whose images stay inside of B(S) under T^i, $1 \le i \le \mu - 1$.

Take the connected component C of $T^{\mu-1}(U) \cap B(S)$ such that the point $T^{\mu-2}(R(\hat{Q}_3))$ is on its the boundary. That boundary must also intersect with the boundary of B(S) (otherwise we are back to case (a)). Denote this set of intersection as K.

Apply T once more on C. $T^{\mu-2}(\hat{Q}_3)$ is mapped into the arc Q_1Q_3, and K is mapped back to either $R(\hat{Q}_3)$ or $R(\hat{Q}_1)$. Therefore T(C) crossly overlap every regular strip on S. Refer to Figure 14.

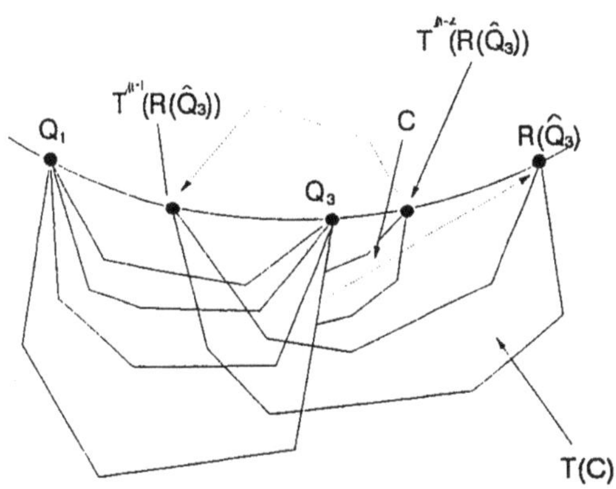

Figure 14: The overlap created by T(C).

One also needs to study the pre-image of C to create the desired horseshoe structure. $D = T^{\mu-2}(R(\hat{Q}_3))$ is on the boundary of C, and $T^{-(\mu-1)}(D) = F^{-1}(\hat{Q}_3) = b_{k+1}$. So the pre-image of the part of C which is close to D under $T^{-(\mu-1)}$ is a sub-strip of U. Refer to Figure 15

Let Ω be the collection of all such sub-strips on S. The image of a sub-strip of Ω stays inside of the Basic strip B(S) under the first $\mu - 1$ iterations of T, then it crossly overlaps all the other sub-strips in Ω under one more iteration of T. Horseshoe structure occurs by applying T^{μ} to Ω.

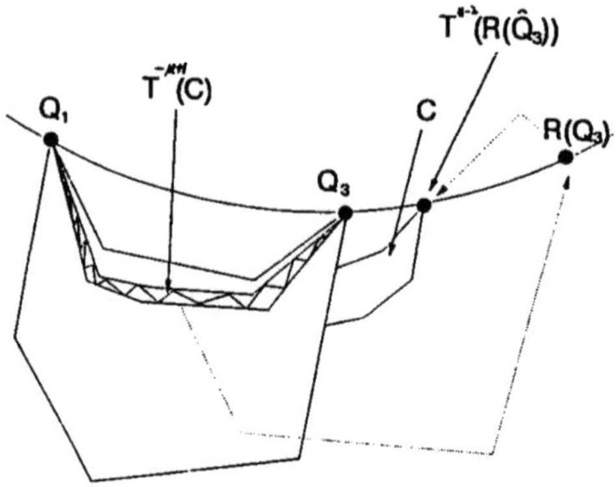

Figure 15: The pre-image of C

3.4. The case of $\mu > 1$: symbolic dynamics

The geometric discussion in last subsection gives a clear indication of a horseshoe structure in the case of $\mu > 1$. However, it is very hard (if not impossible) to form a rigorous proof for the existence of the full symbolic dynamics out of this geometric description. To get the desired symbolic dynamics, one has to trace all the vertical and horizontal paths and their images, and to be sure that these paths overlap each other correctly. The correct overlap of all the paths is guaranteed for the standard horseshoe map by assumption. Here we have a far more complicated situation. Intersections of boundaries of the strips might be very irregular, and the shape of C might be very strange to prevent a correct overlap. It is the goal of this subsection to provide a rigorous proof of the existence of the full symbolic dynamics in the case of $\mu > 1$.

Let $\hat{\Gamma}^+$ be the collection of all finite and infinite sequences of positive integers:

$$\hat{\Gamma}^+ = \{s = (s_0, s_1, ..., s_n, ...) \mid s_n \in Z^+\}$$

We again introduce two additional symbols c, e to augment the set $\hat{\Gamma}^+$. If a given sequence s is finite, we add symbols e or c at the end of s. Denote the concluded set as Γ^+.

For a given sequence $s \in \Gamma$, create a new sequence s_μ according to the given integer μ by inserting $\mu - 1$ of 1's in the middle of every two consecutive symbols of s. Let

$$\Gamma^+_\mu = \{s_\mu \mid s \in \Gamma^+\}.$$

Now for any sequence $s_\mu \in \Gamma_\mu^+$,

$$s_\mu = (s_0, s_1, ...s_n, ...),$$

let

$$\hat{s_\mu} = (...., s_n, ..., s_1, s_0.)$$

We call $\hat{s_\mu}$ the inversion of s_μ. Also, let

$$\Gamma_\mu^- = \{\hat{s_\mu} \mid s_\mu \in \Gamma_\mu^+\}$$

Proposition 10 *Assume $\mu > 1$ for the fictitious flow on the McGehee's collision manifold. For any given sequence $s = (s_0, s_1, ..., s_n, ...) \in \Gamma_\mu^+$, take a regular strip U_{s_0} on S with representation $(s_0 - 1, s_0, s_0)$. For any vertical path l_v of U_{s_0}, there exists at least one point p on l_v, such that the forward solution starting at p corresponds to the sequence s.*

Refer to Theorem 1 for the relation between a given solution and its coding sequence.

Proof: Take any vertical curve l_v, say, $c(t) : (-\infty, +\infty) \mapsto U_{s_0}$. Assume that $\lim_{t\to-\infty} c(t) \in b_{k+1}$ and $\lim_{t\to+\infty} c(t) \in b_k$. F(c(t)) is a regular curve on $\hat{S}$ connecting $\hat{Q}_3$ and $\hat{Q}_1$. i.e. $\lim_{t\to-\infty} F(c(t)) = \hat{Q}_1$, $\lim_{t\to+\infty} F(c(t)) = \hat{Q}_3$,

Define the set K as all the values of t such that $T^i(c(t)) \in B(S)$ $(1 \leq i \leq \mu - 1)$. Since the image of a sufficiently small tails of $F(c(t))$ at $\hat{Q}_3$ and $\hat{Q}_1$ will stay inside of B(S), K is not empty. In fact, there is a sufficiently large constant $M > 0$, such that $(-\infty, -M) \in K$, $(M, +\infty) \in K$.

Let $t_0 = inf\{t \mid t \notin K\}$ and $l_{s_0} = \{c(t) \mid t \in (-\infty, t_0)\}$. $T^\mu(l_{s_0})$ is a regular curve connecting $T^{\mu-1} \circ R(\hat{Q}_3)$ (which is on the arc Q_1Q_3) and $R(\hat{Q}_1)$ (or $R(\hat{Q}_3)$).

Therefore, $T^\mu(l_{s_0})$ crosses all the regular strips on S. For the integer s_1, there exists a regular strip U_{s_1} on S with representation $(s_1 - 1, s_1, s_1)$. $T^\mu(l_{s_0}) \cap U_{s_1}$ at least contains a vertical curve of U_{s_1}. The pre-image of this vertical curve is a closed sub-segment of $c(-\infty, t_0)$. Denote this segment as l_{s_0,s_1}.

Now we observe the image of the concluded vertical curve on U_{s_1}. $T^{2\mu}(l_{s_0,s_1})$ will crossly intersect with all the regular strips on S again. Among all the regular strips, there is one with representation $(s_2 - 1, s_2, s_2)$. The intersection $T^{2\mu}(l_{s_0,s_1}) \cap U_{s_2}$ gives another vertical curve in U_{s_2}. The pre-image of this vertical curve under $T^{2\mu}$ is, again, a closed sub- segment of l_{s_0,s_1}. denote this sub-segment as l_{s_0,s_1,s_2}.

Repeat this process, one obtains a nested sequence of closed sub-segment of c(t):

$$l_v \supset l_{s_0} \supset l_{s_0,s_1} \supset l_{s_0,s_1,s_2}, \cdots$$

The intersection of all these sub-segments is a non-empty set l_s of c(t). The forward solution starts from any point of l_s has the given coding $s = (s_0, s_1, s_2, ...)$.

We also have:

Proposition 11 *Assume $\mu > 1$ for the fictitious flow on the McGehee's collision manifold. For any given sequence $\hat{s} = (...., s_{-n}, ..., s_{-2}, s_{-1}, s_0.) \in \Gamma_\mu^-$, take a regular strip U_{s_0} on S with representation $(s_0 - 1, s_0, s_0)$. For any horizontal path l_h of U_{s_0}, there exist at least one point p on l_h, such that the backward solution starting at p corresponds to the sequence $\hat{s}$.*

Proof: Because the collinear three-body system is reversible, the whole proof for the forward flow can be carried over to the backward flow.

Take two symbolic sequences $s, \hat{s}$. $s = (s_0, s_1, s_2, ..., s_j, ...) \in \Gamma_\mu^+$, $\hat{s} = (..., s_{-j}, ..., s_{-2}, s_{-1}, s_0.) \in \Gamma_\mu^-$. Assume that the starting integer s_0 of s is the same as the terminating integer s_0 of $\hat{s}$. Put s and $\hat{s}$ together to obtain a new sequence

$$y(s, \hat{s}) = (..., s_{-j}, ..., s_{-2}, s_{-1}, s_0; s_1, s_2, ..., s_j, ...)$$

Denote the set of all the symbols $y(s, \hat{s})$ created by $s \in \Gamma_\mu^+$ and $\hat{s} \in \Gamma_\mu^-$ as Γ_μ:

$$\Gamma_\mu = \{y(s, \hat{s}) \mid s \in \Gamma_\mu^+; \hat{s} \in \Gamma_\mu^-\}.$$

Theorem 2 *Assume $\mu > 1$ for the fictitious flow on the McGehee's collision manifold. For any given sequence*

$$y = (..., s_{-j}, ..., s_{-2}, s_{-1}, s_0; s_1, s_2, ..., s_j, ...) \in \Gamma_\mu,$$

there exists at least one solution of the collinear three-body problem which corresponds to the sequence y.

Proof: Combining proposition 10 11 and the next lemma.

Lemma 1 *Take a regular strip U on S. Let C_1, C_2 be two subsets of U. Assume that all the vertical curves of U intersect C_1 and all the horizontal paths of U intersect C_2. Then $C_1 \cap C_2$ is not empty.*

Proof: Since the argument is topological, one only need to prove this lemma on the standard strip V associated with U. Also we can assume a metric and a measure on V.

For a sufficiently small $\varepsilon > 0$, take the ε neighborhood H_ε of C_1 in V. H_ε is open. This open set only has finitely many connected components. One sees that at least one of these components contains a horizontal path. Otherwise, one can draw a vertical curve avoiding all the components of H_ε. This is a contradiction since every vertical curve intersects C_1. Drop all those components of H_ε which do not contain any horizontal path, and rename what left in H_ε as H_ε

Similarly, take the ε neighborhood V_ε of C_2. V_ε must have a connected component containing a vertical curve. Again, drop all the components which do not contain any vertical curve in V_ε, and rename what left as V_ε. One sees that $C_{1\varepsilon} = C_1 \cap H_\varepsilon$, $C_{2\varepsilon} = C_2 \cap V_\varepsilon$ again satisfy the intersection assumptions for C_1, C_2 in the lemma. The closure of H_ε and V_ε intersect with each other, and the set of intersection I_ε is a compact set.

Now take a sequence of positive numbers $\varepsilon_n \to 0$. One can inductively construct $H_{\varepsilon_n} = C_{1\varepsilon} \cap H_{1\varepsilon}$, $V_{\varepsilon_n} = C_{2\varepsilon_{n-1}} \cap V_{2,\varepsilon_{n-1}}$. This gives a nested set sequence of intersections $\{ I_{\varepsilon_n} = \text{closure}(H_{\varepsilon_n}) \cap \text{closure}(V_{\varepsilon_n}) \}$. The intersection of all I_{ε_n} is not empty, and has to be in $C_1 \cap C_2$. This concludes the proof.

4. References

1. R. Devaney. *Singularities in classical mechanical system* Ergodic Theory and Dynamical System I, Ed. A. Kotak, Progress in Math. **10** (1981) Birkhauser, Boston/Basel/Stuttgart, 211-333.
2. J.N. Mather and R. McGehee. *Solution for collinear four-body problem which becomes unbounded in finite time*, Lecture Notes in Physics, Vol. **3**, (1975), Springer-Verlag 573-597.
3. R. McGehee. *Triple collision in the collinear three-body problem*, Invent. Math. **27**, (1974), 191-227.
4. K.R. Meyer and Q.D. Wang. *The global phase structure of the restricted isosceles three- body problem with positive energy*, Tans.Amer. Math. Soc. **338(1)**, (1993), 311-336.
5. K.R. Meyer and Q.D. Wang. *The global phase structure for the three-dimensional isosceles three-body problem with zero energy*, To appear in Hamiltonian Dynamical System II, edited by K.R. Meyer, H.S. Dumas, D.S. Schmidt.
6. K.R. Meyer and Q.D. Wang. *The collinear three-body problem with negative energy*, To appear on J. Diff. Eq., (1994).
7. D. G. Saari and Z.H. Xia. *Oscillatory and super hyperbolic solutions in Newtonian system* J. Diff. Eq. **82** (1988) 342-355.

www.ingramcontent.com/pod-product-compliance
Lightning Source LLC
La Vergne TN
LVHW051053210926
848486LV00002B/93